LIQUID CHROMATOGRAPHY IN PHARMACEUTICAL DEVELOPMENT

An Introduction

LIQUID CHROMATOGRAPHY IN PHARMACEUTICAL DEVELOPMENT

An Introduction

Edited by

Irving W. Wainer

Research Chemist
U.S. Food and Drug Administration
Washington, DC

A S T e R
PUBLISHING
CORPORATION

This book is dedicated to my wife, Pamela A. Zulli, who is a constant source of energy, support, creativity, and enthusiam without whom this work would have been impossible.

Liquid Chromatography in Pharmaceutical Development: An Introduction

ASTER PUBLISHING CORPORATION
320 NORTH A STREET, SPRINGFIELD, OREGON 97477

Library of Congress Cataloging in Publication Data

Liquid chromatography in pharmaceutical development

    Includes index.
    1. Drugs — Analysis.   2. Liquid chromatography.
3. Pharmaceutical technology.   I. Wainer, Irving W.
[DNLM: 1. Chromatography, liquid.   2. Drugs — Analysis.
3. Technology, pharmaceutical.   QV 25 L7673]
RS189.L53 1985   615'.1901   85-18596

ISBN 0-943330-11-4

Printed in the United States of America

10 9 8 7 6 5 4 3 2 1

Published by Aster Publishing Corporation, 320 North A Street,
P.O. Box 50, Springfield, Oregon 97477

CONTENTS

v

LIST OF CONTRIBUTORS

Mark Canales, PhD, Manager, Interface Group, Nelson Analytical, Cupertino, California, USA

Henri Colin, PhD, Vice President, Research and Development, Varex Corporation, Rockville, Maryland, USA

Peter DePhillips, Associate Scientist, Smith Kline & French Laboratories, Philadelphia, Pennsylvania, USA

John Dingerdissen, Associate Senior Investigator, Smith Kline & French Laboratories, Philadelphia, Pennsylvania, USA

Lars-Erik Edholm, PhD, Head, Bioanalytical Section, AB Draco, Lund, Sweden

Robert A. Egli, DiplChem, Director, Analytical Services, Cilag AG, Schaffhausen, Switzerland

Karl Erhard, Senior Medicinal Chemist, Smith Kline & French Laboratories, Philadelphia, Pennsylvania, USA

John Filan, Associate Senior Investigator, Smith Kline & French Laboratories, Philadelphia, Pennsylvania, USA

David E. Games, PhD, DSc, Reader in Chemistry, University College, Cardiff, Wales, UK

Michael P. Henry, PhD, Research Scientist, J.T. Baker Chemical Company, Phillipsburg, New Jersey, USA

William Horwitz, PhD, Scientific Adviser, Food and Drug Administration, Washington, DC, USA

Ante M. Krstulović, PhD, Assistant Director of Analytical Research and Quality Control, Laboratoires d'Etudes et de Recherches Synthelabo, Paris, France

Wolfgang Lindner, PhD, Associate Professor, University of Graz, Graz, Austria

W. John Lough, PhD, Head, Chromatography Laboratory, Physical and Analytical Services Unit, Beecham Pharmaceuticals, Harlow, Essex, UK

Milos Novotny, PhD, Professor of Chemistry, Indiana University, Bloomington, Indiana, USA

Lars Ögren, PhD, Group Leader, Method Development, Bioanalytical Section, AB Draco, Lund, Sweden

Curt Pettersson, PhD, Research Fellow, Department of Analytical Pharmaceutical Chemistry, Biomedical Center, University of Uppsala, Uppsala, Sweden

Willy Roth, PhD, Head, Pharmacokinetics and Metabolism Group, Dr. Karl Thomae GmbH, Biberach, FRG

Eric B. Sheinin, PhD, Chief, Drug Standards Research Branch, Food and Drug Administration, Washington, DC, USA

Robert Sitrin, PhD, Assistant Director, Biopharmaceuticals, Smith Kline & French Laboratories, Philadelphia, Pennsylvania, USA

Donald J. Smith, Research Chemist, Division of Drug Chemistry, Food and Drug Adminstration, Washington, DC, USA

Stanley Stein, PhD, Section Head, Department of Biopolymer Research, Hoffmann–La Roche, Inc., Nutley, New Jersey, USA (Current affiliation: Section Leader, Physical and Analytical Chemistry Research and Development, Schering Corporation, Bloomfield, New Jersey, USA)

Richard M. Venable, Chemist/Computer Specialist, Food and Drug Administration, Washington, DC, USA

Edward S. Yeung, PhD, Professor of Chemistry, Iowa State University and Ames Research Laboratory, U.S. Department of Energy, Ames, Iowa, USA

Morris Zief, PhD, Baker Fellow, Research and Development, J.T. Baker Chemical Company, Phillipsburg, New Jersey, USA

PUBLISHER'S NOTE

Two chapters planned for this text — one on photodiode array detection by A.F. Fell and B.J. Clark, the other on preparation of recombinant proteins by S.J. Tarnowski and E.S. Sharps — were not available in time for inclusion in this volume. The two chapters will appear instead as feature articles — the former in briefer form — in upcoming issues of *LC Magazine: Liquid Chromatography and HPLC*.

INTRODUCTION

Liquid chromatographic (LC) methods are used in every stage of the development of a pharmaceutical preparation, from initial synthesis, isolation of the drug substance, and pharmacological testing to the quality control of the marketed pharmaceutical preparation. This has been especially true since the introduction of modern high performance liquid chromatography (HPLC) in 1969.

The increased use of HPLC and other LC techniques (high performance thin-layer chromatography, affinity chromatography, and others) is attributable to the versatility of the approach. LC can be used on analytical and microanalytical scales to separate and quantitate large or small, polar or nonpolar, and chiral or achiral molecules. The methodology can also be adapted to the preparative scale and is extremely useful in the isolation of small amounts of biologically active molecules, including proteins. In addition, the methods are easily automated, thereby increasing the number of analyses per unit time and reducing costs.

This volume has been designed as a discussion of the utility of LC in the pharmaceutical industry. A full discussion, however, must of necessity be as broad as the method's versatility. Thus, we have had to limit ourselves to a few of the major aspects of the topic. The text begins with an overview of some of the key advances in LC that are having or that should soon have an effect on the industry. These developments include progress in separation and isolation techniques and in detection and identification of the resolved solutes.

The second part of the volume is concerned with the use of LC in the preparative separation of drug substances. Although LC has not as yet become a major production-line technique, it is a major factor in the small-scale separation of both low molecular weight substances and endogenous proteins.

One of the key steps in the development of a new drug substance is pharmacological testing, which includes pharmacokinetic and metabolic studies. The latter studies involve large numbers of biological samples and present the analyst with the challenge of designing methods with high throughput and low cost. An exciting approach to this problem is coupled column chromatography (also known as multiple-column chromatography and column switching), a method that can be readily automated. This approach is discussed in the third part of this volume by two of the leaders in the field.

Once a drug substance has been formulated and introduced to the market, LC (including thin-layer chromatography as well as HPLC) plays an important role in the quality control of the marketed pharmaceutical. The fourth part of this book addresses some of the issues in this area. One of the key questions calls for a definition of a viable LC assay that can be used as a standard assay for the composition of the pharmaceutical. Two of the key points discussed are interlaboratory variations and standardization of HPLC columns.

Finally, the last part presents an overview of present and future trends in the use of LC in the pharmaceutical industry. The discussion includes both sides of the regulatory mechanism — the U.S. Food and Drug Administration and the pharmaceutical industry. The dialogue is instructive, because it points out both the convergence and the divergence of two approaches to the same goal — the development of safe and effective pharmaceuticals.

This volume could not have been completed without the excellent efforts of the contributors who are named elsewhere or without the help and support of my colleagues at INSERM Unit 7, Hôpital Necker, Paris, France, and at the Division of Drug Chemistry of the U.S. Food and Drug Administration. The assistance of the publisher's staff is also appreciated, particularly the efforts of Heather W. Lafferty and James McCloskey in preparing the manuscripts for print. In addition, a special debt of gratitude is owed Professor Gören Schill for his continued inspiration and enthusiasm for the pharmaceutical industry.

—Irving W. Wainer
Washington, DC
September 1985

PART ONE

RECENT ADVANCES IN THE SEPARATION, ISOLATION, AND DETECTION OF DRUG SUBSTANCES

A: SEPARATION AND ISOLATION TECHNIQUES

BIOCHEMICAL AND PHARMACEUTICAL APPLICATIONS OF MICROCOLUMN LIQUID CHROMATOGRAPHY

Milos Novotny

Department of Chemistry
Indiana University
Bloomington, IN 47405

Modern liquid chromatography (LC), the so-called high performance liquid chromatography (HPLC), is often said to be the most successful analytical method of recent times. This notion is supported both by the enormous number of applications associated with this method and by the related success in the manufacture of instrumentation.

Although the trends toward use of packing materials with ever-decreasing particle size (1–3) are to a great extent responsible for the success of today's HPLC, a potential importance of the column's diameter was largely overlooked during earlier investigations. Because many laboratories seemed to experience difficulties in packing narrow columns with small particles, most LC column manufacturers have standardized their products to be of a 4.6-mm internal diameter (i.d.). Correspondingly, most high-pressure pumps, gradient devices, sampling units, and detectors have now been designed for flow rates on the order of milliliters per minute to use with 4.6-mm columns.

The first purposeful attempts to use LC columns with small diameters date to the late 1970s. Scott and Kucera succeeded in packing 1-m segments of 1-mm i.d. columns with very high pressures of slurried packing materials (4–6). They referred to their columns as *microbore columns,* a somewhat misleading name that may, nevertheless, become widely accepted. By joining these 1-m columns together, Scott and Kucera achieved startling overall efficiencies between 10^5 and 10^6 theoretical plates, although the analysis time was very long. The main reasons behind the investigations of Ishii and co-workers (7), who described the use of small plastic columns together with a miniaturized instrument, seemed to be the LC system's simplification and reduced mobile-phase consumption. Finally, Tsuda and Novot-

5

ny (8,9), who sought a substantial departure from the established LC column technologies, reported the first results on the use of capillary columns. Their GC-like column technologies, featuring both the open tubular (9) and packed capillary (8) column varieties, showed that significant instrumental modifications would be needed for flow rates on the order of microliters per minute or less. Thus, while there were initially somewhat different reasons for initiating work on the small-bore LC columns in these three laboratories, the analytical advantages of different microcolumns and their overlapping and complementary features have gradually become recognized.

During the last several years, microcolumn LC has developed into one of the more popular directions taken in the area of analytical separations. In addition, related techniques of capillary supercritical fluid chromatography (10) and capillary zone electrophoresis (11) have also been developed to a fair degree of sophistication. During a period of just one year, three books on the subject have been published (12–14). At present, microcolumn separation techniques are rapidly developing toward new directions, often well beyond the original intent. In addition to the potential of separation performance, unique detection aspects and the abilities to work with small amounts of biological materials hold the primary attraction for newcomers to the field.

Because much has happened in the field of microcolumn LC since the pioneering studies of the late 1970s, there is a need to summarize recent trends and conceptually important applications. The present chapter will review these trends in relation to current and potential applications of microcolumn LC techniques to biochemical and pharmaceutical research.

METHODOLOGICAL SCOPE

In any attempt to analyze biological materials effectively, a researcher is often confronted with unique problems that are seldom shared by other fields: an enormous complexity of most biological samples, the relative instability of biologically important molecules, and a frequent need to perform analytically reliable measurements at extremely low concentrations and in minute sample amounts. When one adds a requirement of rapid determinations, even the most powerful analytical techniques become greatly challenged. Not surprisingly, various chromatographic and electrophoretic methods find enormous application in biochemical and biomedical investigations. Over the years, more primitive separation techniques — conventional LC, thin-layer chromatography (TLC), and packed-column gas chromatography (GC) — are gradually being replaced by more powerful and sensitive analytical options.

In general terms, the miniaturization efforts in chromatography started almost 30 years ago with the development of open tubular (capillary) columns by Golay (15). This trend was predicted by Martin in 1956 (16) and followed by some fascinating projections on the importance of miniaturized analytical methods for biological research during Martin's keynote lecture at the 1962 Hamburg gas chromatography symposium (17). Although one trend emphasized chromatographic performance and the other stressed the capabilities of handling and detecting increasingly small amounts of substances, some parallels of thought are evident. Unfortunately, it took nearly two decades before capillary GC found sufficient recognition in biomedical research; one hopes that microcolumn LC will not follow at the same slow speed.

With most biologically and pharmaceutically important compounds, the presence of various polar groups in the molecules imposes certain limitations on effective use of the highly efficient and sensitive gas-phase analytical techniques. Although many of these compounds can be brought into the gas phase following various chemical derivatizations (18,19), some of these procedures are known to be tedious and nonquantitative and, consequently, are less popular with biomedical researchers. In addition, the lack of volatility of larger molecules, such as peptides or nucleotides, is clearly recognized throughout the pharmaceutical industry where, at least for analyses of uncomplicated mixtures (for example, in quality control situations), HPLC has now gained a significant advantage over GC. For numerous sophisticated analytical problems, however, additional developments must be made before LC techniques will be comparable with the established performance of powerful GC-based techniques such as capillary GC/mass spectrometry, electron capture detection, element-specific detection, multiple-ion detection, or capillary GC/Fourier-transform IR spectrometry. The recently developed field of microcolumn LC appears to provide a potential for such developments.

The incentives for further developments in microcolumn LC include:
- very high column efficiencies for resolution of complex mixtures
- increased *mass sensitivities* with miniaturized concentration-sensitive monitors in combination with very low flow rates
- possibilities of optimizing certain detection techniques as a result of extremely low flow rates as well as the potential for applying new detection principles
- possibilities of using exotic mobile phases and reagents for the sake of improved separations and measurements
- drastically reduced consumption of increasingly expensive or environmentally hazardous mobile phases
- size compatibility with various micromanipulations of modern biology and medicine.

In searching for new analytical capabilities, different types of microcolumns (to be discussed below) might assume different functions. Yet, there is a different set of instrumental problems associated with each column type. Stringent volumetric requirements for sampling and detection devices are prominent among the instrumental difficulties of miniaturized LC. Predominantly for this reason, commercial developments in microcolumn LC seem to follow, at the moment, the woefully slow path of capillary GC immediately after its inception. For example, easy technological fixes seem in short supply for the use of open tubular columns in LC. As the analytical demands increase and additional applications appear, reliable commercial instrumentation will also need to be designed.

Even though the current number of microcolumn LC applications is relatively small, convincing cases can be made for high performance, increased mass sensitivity, and some unique detection capabilities of research micro-LC systems. With the gradual recognition of various analytical advantages, numerous laboratories are now quickly moving into this general area to enrich the field with new ideas and instrumental technology.

Recently developed techniques of capillary supercritical fluid chromatography (10) and capillary zone electrophoresis (11) are viewed as both complementary to and competitive with microcolumn LC, as is discussed in this chapter. With regard to solute mass-transfer kinetics, supercritical fluid chromatography should be superior to LC as far as performance and speed are concerned. Extremely high separation efficiencies encountered in capillary electrophoresis seem to be attributable to the lack of a two-phase system and diffusion-controlled interphase phenomena (11). Although it appears unlikely that the supercritical fluids can readily be found for effectively solvating very polar and large molecules, the general method offers certain unique analytical advantages, such as an easy use of flame detectors, mass spectrometry, and IR spectroscopy. High-resolution electrophoresis is a very competitive approach to ion-exchange chromatography in dealing with the charged biological molecules. Both capillary supercritical fluid chromatography and capillary zone electrophoresis appear to be natural components within the range of microcolumn separation techniques. Further discussion of these techniques falls outside the scope of this chapter; both techniques, however, were recently reviewed (20,21).

MINIATURIZED LC COLUMNS

The column is the site of action to which the other components of a chromatograph must adjust. If the *unique* analytical properties of certain microcolumns demand new instrumentation, every reasonable effort should be made to achieve that adjustment. The opinions as to what is reasonable in the field of microcolumn LC, however, vary considerably.

The major column types under investigation in different laboratories are shown in Figure 1; their typical approximate dimensions and operating parameters are summarized in Table I. The instrumental consequences of using these different microcolumns will be discussed in a following section. At this point, their technologies and analytical characteristics will be briefly described.

Small-bore packed columns, including the already mentioned 1-mm i.d. microbore columns, are a fairly straightforward extension of the previously used general approach — that is, packing with slurries of small-particle materials. Al-

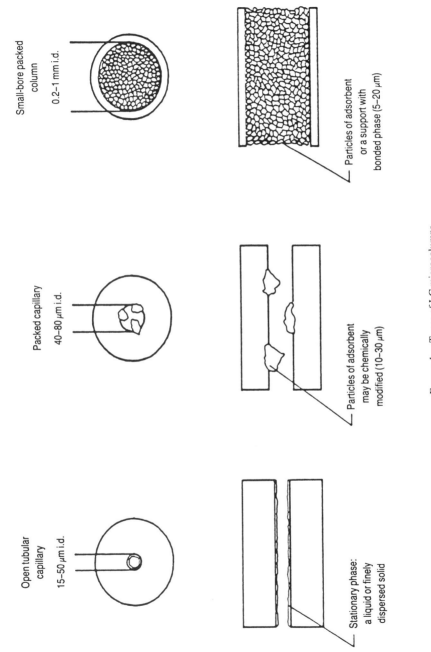

Open tubular
capillary

15–50 μm i.d.

Packed capillary

40–80 μm i.d.

Small-bore packed
column

0.2–1 mm i.d.

Stationary phase:
a liquid or finely
dispersed solid

Particles of adsorbent
may be chemically
modified (10–30 μm)

Particles of adsorbent
or a support with
bonded phase (5–20 μm)

FIGURE 1. Types of LC microcolumns.

TABLE I

Characteristics of Conventional and Microcolumns
for High Performance Liquid Chromatography

Column type	Typical dimensions		Volumetric flow rate	Sample capacity
	i.d.	Length		
Conventional column	4.6 mm	10–25 cm	1 mL/min	10–100 μg
Small-bore packed column	0.2–1 mm	1–10 m	1–20 μL/min	1–10 μg
Packed capillary column	40–80 μm	1–100 m	0.5–2 μL/min	100 ng–1 μg
Open tubular capillary	15–50 μm	1–100 m	<1 μL/min	<100 ng

though there are no obvious theoretical reasons for higher efficiencies of these columns as compared to the conventional HPLC columns (that is, the particle size but not the column's inner diameter is the characteristic dimension controlling kinetic processes in totally packed columns), additional considerations may exist for which further clarification becomes necessary. For instance, the recently studied slurry-packed capillaries made from glass or fused-silica tubing (22–25) yield very different results than the 1-mm i.d. microbore columns.

The open tubular (true capillary) columns should provide results that agree with the Golay theory — that is, extremely high efficiencies (around 10^6 theoretical plates) should result when working with inner diameters below 10 μm (26,27). Although far from being easy, preparation of such columns is considerably less of a technological problem than designing suitable instrumentation in this area because the dead-volume problems are considerably more troublesome than with the other column types. Figure 2, showing a cross-section of a 15-μm i.d. capillary (27), provides an idea of the volumetric requirements of the interconnecting sampling and detection components of the instrument.

In the preparation of narrow-bore open tubular columns, the most popular materials appear to be glass and fused-silica capillaries coated with an outer polymeric protective layer. The inner surfaces of such columns must subsequently be modified to prepare the analytical columns for either adsorption or partition chromatographic modes. Numerous microetching and phase-immobilization procedures of general usage may actually borrow from the knowledge acquired in high-efficiency capillary gas chromatography. Promising results were recently obtained by Tsuda and associates (28) and Jorgenson and co-workers (27,29), demonstrating that a gradual decrease in the column dimensions, coupled with progress in small-volume technology, may actually advance these columns to the point of practical utility sooner than is widely believed.

The properties of semipermeable packed capillary columns are intermediate between those of open tubular and totally packed columns (8). Their sample capaci-

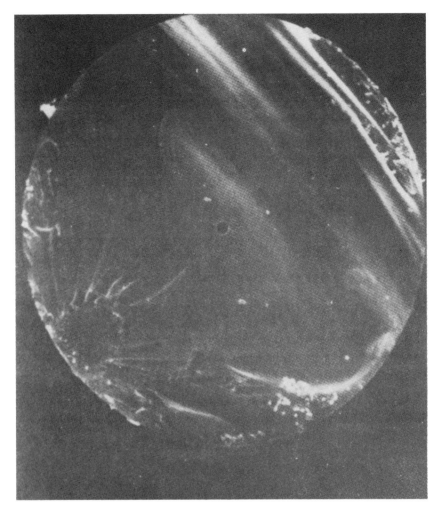

FIGURE 2.   Electron micrograph of the 15-μm capillary column end. (Reproduced, with permission, from Reference 27.)

ties and typical volumetric flow rates fall in the desirable range (Table I), while the column efficiencies are also high (8,30,31). There is a significant rationale for further developments in this area.

The column technology of packed capillaries is somewhat unique. In the first step, small-bore glass tubes are totally filled with particles of a certain size, followed by drawing the glass tube into a capillary. In the initial work on these columns, low packing density was emphasized to maintain good permeability (8,32). Mechanical stability of the packing structure has been maintained by imbedding the particles into the glass column wall. The column evaluation study by McGuffin and Novotny (30) suggested that the column diameter is related to the mobile-phase mass transfer phenomena because the small column radii (at a constant particle size) yielded lower plate-height values. Unusual column hydrodynamic effects were also suggested by the fact that irregular particles provided better efficiencies than spherical particles.

In a recent publication, Tsuda and co-workers (31) prepared packed capillaries with greater packing density; an example is shown in Figure 3. High column efficiencies were realized following their packing procedure despite the fact that particles smaller than 10 μm could not effectively be drawn into glass capillaries. A general disadvantage of this column preparation technology is the fact that, unlike the case with slurry-packed capillaries, selective chemical treatments must be performed *after* the column has been drawn (32). The process is relatively tedious; problems of column deactivation may also exist.

Some confusion still surrounds the analytical merits of the three major types of microcolumns. Apart from their individual compatibilities with available instru-

FIGURE 3.   Cross-sectional photomicrograph of a glass capillary with 10-μm particles drawn inside. (Obtained through courtesy of Dr. Takao Tsuda, Nagoya Institute of Technology, Nagoya, Japan.)

mentation and suitability for various detection techniques, these columns are primarily judged for their separation performance. While the column inner diameter primarily determines column performance in open tubular columns, particle size is the principal factor controlling both column efficiency and speed of analysis in packed columns; with microcolumns, however, the statement is correct in only a semiquantitative sense.

In ordinary packed-column work, the quality of a given column is judged to be superior if the reduced plate-height, h, is equal to the value of twice the particle size, d_p. Numerous columns with such a performance are widely documented in the literature. What is less commonly recognized, however, is that such columns can only be prepared in limited lengths, yielding total plate numbers of less than 10,000–20,000. In different laboratories, combining such columns has been less than satisfactory. In contrast, small-diameter columns were recently prepared in lengths up to several meters, totaling efficiencies well within the order of 10^5 theoretical plates (23,33,34).

The use of open tubular columns in LC creates an unusual situation in that column efficiencies are strongly dependent on solute retention and the capacity ratio. This is a direct consequence of the Golay equation — familiar to the practitioners of capillary GC but less so to the HPLC clientele. Confusion may thus arise from this point. Obviously, long capillary columns that generate plate numbers in excess of 10^6 only for nonretained solutes will be of little use in practical work.

Unusual circumstances are further encountered in work with semipermeable packed capillary columns. As discussed by McGuffin and Novotny (30), neither particle size nor column diameter can be used as adequately as the characteristic column dimensions and the basis for evaluating reduced plate heights.

In assessing the separation potential of various LC microcolumns, appropriate criteria of column performance must be applied. The so-called *separation impedance, E,* introduced by Bristow and Knox (35), appears to be a satisfactory criterion:

$$E = \frac{t_R \cdot \Delta p}{N^2 \, \eta \, (1 + k')} = \frac{H^2}{K^0} = h^2 \, \phi' \qquad [1]$$

The value of separation impedance E takes into consideration the column efficiency N, but also evaluates the time of analysis t_R and the pressure gradient Δp required to achieve the separation. The pressure gradient is, basically, the technological price paid for the efficiency and speed of analyses. The solvent viscosity η and the capacity ratio k' are also considered. It can further be shown that the separation impedance is equal to the square of the plate height H divided by the column permeability K^0, or, alternatively, is equal to the square of the reduced plate height h multiplied by a column resistance factor, $\phi' = d^2/K^0$; d is the particle size or the inner diameter of an open tubular column. Using these criteria, an actual column performance can be easily compared with theoretically predicted values.

TABLE II

**Comparison of the Theoretical Performance of
Conventional and Microcolumns in Liquid
Chromatography**

Column type	h_{min}	ϕ'	E_{min}
Conventional or small-bore packed column	2	500–1000	2000
Packed capillary column	2	~150	600
Open tubular capillary	0.8	32	20

Table II shows the theoretical minimum values for conventional columns and microcolumns, as predicted by Knox (36). For conventional and small-bore columns, the minimum reduced plate height is 2, the column resistance factor between 500 and 1000. The minimum separation impedance is 2000, although the actual experimental values are usually higher. Clearly, open tubular columns exhibit best theoretical values (separation impedance as low as 20), while semipermeable packed capillary columns are intermediate between conventional and open tubular columns.

There are differences between the values in Table II and current experimental results obtained with various microcolumns. Although open tubular columns are potentially the most efficient of the microcolumn family, their potential will not be realized until stringent technological requirements are met to work with the inner diameters of less than 10 μm (26,27), although totally packed microcolumns (for example, 1-mm i.d. microcolumns) have been successfully prepared and used in a variety of applications.

The best column efficiencies have thus far been obtained with the slurry-packed fused-silica capillaries (22–25). Values of reduced plate height that are close to theoretical are achieved even for relatively long columns (23,33,34). It is of some interest that such columns may occasionally exhibit separation impedance values well below the theoretical E_{min}(37), suggesting their loose packing structure. Similarly, some drawn semipermeable packed capillary columns may yield better results than are expected from theory (30). Further research in this area is desirable to improve analytical separations, in addition to improving the understanding of the separation processes involved. Geometrical features and certain analytical attributes of packed capillary columns are intermediate between those of microbore columns and open tubular columns. Their ease of preparation, high efficiencies, adequate sample capacities, and easy coupling with various unique detectors place these columns among the most promising tools in modern separation science.

INSTRUMENTAL CONSIDERATIONS

Although the use of microcolumns in modern LC offers a number of significant advantages, the instrumentation associated with such columns can be quite demanding. During the last several years, much effort has been spent on designing suitable systems for various microcolumn types. Such activities have ranged from various attempts to modify existing commercial equipment to investigations aimed at radically new ways of sample introduction and solute detection.

The overall system performance in liquid chromatography can be critically influenced by extracolumn contributions to band dispersion. The very small dimensions of microcolumns and correspondingly low flow rates make the reduction of these contributions particularly critical. Improperly designed components of a micro-LC system, in a volumetric or hydrodynamic sense, can entirely negate performance achieved by a column.

The band dispersion observed in liquid chromatography may be volumetric in nature (contributions from dead volume, for example, in the injector, detector, or connecting tubes), or it may be temporal, originating in the slow response of detecting or recording devices. If these factors operate independently of each other, the total peak dispersion is the sum of individual contributions from different sources:

$$\sigma^2_{\text{total}} = \sigma^2_{\text{column}} + \sigma^2_{\text{conn}} + \sigma^2_{\text{inj}} + \sigma^2_{\text{det}} + \sigma^2_{\text{temporal}} \qquad [2]$$

where the variance (σ^2) is used as a measure of band dispersion. The volumetric variance of the column itself is given as

$$\sigma^2_{\text{column}} = (\pi r^2 \varepsilon_T)^2 \, HL \qquad [3]$$

where
r = the column radius
L = the length
H = the plate height
ε_T = the total porosity.

Understanding the origin and magnitude of band broadening in microcolumn LC is indeed crucial to designing adequate analytical systems. Sternberg's classical theory of extracolumn contributions (38) has been discussed explicitly for microcolumns by Yang (39) and more recently reviewed by Gluckman and Novotny (40). Obviously, open tubular columns of 10-μm i.d. impose the most stringent requirements on the system; the necessary volumes for the detectors and sampling devices are in the low nanoliter and subnanoliter range. These requirements are considerably more relaxed for various packed capillary columns (low fractions of a microliter), while 1-mm i.d. microbore columns can already tolerate volumes greater by almost an order of magnitude. Consequently, much currently available commercial equipment can be modified for work with 1-mm i.d. microbore columns.

Besides strictly volumetric requirements, there may be other needs for the instrumental design that must aim at the elimination of other less visible sources of band dispersion, such as discontinuity of tube cross-sections, occurrence of stagnant zones, mixing chambers, and so forth. Some examples of these sources are described in a recent review by Ishii and Takeuchi (41). Ideally, the connecting tubes and unions should be eliminated altogether. Although certain solutes can now be detected on-column, eliminating connecting tubing is considerably more difficult for the sampling techniques.

In general terms, the injected volume must be sufficiently small to minimize extracolumn dispersion, yet the sample should be sufficiently large to detect the sample components. The resulting compromise between the loss of chromatographic resolution and the gain in detection sensitivity with increasing injection volume is particularly important for high-speed and high-efficiency separations.

The maximum permissible injection volume (V_{inj}) that will produce a fractional (θ^2) increase in the volumetric variance of a nonretained peak (42) is given by Equation 4:

$$V_{inj}^2 = (\theta K \pi r^2 \varepsilon_T)^2 HL = K^2 \sigma_{inj}^2 \qquad [4]$$

The maximum injection volume decreases rapidly with the column radius r, but it is also dependent on the column length L and the plate height H. Moreover, the above equation also emphasizes the importance of the input function, because the constant K^2 is characteristic of the injection profile (42).

Table III lists various column types and their typical geometric characteristics together with volumetric requirements for the sampling process. The magnitude of the general problem is easily evident from the table. At present, the requirements for the size of sampling loops in ordinary HPLC are easily within the cur-

TABLE III

Maximum Injection Volume and Variance for Conventional and Microcolumns

Column type	i.d. (mm)	Particle size (μm)	Length (m)	σ_{column}^2 *	σ_{inj}^2 †	V_{inj} †
Conventional	4.6	5	0.25	500 μL²	25 μL²	17 μL
Small-bore	1	5	1	4.5 μL²	0.22 μL²	1.6 μL
Packed capillary	0.1	30	10	0.062 μL²	3100 nL²	190 nL
	0.07	30	10	0.011 μL²	530 nL²	80 nL
Open tubular	0.03	–	10	120 nL²	6 nL²	8 nL
capillary	0.01	–	5	2 nL²	0.1 nL²	0.4 nL

* Optimum velocity and plate height are assumed; total porosity (ε_T) was 0.85 for conventional and small-bore packed columns, and 1.0 for packed and open tubular capillaries.
† Ideal injection profile is assumed ($K^2 = 12$); 5% increase in column variance permitted ($\theta^2 = 0.05$).

rent technological means, but the small-bore columns (50–200 nL) are pushing the limits while open tubular columns can operate only under the conditions of indirect sample introduction, such as splitting injectors (8,27) or a heart-cutting device (43). Several workers have also reported injecting small volumes using temporary (44,45) or moving injection techniques. Split injections are convenient and relatively precise, but the technique is wasteful of the available sample and solvent and unsuitable for trace analyses.

The detector to be used in microcolumn LC must also pass certain stringent requirements:

• the detector volume and time constants must be quite low
• sensitivities (signal-to-noise ratio) of miniaturized detectors should be equal to or better than those for conventional LC
• the detector should permit convenient and reliable measurements.

Volumetric requirements will, again, depend on flow cell radius r and length L:

$$\sigma^2_{det} = \frac{V_{det}^2}{12} = \frac{(\pi r_{cell}^2 L_{cell})^2}{12} \qquad [5]$$

Although the calculated values are somewhat less stringent than those for the input volumes discussed earlier, one still must deal with submicroliter volumes for the small-bore columns and low nanoliter volumes for the open tubular columns. The sources of band dispersion, including those related to flow irregularities, have been discussed elsewhere (40). With careful attention to the design and construction of the detection cell and connecting unions, band dispersion can effectively be minimized for small-bore columns. Ideally, on-column detection is preferable, as has been shown with the examples of UV absorbance (10,22), laser fluorimetry (46,47), and electrochemical measurements (48,49).

The system's time constant is important for fast separation. At present, only a minority of microcolumn applications are encountered in this area. As pointed out previously (50–52), the detector time constant for such separations should be no more than a few tenths of a second. In view of the fact that some commercial spectroscopic detectors may have time constants as large as 3 s, the magnitude of this problem must be investigated for individual cases; obviously, difficulties can be encountered for separations that occur on the scale of seconds. As shown by Scott and co-workers (51), even conventional photocells of spectroscopic detectors can be too slow for the purpose of fast separations. A recent review by Hartwick and Dezaro is recommended for more detailed information on this subject (52).

Chemical derivatization in HPLC is frequently sought to overcome general detection problems. In conjunction with microcolumn chromatography, solute precolumn derivatization will obviously be preferred over postcolumn derivatization that requires instantaneous reactions and mixing volumes. Nevertheless, postcolumn derivatization techniques were demonstrated as being feasible with 1-mm i.d. columns (53,54). Work with smaller column diameters may require dramatic miniaturizations of the postcolumn reactors. Precolumn derivatization methods are becoming increasingly popular, primarily because of a wealth of solute tagging approaches available in modern chemistry; here, time of derivatization is relatively unimportant. UV absorbance, fluorimetric, and electrochemical detectors can all be used for quantitative measurements of derivatized molecules. The often-raised objection against precolumn derivatization — that the chemical reactions tend to mask the solute properties beneficial for separations — does not appear particularly serious. It is even less serious with significantly enhanced column efficiencies that are now available with certain microcolumns. An example of successful use of precolumn derivatization (benzoylation of model hydroxysteroids) is illustrated in Figure 4 (55); this chromatogram demonstrates an adequate resolution of isomeric substances.

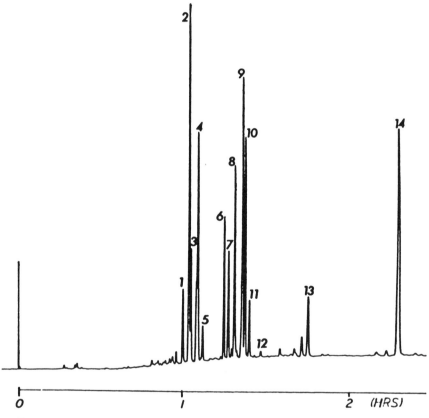

FIGURE 4.   Separation of 14 benzoylated steroid standards. Stepwise gradient conditions: 80% acetonitrile (ACN)/H_2O (15 min); 85% ACN/H_2O (14 min); 90% ACN/H_2O (15 min); 95% ACN/H_2O (18 min); 100% ACN. (Reproduced, with permission, from Reference 55; copyright 1983 American Chemical Society.)

There are certain auxiliary techniques of modern LC that allow more effective uses of the general method. These include solute precolumn concentration, multidimensional chromatography, and recycling. They, too, require significant volumetric miniaturization if they are to be used effectively in microcolumn chromatography. Once again, these techniques are relatively easy to implement with 1-mm i.d. columns, while reducing the column diameter down to the desirable 0.2-mm i.d. presents difficulties of dead-volume–free connections and the availability of small-volume valves. The early successful studies of precolumn concentration schemes in conjunction with microcolumns were demonstrated by Ishii and co-workers, who applied these techniques to the analysis of corticosteroids in serum (56,57), catecholamine metabolites in urine (58), trace organic components of water (59), and bile acids in biological media (60).

For a variety of trace determinations, it is essential that trace components first be enriched from larger volumes and subsequently analyzed by microcolumn LC. The *peak compression* technique can often be used for injecting large volumes directly onto the column. Under conditions of preconcentration, the conditions must be chosen to favor large distribution coefficients, such as are present in the preconcentration of organics from an aqueous solution on a reversed-phase column. Scott and Kucera (61) and, more recently, Krejci and co-workers (62) demonstrated that the principle is applicable with microcolumns. Figure 5 demonstrates preconcentration of catecholamines by peak focusing with a coeluting additive (62).

The peak compression technique will also be important in certain multicolumn schemes (column-switching techniques). In these cases, the selected solutes, originally introduced in a complex matrix onto the first column, are more effectively analyzed with a second column. An example has been described (63) in which the excess derivatization agent was removed while the components of interest — derivatized prostaglandins — were introduced onto a second column. Given the enormous complexities of most biological mixtures, column-switching methods may indeed be essential. To date, however, no serious efforts have been made to develop such systems for microcolumn LC.

While recycling methods have been fairly common in conventional LC, Kucera and Manius also explored this approach with microbore columns of overall high efficiency and demonstrated resolution of certain isomeric compounds (64). The separation of the *cis/trans* isomers of retinoic acid, shown in Figure 6, is one example (64). Once again, minimizing the extracolumn sources of volumetric band dispersion was crucial to the success of this approach.

Although an increasing number of chromatographers have shown interest in microcolumn LC over the past several years, the commercial aspects of this method are still in their infancy. Only recently have miniaturized solvent delivery systems and detectors appeared on the market. The technical requirements of microcolumn LC must be met by such equipment to ensure further applications of this powerful method.

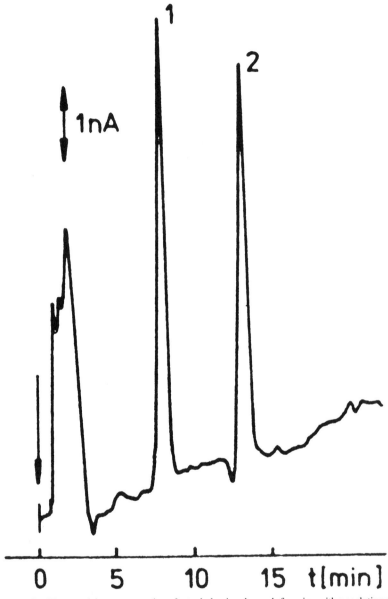

FIGURE 5.   Trace-enrichment separation of catecholamines by peak-focusing with a coeluting additive. Column: 150 mm × 0.7 mm i.d. packed with 10-μm Separon Si-C_{18}; mobile phase: 10^{-1} M $NaClO_4$ + 10^{-3} M $HClO_4$ + 10^{-3} M disodium ethylenediamine tetraacetate (EDTA) in water; flow rate, 0.8 μL/s; amperometric detection. Sample injected: 0.1 mL of a solution of 4.10^{-7} M noradrenaline + 4.10^{-7} M adrenaline + 2.10^{-2} M $C_{10}H_7SO_3Na$. Peaks: 1 = noradrenaline: 2 = adrenaline. (Reproduced, with permission, from Reference 62.)

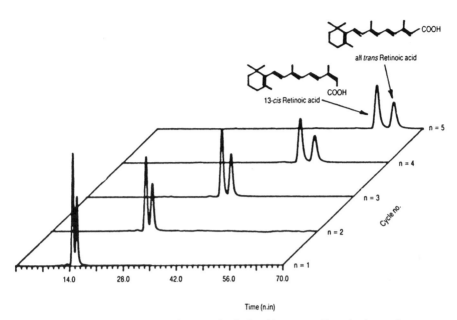

FIGURE 6.   Separation of *cis/trans* isomers of retinoic acid on a recycling microbore column system. Columns: two ODS columns, 50 cm × 1 mm each; eluent: chloroform/methanol (5:95) containing 0.3% (w/v) ammonium acetate; flow rate: 40 μL/min; detection: 350 nm. (Reproduced, with permission, from Reference 64.)

HIGH-EFFICIENCY SEPARATIONS

As discussed already in the section on columns, various microcolumns differ in their potential for component resolution. While open tubular columns are potentially most efficient, little practical use has thus far been made of them. Efficiencies of the open tubular columns with 40-μm i.d. or above are hardly more efficient than conventional columns. The efficiencies reported on nonretained components, no matter how impressive, do not count as far as the practical uses of open tubular columns are concerned. With the exception of some recent studies (27,49), open tubular columns have not been applied thus far to what could be called a difficult sample.

The vast majority of complex nonvolatile samples have been successfully chromatographed with microbore or slurry-packed capillary columns. In their pioneering studies, Scott and Kucera (4,5) demonstrated high efficiencies — in excess of 500,000 theoretical plates — through coupling 1-m long, 1-mm i.d. columns. High efficiencies were, however, attained after very long analysis times. These results were extended to an extreme in a recent study (65) in which more than a million theoretical plates were achieved. Component resolution was achieved for α-values as low as 1.02.

It should be pointed out that order-of-magnitude increases in column efficiency are significant in efforts to analyze complex mixtures. The practical lesson of capillary versus packed-column GC is fairly clear, but which analysis times in LC shall be tolerated? The answer to that question is less straightforward, for it will be strongly dependent upon sample type. As expressed by Scott, "today, such efficiencies are novel and are required only in a minority of circumstances. As the technique develops and expands into the field of biological materials, very high efficiency columns will no longer be a luxury but a necessity" (5).

In biological and environmental samples, and perhaps less commonly with technologically important samples, adequate component resolution within a complex mixture is often the key to successful investigations. Two examples from different areas will now be presented.

Various fossil fuels — petroleum or coal-derived liquids, for example — and their fractions are among the most complex mixtures of organic compounds. To understand their composition is an exceedingly important task for both technological and environmental reasons. Although capillary GC is now a well-proved separation technique for relatively small molecules, less volatile compounds may best be resolved by microcolumn LC (33,34,66). Figure 7 demonstrates a separation of a coal-tar fraction in which a slurry-packed capillary column was used to resolve the individual components for subsequent mass-spectral investigations (34). Similarly, 5- to 9-ring aromatic compounds, including various isomers, were resolved from an extract of carbon black at column efficiencies close to 250,000 theoretical plates (33). Such separations typically take 5–10 h, but for more rapid sample profiling less resolution may be adequate (66).

Chromatographic analyses of complex biological samples are frequently hampered both by the lack of resolution and by detection problems. Because various compounds within a chemical class (peptides, steroids, prostaglandins, and so forth) possess a characteristic functional group, derivatization methods can be used for enhanced detection. An example is seen in Figure 8, in which two samples of urinary hydroxysteroids are displayed as distinct metabolic profiles after these metabolities were benzoylated to facilitate UV detection and separated by microcolumn LC (55).

In both separation examples mentioned above, slurry-packed capillary columns were used. Could these samples be similarly resolved if sufficiently long *conventional* columns were used? In principle, yes. There are no theoretical reasons to suggest that *totally packed* columns of different inner diameters will have different efficiencies. In practice, however, it has been exceedingly difficult to couple small, conventional columns in series to yield such efficiencies. Such a procedure would be tedious and expensive. Conversely, preparation of slurry-packed capillaries is easy and reproducible (23).

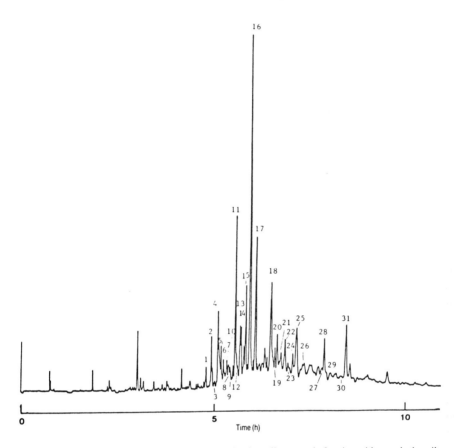

FIGURE 7.   High-efficiency separation of neutral polycyclic aromatic fraction with a packed capillary column. Column: 1.8 m × 200 μm capillary packed with 3-μm C_{18} Spherisorb; mobile phase: 90% acetonitrile/10% water for 3.7 h, followed by 100% acetonitrile for 4.2 h, and 95% acetonitrile/5% ethyl acetate thereafter. (Reproduced, with permission, from Reference 34; copyright 1984 American Chemical Society.)

Semipermeable packed capillary LC columns provide yet another route to achieve high column efficiencies and resolution of complex mixtures (8). Promising results in pursuit of this goal were recently presented by Tsuda and co-workers (31).

FAST ANALYSES

Speed of separation has been of continuing interest to certain chromatographic applications in process control, drug monitoring, and kinetic measurements. In gas chromatography, remarkable separation speeds were achieved in the early

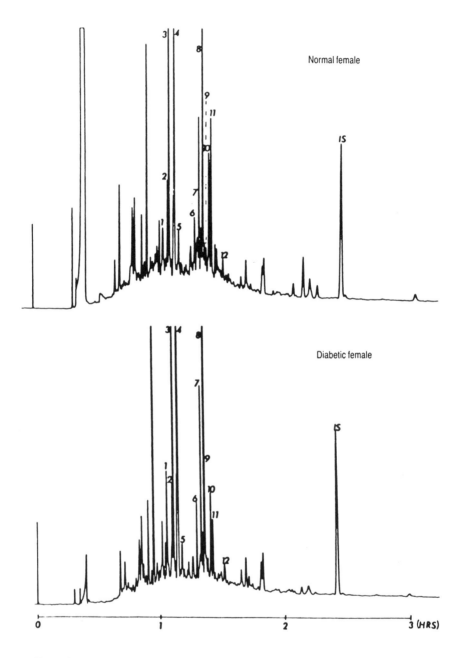

FIGURE 8.   Comparison of the representative metabolic profiles of urinary steroids from normal and diabetic human females. (Reproduced, with permission, from Reference 55; copyright 1983 American Chemical Society.)

years of GC (67). In modern liquid chromatography, a gradual decrease in particle size — down to 3 μm — has generated some interest in the subject (68). Although less efficient than GC columns, the modern small-particle columns are now capable of generating approximately 100–500 theoretical plates/s.

Why are the microcolumns of interest in high-speed separations? Once again, there is no obvious advantage here in reducing column diameter, at least in theory; some hidden, potential advantages might, however, be found in association with a reduction in the wall effect (69) and the heat of friction (70,71). With further possibilities of particle size reduction to 1 μm (72), these aspects warrant further exploration. The most prevalent reason for using microcolumns in high-speed LC is, however, a very practical one: a significantly reduced consumption of the mobile phase.

Scott and co-workers were the first to explore uses of microcolumns for fast analyses (51). In their work, seven model substances were resolved in less than 30 s. Since then, the subject has been reviewed by Hartwick, Dezaro, and Meyer (52,73), with numerous new examples described. Figure 9, showing a rapid sepa-

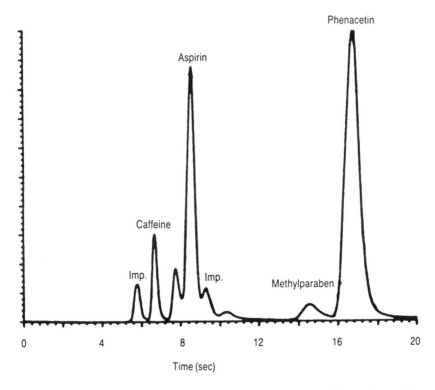

FIGURE 9. Rapid ion-pairing separation of common analgesics. Column: 10 cm $\times$ 1 mm column packed with 3-μm reversed-phase C_{18} material; mobile phase: 18% ACN with 5 mM sodium octylsulfate. (Reproduced, with permission, from Reference 52.)

ration of common analgesics, is quite representative of such efforts (52). The separation medium used in this application was a 10 cm × 1 mm stainless steel column packed with 3-μm C_{18} material. Short, fused-silica capillary columns, packed with small particles, were also found by Ishii and co-workers to achieve similar results (74).

Although the rationale for high-speed work is relatively straightforward, the instrumental aspects — including primarily injection volumes and the modes of sampling as well as the time constants of corresponding detectors and recording devices — will undoubtedly require further attention.

IMPROVED DETECTION CAPABILITIES

As previously discussed, the detectors used in microcolumn LC must have a minimal volume to avoid band broadening. Conventional detectors are too large for the purpose because their volumes are typically at the microliter level. Depending on the type of microcolumn used, the volumetric requirements may range from a few tenths of a microliter to fractions of a nanoliter. Following the initial miniaturization efforts for UV absorbance (7), fluorimetric (75), and electrochemical (76) detectors, numerous small-volume designs appeared in the literature. Two instances can readily be recognized in microcolumn LC detection: the use of miniaturized conventional (mainly concentration-sensitive) detectors, and the use of unique detection principles that take advantage of extremely low flow rates.

Concentration-sensitive devices, when miniaturized and coupled with microcolumns, exhibit higher mass sensitivities (61,77). This unique advantage has now been recognized for various sample-limited applications. In fact, a great deal of interest in microcolumns will likely be spurred by this very straightforward fact in the areas of pharmacological research, biotechnology, and modern medicine, all of which are frequently concerned with an optimum analytical use of extremely small samples. With an increasing emphasis on analyses at the cellular level in the future, microcolumn techniques are likely to be popular.

Despite drastic decreases in the optical path length following detector miniaturization, UV detectors still possess reasonable sensitivities because of the above-mentioned mass sensitivity enhancement. Although their concentration sensitivities are impaired because of the increased noise level, detection of nanogram quantities is still feasible (55,77,78). For slurry-packed capillary columns (23), detection volumes of 100–200 nL appear practical. When fused-silica tubing is used as a column material, it is advantageous to use on-column detection, as is illustrated in Figure 10 (22).

A substantial amount of miniaturization appears feasible with electrochemical detectors. As shown by Slais and Krejci, a wall-jet electrochemical detector could be constructed with an effective volume of 1 nL and, correspondingly, a very high sensitivity (79). Manz and Simon used an ion-selective detector with a liquid

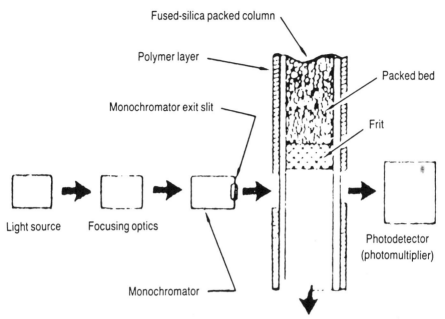

FIGURE 10. Schematic diagram for on-column detection using narrow-bore fused-silica microparticle-packed column. (Reproduced, with permission, from Reference 22.)

membrane that could be positioned directly at the column exit; under favorable conditions, sensitivities in the femtomole range were achieved for monovalent cations (48). Finally, Knecht and co-workers inserted a single-fiber carbon electrode inside the end-part of a capillary column (49). They obtained nearly 100% coulometric efficiencies with this detector. As suggested by Figure 11, such a detector could be very useful in the low-level detection of neurochemically important compounds and drug metabolites (49). Further advances in the technology of microelectrodes and other small sensors can conveniently combine with the modern trends in column design, permitting some unrealized opportunities in ultrahigh-sensitivity detection.

Yet another class of unique devices that combine naturally with microcolumns are various laser-based detectors, as reviewed recently by Yeung (80). The highly collimated nature of laser beams makes them nearly ideal for the small volumes associated with microcolumn techniques. Furthermore, lasers readily yield high photon densities, which are beneficial in certain high-sensitivity measurements. Although thermal-lens calorimetry (81,82), light-scattering phenomena (83,84), and coherent anti-Stokes Raman spectroscopy (85) have been explored in HPLC detection, laser-induced fluorescence and optical rotation measurements are perhaps the most attractive applications of laser-based detectors in the areas of biochemical and pharmaceutical analysis.

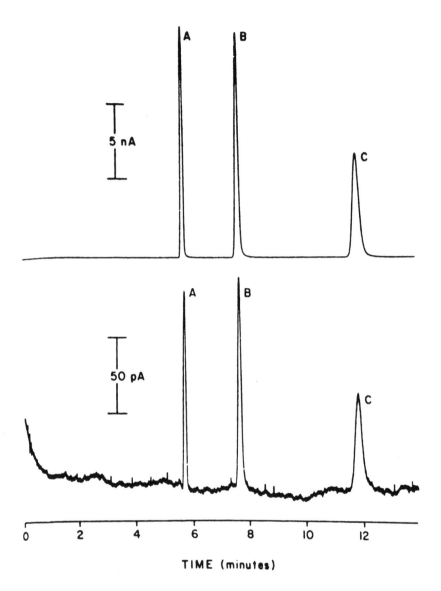

FIGURE 11. Chromatograms of equimolar mixtures of ascorbic acid (A), catechol (B), and 4-methylcatechol (C) obtained with a miniature electrochemical detector. Concentrations: (upper) 1×10^{-4} M; (lower) 1×10^{-6} M. (Reproduced, with permission, from Reference 49; copyright 1984 American Chemical Society.)

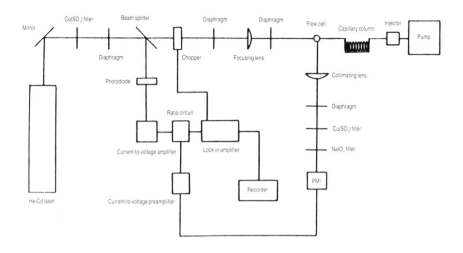

FIGURE 12.   Schematic diagram of chromatographic system, laser-induced fluorescence detector, and ancillary electronic equipment. (PMT = photomultiplier tube; reproduced, with permission, from Reference 47.)

Folestad and co-workers (86) and Hershberger and associates (87) were first to point out that laser-based fluorescence detectors combine well with miniaturized cells. Various laser systems were subsequently used (30) — but a crucial question remains whether chromatographers are sufficiently dedicated to become involved with expensive, high-power lasers that could occupy a sizable part of their laboratories. Simpler systems may thus be a key to widespread acceptance of the technique. The use of a helium-cadmium laser, as originally proposed for LC detection by Diebold and Zare (88), is an example of such a relatively simple device.

   Recent applications of a helium-cadmium laser detector have shown considerable promise for ultrahigh-sensitivity measurements (46,47). An example of a pertinent detection system is shown in Figure 12; in this form of detection, the 325-nm laser line is used, while the 442-nm blue line is filtered away (47). The ultraviolet line is useful for many molecules with native fluorescence as well as for biological compounds that can be derivatized to form suitable molecules for detection (89,90). Alternatively, the blue line, which is considerably more intense, can also be used in detection, but suitable derivatization agents must first be developed. As shown recently by the author's laboratory (47) and Guthrie and co-workers (46), low femtogram sensitivities can be achieved with such a detector.

Figure 13 shows a chromatogram obtained for steroid metabolites extracted from human blood at low picogram levels and tagged with a coumarin moiety before detection (47).

Recent studies of Yeung and associates (91–93) have convincingly shown that a laser is the crucial component of a detection system that is capable of measuring small changes in optical rotation. This optical activity detector can also be used to monitor ordinary solutes as vacancy peaks when an optically active mobile phase is used (93).

Whereas adequate detection of chromatographic peaks in miniaturized LC is important for further advances of these techniques, powerful ancillary methods for identification of the individual solutes that are separated through high-

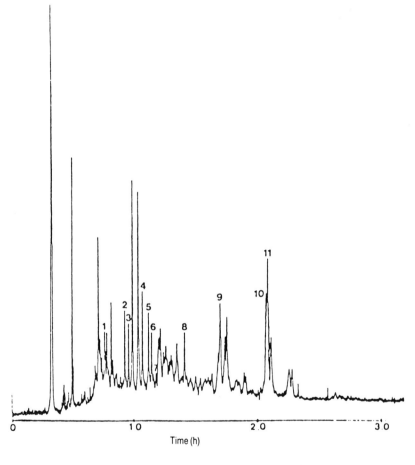

FIGURE 13.   Chromatogram of solvolyzed plasma steroids. Column: 2.25 m × 220 μm. Approximately 50 pg of each steroid was injected. (Reproduced, with permission, from Reference 47.)

efficiency columns are also needed. Traditionally, mass spectrometry surpasses such ancillary methods as far as the power of identification is concerned. Combining liquid chromatography with mass spectrometry (LC/MS) in general — and microcolumn LC in particular — has been among the most interesting and challenging tasks of the field. After more than a decade of intensive research in LC/MS, several important trends seem to emerge. Some take advantage of very low flow rates associated with microcolumns. As far as transport LC/MS interfacing is concerned, Scott (94) emphasized long ago the advantages of the total sample usage that can easily be done with small-bore columns. On the one hand, in the competitive approach of the direct inlet interface, microcolumns once again are preferable, as is emphasized in the recent reviews by Henion (95,96) and Tsuge (97). On the other hand, the recent important technique of thermospray LC/MS has been primarily designed for high flow rates (98,99).

While further development of LC/MS techniques seems both popular and necessary, some of the current trends will undoubtedly become modified. It is likely that microcolumns will continue to be of much interest in LC/MS. One particular advantage, emphasized by Henion, rests in the use of microcolumns in sample-limited situations, such as the detection of drug metabolites or of low-level pollutants (95). Figure 14, demonstrating the detection of 20-ng quantities of cortisone and dexamethasone using LC/MS, appears representative of current sensitivities using microcolumns with this general method (96).

Other ancillary methods that are complementary to mass spectrometry must also be developed. Coupling LC with infrared spectroscopy (LC/IR) remains a difficult task because of mobile-phase interference. Although far from ideal, improvements in LC/IR were recently shown using microcolumns and deuterated solvents (100).

Additional optical identification methods in combination with microcolumn LC have recently surfaced. Based on various photodiode array technologies, miniaturized imaging detectors based on UV-absorbance measurements (101) and fluorimetric principles (102,103) are becoming feasible. An application of the miniaturized UV photodiode array detector, as shown in Figure 15, has some relevance to pharmaceutical research (101). The intensified photodiode array detector (102,103) used in fluorescence measurements exhibits very high sensitivity (low picogram levels) even with a conventional source; combining this device with a laser excitation source for enhanced sensitivities remains a distinct possibility. Further developments of miniaturized optical imaging detectors appear desirable and are likely in the near future.

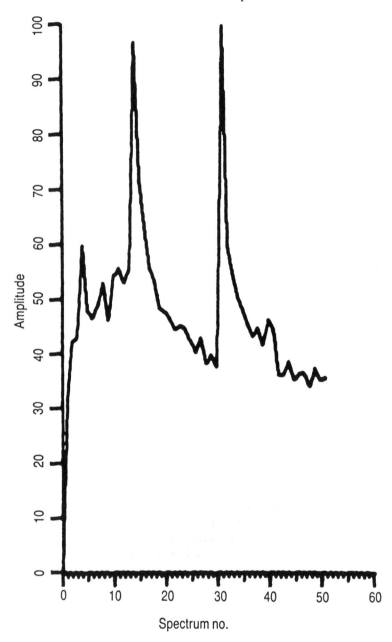

FIGURE 14:   Total ion current profile plot for 20 ng cortisone and 20 ng dexamethasone. The separation and micro-LC/MS analysis was achieved using a 7 cm × 0.5 mm SC-01 column with 8 μL/min
CH$_3$CN/H$_2$O (40:60) as micro-LC eluent/CI reagent gas. (Reproduced, with permission, from Reference 96.)

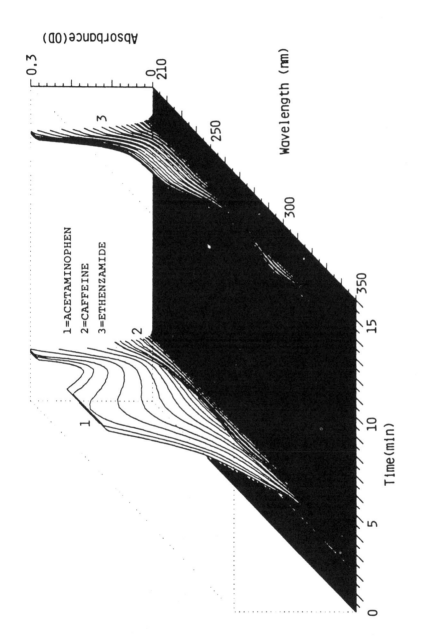

FIGURE 15: Three-dimensional spectrochromatogram of components in a medicine. (Reproduced, with permission of Huthig Verlag, from Reference 101.)

REFERENCES

(1) Cs. Horváth, B. Preiss, and S.R. Lipsky, *Anal. Chem.* **39**, 1422 (1967).

(2) J.H. Knox and M. Saleem, *J. Chromatogr. Sci.* **7**, 614 (1969).

(3) J.F.K. Huber, in *Column Chromatography,* Chimia Supplement, E. Kovats, ed. (Sauerlander, Aarau, 1970), p. 24.

(4) R.P.W. Scott and P. Kucera, *J. Chromatogr.* **125**, 251 (1976).

(5) R.P.W. Scott, *Analyst* (London) **103**, 37 (1978).

(6) R.P.W. Scott and P. Kucera, *J. Chromatogr.* **169**, 51 (1979).

(7) D. Ishii, K. Asai, H. Hibi, T. Jonokuchi, and M. Nagaya, *J. Chromatogr.* **144**, 157 (1977).

(8) T. Tsuda and M. Novotny, *Anal. Chem.* **50**, 271 (1978).

(9) T. Tsuda and M. Novotny, *Anal. Chem.* **50**, 632 (1978).

(10) M. Novotny, S.R. Springston, P.A. Peaden, J.C. Fjeldsted, and M.L. Lee, *Anal. Chem.* **53**, 407 (1981).

(11) J.W. Jorgenson and K.D. Lukacs, *Anal. Chem.* **53**, 1298 (1981).

(12) P. Kucera, ed., *Microcolumn High-Performance Liquid Chromatography* (Elsevier, Amsterdam, 1984).

(13) R.P.W. Scott, ed., *Small Bore Liquid Chromatography Columns* (Wiley-Interscience, New York, 1984).

(14) M. Novotny and D. Ishii, eds., *Microcolumn Separation Methods* (Elsevier, Amsterdam, 1985).

(15) M.J.E. Golay, in *Gas Chromatography 1958,* D.H. Desty, ed. (Butterworths, London, 1958), p. 36.

(16) A.J.P. Martin, in *Vapour Phase Chromatography,* D.H. Desty, ed. (Academic Press, New York, 1957), p. 1.

(17) A.J.P. Martin, in *Gas Chromatography 1962,* M. van Swaay, ed. (Butterworths, London, 1962), p. xxvii.

(18) E.C. Horning and M.G. Horning, *Methods in Medical Research* **12**, 369 (1970).

(19) M. Novotny and D. Wiesler, in *Separation Methods: New Comprehensive Biochemistry,* Volume 8, Z. Deyl, ed. (Elsevier, New York, 1984), p. 41.

(20) J.C. Fjeldsted and M.L. Lee, *Anal. Chem.* **56**, 619A (1984).

(21) J.W. Jorgenson and K.D. Lukacs, *Science* **222**, 266 (1983).

(22) F.J. Yang, *J. Chromatogr.* **236**, 265 (1982).

(23) J.C. Gluckman, A. Hirose, V.L. McGuffin, and M. Novotny, *Chromatographia* **17**, 303 (1983).

(24) Y. Hirata and K. Jinno, *J. High Resolut. Chromatogr.* **6**, 196 (1983).

(25) D. Ishii and T. Takeuchi, *J. Chromatogr.* **255**, 349 (1983).

(26) J.H. Knox and M.T. Gilbert, *J. Chromatogr.* **186**, 405 (1979).

(27) J.W. Jorgenson and E.J. Guthrie, *J. Chromatogr.* **255**, 335 (1983).

(28) T. Tsuda, K. Tsuboi, and G. Nakagawa, *J. Chromatogr.* **214**, 283 (1981).

(29) J.W. Jorgenson, E.J. Guthrie, R.L. St. Claire III, P.R. Dluzneski, and L.A. Knecht, *J. Pharm. Biomed. Anal.* **2**, 191 (1984).

(30) V.L. McGuffin and M. Novotny, *J. Chromatogr.* **255**, 381 (1983).

(31) T. Tsuda, I. Tanaka, and G. Nakagawa, *Anal. Chem.* **56**, 1249 (1984).

(32) Y. Hirata, M. Novotny, T. Tsuda, and D. Ishii, *Anal. Chem.* **51**, 1807 (1979).

(33) A. Hirose, D. Wiesler, and M. Novotny, *Chromatographia* **18**, 239 (1984).

(34) M. Novotny, A. Hirose, and D. Wiesler, *Anal. Chem.* **56**, 1243 (1984).

(35) P.A. Bristow and J.H. Knox, *Chromatographia* **10**, 279 (1977).

(36) J.H. Knox, *J. Chromatogr. Sci.* **18**, 453 (1980).

(37) D.C. Shelly, J.C. Gluckman, and M. Novotny, *Anal. Chem.* **56**, 2990 (1984).

(38) J.C. Sternberg, *Adv. Chromatogr.* **2**, 205 (1966).

(39) F.J. Yang, *J. Chromatogr. Sci.* **20**, 241 (1982).

(40) J.C. Gluckman and M. Novotny, in *Microcolumn Separation Methods*, M. Novotny and D. Ishii, eds. (Elsevier, Amsterdam, 1985), p. 57.

(41) D. Ishii and T. Takeuchi, *ibid.*, p. 73.

(42) M. Martin, C. Eon, and G. Guiochon, *J. Chromatogr.* **108**, 229 (1975).

(43) V.L. McGuffin and M. Novotny, *Anal. Chem.* **55**, 580 (1983).

(44) B. Coq, G. Cretier, J.L. Rocca, and M. Porthault, *J. Chromatogr. Sci.* **19**, 1 (1981).

(45) M.S. Harvey and S.D. Stearns, in *Liquid Chromatography in Environmental Analysis,* J.F. Lawrence, ed. (Humana Press, Clifton, New Jersey, 1982).

(46) E.J. Guthrie, J.W. Jorgenson, and P.R. Dluzneski, *J. Chromatogr. Sci.* **22**, 171 (1984).

(47) J.C. Gluckman, D.C. Shelly, and M. Novotny, *J. Chromatogr.* **317**, 443 (1984).

(48) A. Manz and W. Simon, *J. Chromatogr. Sci.* **21**, 326 (1983).

(49) L. Knecht, E.J. Guthrie, and J.W. Jorgenson, *Anal. Chem.* **56**, 479 (1984).

(50) M. Krejci, K. Tesarik, M. Rusek, and J. Pajurek, *J. Chromatogr.* **218**, 167 (1981).

(51) R.P.W. Scott, P. Kucera, and M. Munroe, *J. Chromatogr.* **186**, 475 (1979).

(52) R.A. Hartwick and D.D. Dezaro, in *Microcolumn High-Performance Liquid Chromatography,* P. Kucera, ed. (Elsevier, Amsterdam, 1984), p. 75.

(53) P. Kucera and H. Umagat, *ibid.*, p. 154.

(54) H. Umagat, P. Kucera, and L.-F. Wen, *J. Chromatogr.* **239**, 463 (1982).

(55) M. Novotny, M. Alasandro, and M. Konishi, *Anal. Chem.* **55**, 2375 (1983).

(56) D. Ishii, K. Hibi, K. Asai, M. Nagaya, K. Mochizuki, and Y. Mochida, *J. Chromatogr.* **156**, 173 (1978).

(57) T. Takeuchi and D. Ishii, *J. Chromatogr.* **218**, 199 (1981).

(58) M. Goto, T. Nakamura, and D. Ishii, *J. Chromatogr.* **226**, 33 (1981).

(59) T. Takeuchi and D. Ishii, *J. High Resolut. Chromatogr.* **6**, 310 (1983).

(60) T. Takeuchi and D. Ishii, *J. High Resolut. Chromatogr.* **6**, 571 (1983).

(61) R.P.W. Scott and P. Kucera, *J. Chromatogr.* **185**, 27 (1979).

(62) M. Krejci, K. Slais, D. Kourilova, and M. Vespalcova, *J. Pharm. Biomed. Anal.* **2**, 197 (1984).

(63) M. Novotny, *J. Pharm. Biomed. Anal.* **2**, 207 (1984).

(64) P. Kucera and G. Manius, in *Microcolumn High-Performance Liquid Chromatography,* P. Kucera, ed. (Elsevier, Amsterdam, 1984), p. 111.

(65) H.G. Menet, P.C. Gareil, and R.H. Rosset, *Anal. Chem.* **56**, 1770 (1984).

(66) M. Novotny, M. Konishi, A. Hirose, J. Gluckman, and D. Wiesler, *Fuel* **64**, 523 (1985).

(67) D.H. Desty and A. Goldup, in *Gas Chromatography 1960,* R.P.W. Scott, ed. (Butterworths, Washington, DC, 1960), p. 162.

(68) J.L. DiCesare, M.W. Dong, and L.S. Ettre, *Chromatographia* **14**, 257 (1981).

(69) J.H. Knox, G.R. Laird, and P.A. Raven, *J. Chromatogr.* **122**, 129 (1976).

(70) I. Halasz, R. Endele, and J. Asshauer, *J. Chromatogr.* **122**, 37 (1975).

(71) H. Poppe, J.C. Kraak, J.F.K. Huber, and J.H.M. van den Berg, *Chromatographia* **14**, 515 (1981).

(72) M. Verzele, lecture presented at the Eastern Analytical Symposium (New York City, 1984).

(73) R.A. Hartwick and R.F. Meyer, in *Microcolumn Separation Methods,* M. Novotny and D. Ishii, eds. (Elsevier, Amsterdam, 1985), p. 87.

(74) D. Ishii, M. Goto, and T. Takeuchi, *J. Pharm. Biomed. Anal.* **2**, 223 (1984).

(75) Y. Hirata and M. Novotny, *J. Chromatogr.* **186**, 521 (1979).

(76) Y. Hirata, P.T. Lin, M. Novotny, and R.M. Wightman, *J. Chromatogr.* **181**, 787 (1980).

(77) M. Novotny, in *Microcolumn High-Performance Liquid Chromatography,* P. Kucera, ed. (Elsevier, Amsterdam, 1984), p. 194.

(78) F.J. Yang, *J. High Resolut. Chromatogr.* **4**, 83 (1981).

(79) K. Slais and M. Krejci, *J. Chromatogr.* **235**, 21 (1982).

(80) E.S. Yeung, in *Microcolumn Separation Methods,* M. Novotny and D. Ishii, eds. (Elsevier, Amsterdam, 1985), p. 135.

(81) R.A. Leach and J.M. Harris, *J. Chromatogr.* **218,** 15 (1981).

(82) C.E. Buffett and M.D. Morris, *Anal. Chem.* **54,** 1824 (1982).

(83) J.W. Jorgenson, S.L. Smith, and M. Novotny, *J. Chromatogr.* **142,** 233 (1977).

(84) A. Stolyhwo, H. Colin, M. Martin, and G. Guiochon, *J. Chromatogr.* **288,** 253 (1984).

(85) L.A. Carreira, L.B. Rogers, L.P. Goss, G.W. Martin, R.M. Irwin, R. von Wondruszka, and D.A. Berkewitz, *Chem. Biochem. Environ. Instrum.* **10,** 249 (1980).

(86) S. Folestad, L. Johnson, B. Josefsson, and B. Galle, *Anal. Chem.* **54,** 925 (1982).

(87) L.W. Hershberger, J.B. Callis, and G.D. Christian, *Anal. Chem.* **51,** 1444 (1979).

(88) G.J. Diebold and R.N. Zare, *Science* **196,** 1439 (1977).

(89) K.-E. Karlsson, D. Wiesler, M. Alasandro, and M. Novotny, *Anal. Chem.* **57,** 229 (1985).

(90) M. Novotny, K.-E. Karlsson, M. Konishi, and M. Alasandro, *J. Chromatogr.* **292,** 159 (1984).

(91) E.S. Yeung, L.E. Steenhoek, S.D. Woodruff, and J.C. Kuo, *Anal. Chem.* **52,** 1399 (1980).

(92) D.R. Bobbitt and E.S. Yeung, *Anal. Chem.* **56,** 1577 (1984).

(93) E.S. Yeung, *J. Pharm. Biomed. Anal.* **2,** 255 (1984).

(94) R.P.W. Scott, in *Trace Organic Analysis: A New Frontier in Analytical Chemistry* (National Bureau of Standards, Special Publication 519, U.S. Government Printing Office, Washington, DC, 1979), p. 637.

(95) J. Henion, in *Microcolumn High-Performance Liquid Chromatography,* P. Kucera, ed. (Elsevier, Amsterdam, 1984), p. 260.

(96) J. Henion, in *Microcolumn Separation Methods,* M. Novotny and D. Ishii, eds. (Elsevier, Amsterdam, 1985), p. 243.

(97) S. Tsuge, *ibid.,* p. 217.

(98) C.R. Blakley and M.L. Vestal, *Anal. Chem.* **55,** 750 (1983).

(99) M.L. Vestal, *Science* **226,** 275 (1984).

(100) R.S. Brown and L.T. Taylor, *Anal. Chem.* **55,** 1492 (1983).

(101) T. Takeuchi and D. Ishii, *J. High Resolut. Chromatogr.* **7,** 151 (1984).

(102) J.C. Gluckman, D.C. Shelly, and M. Novotny, *Anal. Chem.* **57,** 1546 (1985).

(103) J.C. Gluckman and M. Novotny, *J. High Resolut. Chromatogr.,* in press.

LARGE-PORE BONDED PHASES IN THE CHROMATOGRAPHY OF PHARMACEUTICALS

Michael P. Henry

Research & Development Laboratories
J.T. Baker Chemical Company
Phillipsburg, NJ 08865

There is a large body of research relating the chromatographic behavior of large molecules (especially polymers) to pore size. In fact, in types of size-exclusion chromatography (SEC), pore size is the only parameter operating to achieve selectivity. The work of Porath (1) and Moore (2) in gel-filtration (GFC) and gel-permeation chromatography (GPC), respectively, laid the groundwork for these techniques, which are still undergoing important advances today (3). Size-exclusion chromatography is the technique that encompasses both GFC and GPC; it occurs without major solute/stationary-phase interactions and consequently has limited resolving power. Although SEC has been demonstrated to resolve pharmaceuticals and other substances of low molecular weight (4,5), other separation techniques (liquid-solid, reversed-phase, and gas chromatography) are considered more applicable for separations of small molecules.

There is a small but growing body of both theoretical (6,7) and empirical (8–10) work implicating the importance of pore size in determining certain parameters in the interactive liquid chromatography of a wide range of substances of low molecular weight. Some of these materials are of pharmaceutical interest in view of their useful biological and biochemical properties. It is the objective of this chapter to bring together representative research work from the literature and elsewhere that deals with those modes of chromatography in which pore size is one of several parameters determining resolution. Emphasis will be placed on the chromatography of small molecules, because the current state of research with large molecules on large-pore stationary phases has been adequately described in the literature (10,11) and to some extent elsewhere within this book.

It is important to define the terms large pore and large, medium-size, and small molecules in the context of this chapter. *Large pore* covers a portion of two ranges of pore size defined by Unger (12): 2–50 nm, the mesopore range; and greater than 50 nm, the macropore range. In this chapter large pore will refer to stationary phases with mean pore diameters that are greater than 15 nm. *Small* molecules have a molecular weight of less than 2000; *medium-size* molecules have a molecular weight between 2000 and 15,000; and *large* molecules have a molecular weight of greater than 15,000.

Pore size generally refers to the mean pore diameter, D. The difficulties in assigning a coherent physical meaning to this term have been described by Unger (12). Although values of D may be measured in several ways (Equations 1 and 2), its relation to chromatographic behavior can be difficult to predict (see the section on theory, below).

$$D_h = 2V_p/S_i \tag{1}$$

where
$\qquad D_h$ = hydraulic pore diameter
$\qquad V_p$ = pore volume
$\qquad S_i$ = specific internal surface area (13)
or

$$D = (1.32 - \psi)4\psi/S_i \tag{2}$$

where
$\qquad \psi$ = fraction of the particle volume that is porous (14).
The control over pore size and pore-size distribution that can be exerted during the synthesis of porous silica and its chemically modified forms (bonded phases) may explain the emphasis that is placed on this material in published research on pore-size effects. This chapter will be largely confined to silica-based bonded phases, but the reader is referred to the work of Cope (15) and Kato and co-workers (16), which describes pharmaceutical-related chromatography with synthetic polymer-based supports.

THEORY

A number of important properties of bonded phases are determined in part by pore size. These properties will in turn determine the fundamental chromatographic behavior of bonded phases:
• adsorption characteristics
• size exclusion properties
• rate of mass transfer of solute molecules between the mobile and stationary phases.
Adsorption will determine stationary phase selectivity and peak symmetry; size exclusion may influence retention and band width; rates of mass transfer will determine peak width or efficiency.

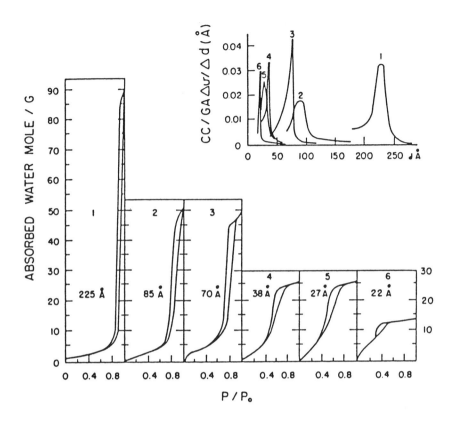

FIGURE 1.   Adsorption isotherms of water on various silica gels and corresponding pore-size distribution curves. Peak pore diameters given in angstroms. (Reprinted, with permission, from Reference 17.)

Pore Size and Adsorption

In summarizing the work of Barby (17), Iler (18) has illustrated the dependence upon pore size of the characteristic adsorption/desorption isotherms for gases or vapors (Figure 1). The isotherms can give the approximate pore-size distribution of a stationary phase particle according to the Kelvin equation (19):

$$\bar{r} = -2\sigma V \cos\theta / 2.303RT \log(p/p_0) = \bar{d}/2 \qquad [3]$$

where

σ = surface tension of adsorbed liquid at its boiling point

θ = angle of wetting (taken as zero)
V = molar volume of liquid
$\bar{d}$ and $\bar{r}$ = average diameter and radius, respectively, of those pores that are filled at a partial pressure p/p_0 for the liquid.

For liquid nitrogen,

$$\bar{r} = 4.146/\log(p_0/p) \qquad [4]$$

Because $\bar{d}$ (average pore diameter) $= 2\bar{r}$, then

$$\log(p_0/p) = 8.292/\bar{d} \qquad [5]$$

Adsorption isotherms relate the volume (or mass) of an adsorbate (for example, liquid nitrogen) adsorbed per gram of adsorbent to the partial pressure (p/p_0) of the adsorbate (Figure 1). If it is assumed that negligible adsorption occurs outside the pores, then the volume of nitrogen adsorbed at a given partial pressure is the partial pore volume (v) at that pressure. Thus, from Equation 5,

$$v = f(p/p_0) = f(\bar{d}) \qquad [6]$$

When $p/p_0 = 1$, then $v = v_p$; that is, the specific pore volume of the solvent is defined when $p/p_0 = 1$.

Differentiating Equation 6 with respect to $\bar{d}$ gives:

$$dv/d\bar{d} = f'(\bar{d}) \qquad [7]$$

A plot of $dv/d\bar{d}$ (the slope of the adsorption isotherm curve) versus $\bar{d}$ is the usual form of a pore-size distribution function.

In liquid chromatographic processes, the nature of the adsorption isotherms of solute molecules in stationary phases may similarly depend in part upon pore size and pore-size distribution. For example, small-pore silicas will generally have a higher number of micropores (less than 2-nm diameter) compared to large-pore silicas. Molecules of any size will be sterically excluded from a proportion of these micropores. In addition, the pore may be more or less permanently filled with highly structured mono- and multilayers of solvent (10,18), consequently giving nonlinear isotherms at low solute concentrations. Larger-pore silicas, by contrast, may have a larger proportion of their surface accessible for binding — especially for larger molecules — and may exhibit higher capacities. Isotherms remain linear at higher concentrations, peaks remain symmetrical at higher loadings, and resolution is maintained under these conditions — assuming, of course, that the small-pore and large-pore materials have the same surface area.

Steric Exclusion Effects

Only solute molecules that enter the pores of bonded phases will contribute to their binding properties, because less than 1% of surface area is located on the particle surface. Thus, in size-exclusion chromatography, all molecules spend more of their time outside the particle than solvent molecules, and capacity is very low. With a choice of pore size inappropriate to the size of the molecule being adsorbed, it has been estimated that as little as 5% of the bonded-phase surface may be involved in the binding process (20).

Snyder and co-workers (21) have determined that polystyrenes of molecular weight 2000 are excluded slightly from 30-nm pores, reducing their retention by 14% (relative to an entirely included molecule) under purely size-exclusion conditions. These polymer molecules exhibit a 66% size-exclusion effect in 6-nm pores. Clearly, much smaller molecules will also be retained less on small-pore supports because of this exclusion effect. Multiple separation mechanisms will not only contribute to altered retention but will also result in peak broadening.

Hearn and Grego (7) have developed a theory that relates k', the capacity factor, to an exclusion effect, a solvophobic (reversed-phase) effect, and a polar retention effect for polypeptides on nonpolar bonded phases. These authors estimate that in cases in which a solute molecule is one-tenth the size of the pore or less, it will have complete access to the pore surface. Thus for a 6-nm pore, a 6-Å molecule has complete access; and for a 30-nm pore, a 30-Å molecule has complete access to the pore surface.

Hearn and Grego (7) also noted that, for small pores (7.5-nm diameter) and small molecules (up to 530 MW), retention is reduced compared to that on medium-pore material (10-nm diameter). Furthermore, the effect appears to be greater for smaller molecules. These authors explain this phenomenon by suggesting that there is significant lack of access to surface area in small-pore material, which is most likely attributable to steric exclusion even for very small molecules. The molecule will be partially excluded from the pore entrance (any size molecule will miss a proportion of pores); but, once inside, that molecule may also be excluded from low-volume regions in the pore.

A systematic study of the influence of pore size on selectivity has been carried out by Sander and Wise (22). In examining the chromatographic properties of a series of octadecyl bonded silicas of various pore sizes, they found that selectivity of monomeric phases toward a group of polyaromatic hydrocarbons (PAHs) does not change greatly with pore size (Table I). Nonetheless, there *are* significant changes in PAH selectivity with pore size for polymeric phases. This observation was explained on the basis of the greater restriction of access of polymeric silanes to small pores relative to large pores. Monomeric silanes were equally accessible to pores varying in size from 6 nm to 100 nm. Another significant trend observable in this work was that monomeric bonded phases showed better selectivity than the polymeric phases.

TABLE I

Variation of Selectivity with Pore Size for
Monomeric and Polymeric Octadecyl Bonded Phases*

Silica	Phase type	Pore diameter (Å)	α (TBN/BAP)	k' (BAP)
Zorbax	Monomeric	300	2.12	3.46
Vydac TP	Monomeric	300	2.00	1.57
Zorbax	Monomeric	100	1.97	4.56
Zorbax	Monomeric	150	1.32	2.34
Zorbax	Monomeric	60	1.78	3.18
Zorbax	Polymeric	60	1.33	8.42
Partisil	Polymeric	85	0.97	11.11
Lichrosorb	Polymeric	100	0.97	9.41
Hypersil	Polymeric	120	0.79	11.73
Lichrospher	Polymeric	300	0.71	6.57
Vydac TP	Polymeric	330	0.62	6.77

* Adapted from Reference 22.

There were no simple relationships between pore size on the one hand and surface area, surface coverage, selectivity, and retention (22) on the other hand for the monomeric or polymeric phases. For the latter phases, although selectivity varies over a greater range, there is no significant correlation with pore size of a variety of silicas. Within the polymeric Zorbax (Du Pont Company, Wilmington, Delaware) series (6-, 10-, 15-, and 30-nm pores), however, there were some trends observable with increasing pore size. In this series, as pore size increases from 6 nm to 30 nm, carbon loading decreases from 17.7% to 4.28% and coverage (μmol/m^2) increases from 2.57 to 5.35. The increased loading is likely to be a result of increased layer thickness on the silica surface (23).

At this point the meaning of a pore-size effect should be explained. From the previous discussion, these effects are seen to be fundamentally steric in origin and thereby influence adsorption capacity, peak width, and peak symmetry. The smaller k' values commonly seen with small molecules on larger pore silicas is not a pore-size effect as Karch and associates postulate (24), but stems from the lower carbon content, which in turn results from the lower surface area. Wide-pore silicas, however, do not necessarily have low surface area, because surface area also depends upon pore volume (see Table I, Reference 22, and Equation 1).

Clearly, there are many physicochemical factors influencing the adsorption characteristics of a given bonded phase. These parameters include:

• silica pretreatment protocol
• type of silane (monomeric or polymeric)
• pore diameter
• surface area
• carbon loading
• surface coverage

- pore volume
- mobile phase
- temperature.

Correlating the individual contributions of these parameters to chromatographic selectivity is likely to be a very challenging task, especially in view of the difficulty in obtaining the above data from commercial suppliers. Large-pore silicas have several well-characterized applications in small molecule analysis (see later section), but finding more of these applications is likely to be challenging.

PORE SIZE AND EFFICIENCY

There is ample experimental evidence that large-pore bonded phases are generally more efficient than small-pore bonded phases when used to separate mixtures of high molecular weight components (greater than 15,000 daltons) (Figure 2). There is, however, comparatively little fundamental data available to determine whether pore size influences efficiency for small molecules. Snyder, Stadalius,

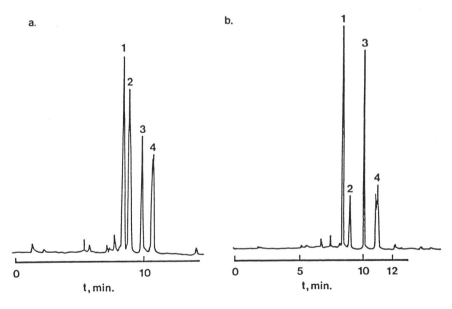

FIGURE 2.   Separation of proteins by reversed-phase HPLC. Gradient from 0.1 % trifluoroacetic acid to 0.1 % trifluoroacetic acid in acetonitrile, over 30 min; flow rate: 1 mL/min; pressure: 1200 psi; detection: 280 nm, 0.08 AUFS; injection volume: 10 μL. Sample: 1 = cytochrome c (6 μg), 2 = α_1-acid glycoprotein (20 μg), 3 = human serum albumin (20 μg), 4 = β-lactoglobulin A and B (20 μg). Column: (a) Bakerbond Octyl (C_8), 5 μm, 250 mm × 4.6 mm, experimental phase, 5-nm pores; (b) Bakerbond Wide-Pore Octyl (C_8), 5 μm, 250 mm × 4.6 mm, 30-nm pores.

and Quarry (21), Stout, DeStefano, and Snyder (6), and Hearn and Grego (7) have derived quantitative relationships indicating the extent of the contribution of pore size effects to peak broadening with both small and large molecules. The following is a summary of aspects of the theoretical discussions of the above authors.

Parameters affecting peak width have been related in the Knox equation (25):

$$h = Av^{1/3} + B/v + Cv \qquad [8]$$

where

h = reduced plate height = H/d_p
H = height equivalent to a theoretical plate
d_p = particle diameter
v = reduced mobile phase velocity = ud_p/D_m
u = linear velocity
D_m = diffusion coefficient of solute in mobile phase.

It is desirable to keep the value of h small for greatest efficiency. A, B, and C are variables that are a measure of static and dynamic (eddy) radial diffusion in the interparticle mobile phase (A), longitudinal diffusion of the solute (B), and mass transfer within the porous bonded-phase particles (C).

Stout and co-workers (6) have postulated a physical model describing distinct regions within a typical pore in which diffusion may occur at rates different from the rate found in the mobile phase. If we further postulate a double layer on the surface within a pore, there are three types of diffusion possible as illustrated in Figure 3. D_p' is the diffusion coefficient in the bulk of the intrapore liquid, D_p'' is the diffusion coefficient along the pore surface when there is little steric hindrance, and D_p''' is a restricted diffusion coefficient attributable to a steric impediment to free movement of a solute molecule at the junction of two microspheres that comprise part of the structure of the porous particle. The extent of reduction in diffusion rates caused by this effect is measured by ϱ — a quantity less than 1 — also known as the restriction factor (6); when $\varrho = 1$ there is no restriction to diffusion. This factor will influence C directly as given in the following equation:

$$C = [1/30\gamma(1-x)][(1+k'-x)/(1+k')]^2 D_m/D_p\varrho \qquad [9]$$

C is thus seen to vary with the tortuosity factor γ (nonsmooth surfaces), with x (fraction of mobile phase outside the particle), with k', the solute capacity factor, with D_m, the diffusion coefficient in the extraparticle mobile phase, and with $D_p\varrho$, the effective diffusion coefficient for all parts of the pore. Values of ϱ have been calculated by Stout and co-workers for small phthalate esters and anilides in packings of various particle diameters and pore sizes (6). The results are given in Table II. The carbon chain length in the bonded phase also influences the value of ϱ. Table III gives selected ϱ values obtained by Snyder as they relate to solute size, mobile phase composition, and pore size.

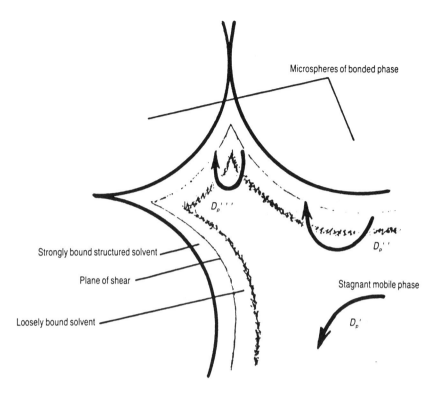

FIGURE 3.   Diffusion of solute molecules within the pore structure of a particle. (Adapted, with permission, from Reference 6.)

Thus, the following general trends are observed in this limited series of parameters:
• values of ϱ increase with increasing pore size for both bare silica and bonded phases
• the larger the solute molecules, the smaller ϱ becomes

TABLE II

Values of the Restriction Factor ϱ for Various Columns*

Column	Pore diameter (nm)	ϱ (Silica)	ϱ (C$_8$)	ϱ (C$_{18}$)
Zorbax (3 μm)	6	0.58	0.47	0.21
Zorbax (5–7 μm)	7.5	0.75	0.45	–
Zorbax (4 μm)	17	–	0.87	0.35
Perkin-Elmer (3 μm)	9	–	–	0.27
Supelco (5 μm)	11	–	0.42	0.37

* Data taken from Reference 6.

TABLE III

Values of ϱ for Various Chain Lengths,
Pore Sizes, Solutes, and Mobile Phases*

Chain length	Pore size	Solute	Mobile phase	ϱ	Particle diameter (μm)
Zero	6 nm	Benzanilide	CH_2Cl_2	0.71	3
	7.5 nm	Benzanilide	CH_2Cl_2	0.91	5.7
C_8	6 nm	Pentyl phthalate	CH_3CN/H_2O (85:15)	0.56	3
	7.5 nm	Pentyl phthalate	CH_3CN/H_2O (85:15)	0.54	5.7
	17 nm	Pentyl phthalate	CH_3CN/H_2O (85:15)	1.01	4.0
C_{18}	17 nm	Butyl phthalate	CH_3CN/H_2O (85:15)	1.05	4.0
	17 nm	Pentyl phthalate	CH_3CN/H_2O (85:15)	1.01	4.0
	17 nm	Octyl phthalate	CH_3CN/H_2O (85:15)	0.5	4.0

* Data from Reference 6.

- the effect is observed in both reversed- and normal-phase systems
- longer chain length results in reduced ϱ values
- bare silica apparently exhibits the best ϱ value.

Stout and co-workers concluded that wide-pore particles and shorter-chain bonded phases can provide greater efficiency, particularly at higher flow rates and with faster separations. This work was based upon the use of well-characterized spherical supports consisting of fused-silica microspheres (Zorbax). Data presented using other columns is not sufficiently detailed to determine whether the phenomenon of the restriction factor is consistent with the proposed model for all bonded phases.

Hearn and Grego have raised other issues that begin to indicate the difficulty of attempting to relate pore-size effects using too simple a model (10). For example, data these authors have compiled indicate that a very complex relationship exists between h, k', and pore size. These authors suggest that pore-size effects may be less important than better selectivity of wide-pore silicas in some cases. Resolution differences occur between silicas that are of the same pore size but that have been synthesized or prepared for bonding in a different fashion (26). Results from the author's laboratories, however, strongly implicate pore size in efficiency and resolution even when selectivity may be worse. Substantial decreases in plate height have been found for the isocratic separation of β-lactoglobulins A and B using a wide-pore weak anion exchanger relative to a narrow-pore (11-nm diameter) exchanger with the same bonded functional group (Figure 4).

It seems that the elucidation of genuine pore-size effects remains to be determined by the use of a suitable model in the first instance, by the choice of appropri-

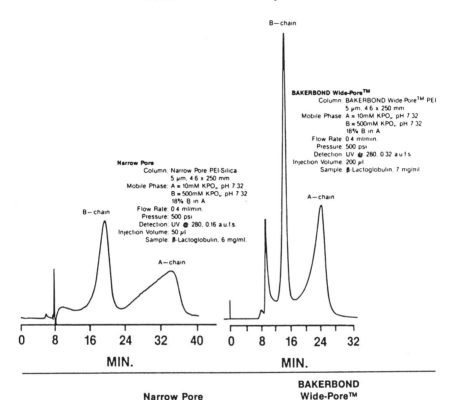

FIGURE 4.   Isocratic separation of A and B forms of β-lactoglobulin. Narrow-pore: 11 nm; wide-pore: 30 nm.

	Narrow Pore Efficiency*	A_f	BAKERBOND Wide-Pore™ Efficiency*	A_f
β-Lactoglobulin B	1300	0.83	4100	0.95
β-Lactoglobulin A	529	0.25	1385	0.35

*Calculated at flow rate 0.4 ml/min., 18% solvent B, isocratic, N/M.

ate silicas whose macro- and microscopic natures are similar for a range of pore sizes, and by sufficient separation of all physicochemical effects that contribute to selectivity and band broadening. Stout and associates (6) have done this, although a great deal more data need to be collected to validate their approach in a more general fashion.

SEPARATION OF SMALL AND MEDIUM-SIZE MOLECULES ON LARGE-PORE SUPPORTS

The applications of wide-pore bonded phases in the analysis of low molecular weight pharmaceuticals described below are provided without detailed explanations of the reason the pore size was considered important. In general, most analysts use large-pore packings simply because of the improvements in resolution and recovery they offer in certain instances. The actual explanation for the observed chromatography is rarely offered and in any case would require a fundamental study of the type carried out by Stout and co-workers (6). Only in a few instances, some of which were described in the previous section, are any sound arguments proposed to explain pore diameter–related chromatographic effects.

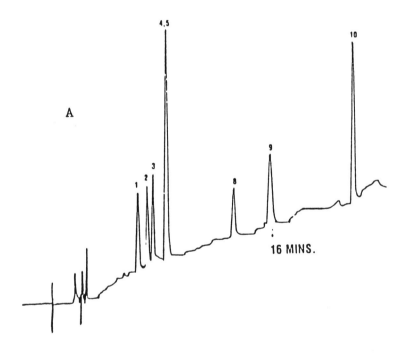

FIGURE 5. Separation of low/medium molecular weight peptides on small-pore and wide-pore bonded phases. Column: (a) Vydac octadecyl (C_{18}), 5 μm, 8-nm pore; (b) Vydac octadecyl (C_{18}), 5 μm, 30-nm pore; (c) Vydac butyl (C_4), 5 μm, 30-nm pore. Mobile phase: A = 0.1% TFA; B = 70:30 0.1% TFA in acetonitrile/water. Gradient: 5% B to 100% B over 30 min; flow rate: 1.5 mL/min, detection at 220 nm. Peaks: 1 = oxytocin, 2 = bradykinin, 3 = angiotensin II, 4 = neurotensin, 4 = eledosin, 6 = impurity from glucagon, 7 = impurity from glucagon, 8 = glucagon, 9 = corticotropin releasing factor (CRF), 10 = gastric related peptide. (Reprinted, with permission, from Reference 27.)

Biologically Active Peptides

A group of relatively small peptides has been chromatographed (27) on three bonded phases made from Vydac silica (The Separations Group, Hesperia, California) (Figure 5). Two pore sizes (8 nm and 30 nm) were chosen, and two octadecyl bonded phases were compared. Peak shapes (sharpness and symmetry) are clearly superior on the larger-pore C_{18} material. Elution order was identical on all three packings, but substantial selectivity differences were observed on the different packings.

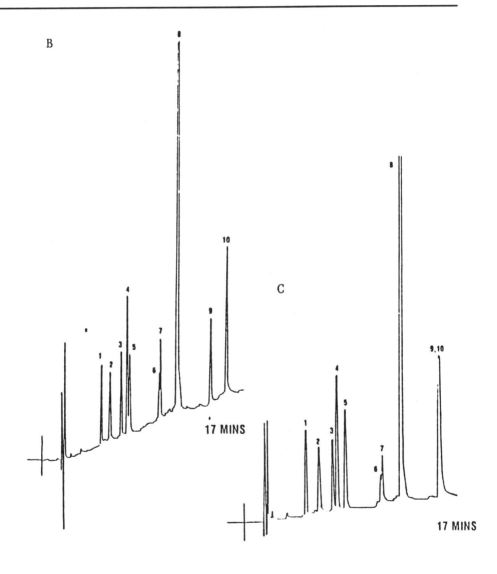

Small peptides are most commonly analyzed and purified using small-pore bonded phases of less than 15-nm diameter (28,29). Rivier and co-workers (30–32), however, found that substantial improvements in peak symmetry and recovery could be obtained with larger-pore (30 nm) supports in the chromatography of small and large synthetic peptides. Furthermore, the manipulation of the bonded-phase functional group (alkyl or aryl) may often introduce significant changes in selectivity.

Insulins

Reversed-phase HPLC has been used extensively in the analysis of insulins (33–37). The similarity of the hydrophobic properties of insulins from animal species (bovine, porcine, or human) prevents a simple resolution of the various forms from each other. The majority of work in this area involves the use of small-pore supports, very shallow gradients or isocratic mobile phases, moderate ionic strengths, and low pH values. Rivier (34), however, has obtained resolution of bovine, human, and porcine insulins on a wide-pore (30-nm) butyl column using low and moderate ionic strength mobile phases containing trifluoroacetic acid and triethylamine phosphate (Figure 6). A wide variety of buffer systems have been used in this analysis, so a comparison of the effect of pore size is difficult. There appears to be no advantage, however, in using a wide-pore nonpolar bonded phase in this application.

Vitamins

Metabolites of the fat-soluble vitamin A have been separated on a C_{18} wide-pore silica (27). This bonded phase is generally suitable for very hydrophobic small molecules such as the vitamins A and their metabolites.

The water-soluble vitamins have been analyzed isocratically using a Bakerbond Wide-Pore octyl-bonded 30-nm pore silica (J.T. Baker Chemical Co., Phillipsburg, New Jersey) and ion-pairing reagents (38). A small-pore (11-nm) Bakerbond octyl silica gave comparatively poor peak shape and selectivity under similar conditions (Figure 7). Equivalent resolution of these and other B vitamins on a small-pore support requires gradient elution (27).

A major improvement in selectivity in the baseline separation of vitamins D_2 and D_3 is obtained on a wide-pore octadecyl bonded packing (27). The equivalent small-pore material is unable to resolve these closely related vitamins.

Koskinen and Valtonen have illustrated a mediocre separation of vitamin D_3 and several of its more than 20 metabolites on a large-pore C_{18} bonded silica (39). The method of choice in this application is normal phase on 3-μm unbonded silica because the metabolites are quite polar.

Tricyclic Antidepressants

A standard mixture of eight tricyclic antidepressant pharmaceuticals has been

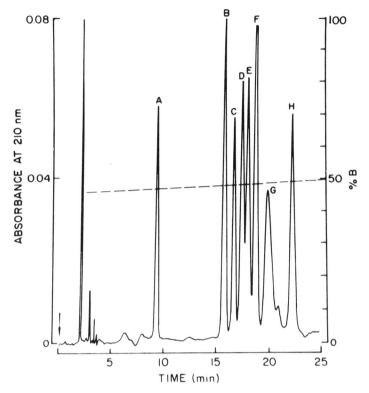

FIGURE 6.   Separation of insulins. Column: Vydac C₄; flow rate: 2 mL/min; pressure: 2900 psi; mobile phase A: 0.1% TFA in water; mobile phase B: 0.1% TFA in acetonitrile/water (60:40); gradient: from 45% B to 50% B in 25 min; detection: 210 nm, 0.08 AUFS. Sample size: 5 μg each insulin. Peaks: A = chicken insulin, 9.23 min; B = bovine insulin, 15.70 min; C = ovine insulin, 16.66 min; D = rabbit insulin, 17.43 min; E = human insulin, 18.00 min; F = porcine insulin, 18.76 min; G = rat I insulin, 19.96 min; H = rat II insulin, 22.36 min. (Reprinted, with permission, from Reference 34.)

resolved on an experimental 10-μm large-pore (30-nm) cyanopropyl bonded phase in the author's laboratories. In comparison to a Bakerbond 5-μm small-pore support bearing the same functional group, the order of elution is the same — except for protriptyline, which elutes last on the small-pore packing but second to last on the wide-pore packing (Figure 8). Resolution is relatively poor on the 10-μm bonded phase, mainly because of the larger particle diameter.

Resolution of Racemic Drugs

In a recent series of articles, Hermansson (40–43) has shown that chromatographic resolution of underivatized basic drugs is possible using a chiral bonded phase. The packing consists of a 10-μm, spherical, 30-nm pore silica to which the protein α₁-acid glycoprotein is covalently bound (35 mg/g silica). A simple 50-

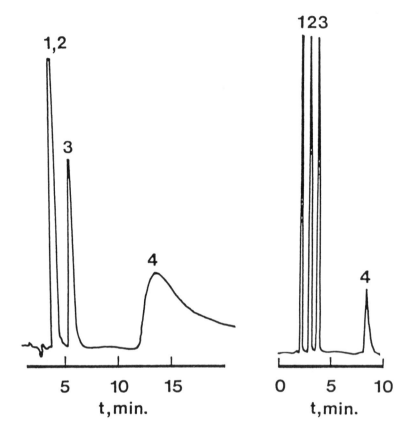

FIGURE 7.   Separation of water-soluble vitamins. (a) Column: Bakerbond Octyl (C₈), 250 mm × 4.6 mm, 5 μm, 11-nm pores; flow rate: 1 mL/min; (b) column: Bakerbond Wide-Pore Octyl (C₈), 250 mm × 4.6 mm, 5 μm, 30-nm pores; flow rate: 3 mL/min; mobile phase: 90:10 sodium heptanesulfonate (4 mM) in acetic acid/water (2:98)/methanol; pressure: 1400 psi; detection: 280 nm, 0.08 AUFS. Peaks: 1 = niacinamide, 2 = pyridoxine, 3 = riboflavin, 4 = thiamine.

mM phosphate buffer is used, with retention of a number of basic drugs increasing almost linearly with hydrogen ion activity. Table IV lists some of the drugs resolved and the selectivity factors (α) obtained at the noted pH.

Hermansson has postulated that the wide pore is necessary for access of the α_1-acid glycoprotein to the interior of the particle (10-μm Lichrospher, E. Merck, Darmstadt, West Germany). Only 2.5 mg/g of the protein could bind to a 10-nm pore silica, compared to 35 mg/g bound to the 30-nm pore silica. This high density of protein groups on the silica surface thus produces the retention (albeit quite weak) of the drugs on the column that is required for resolution.

TABLE IV

Selectivity Factors (α) for
the Resolution of Racemic Drugs
on Covalently Bound α_1-Acid Glycoprotein*

Drug	α	pH	Mobile phase organic solvent
Disopyramide	3.0	6.5	–
Desisopropyl disopyramide	1.3	6.5	–
Propiomazine	1.6	6.52	0.5% propanol
Promethazine	0	–	–
Bupivacaine	1.6	7.06	–
Mepivacaine	1.2	7.06	–
RAC 109	2.1	6.55	–
RAC 109	2.6	7.01	–
Oxyphencyclimine	1.6	7.6	–
Mepensolate bromide	1.2	7.6	–

* Data obtained from Reference 40.

In related work, Hermansson used a chiral mobile phase containing the same α_1-acid glycoprotein that binds to the drug molecule, this time in solution (44). The stationary phase was a diol-bonded packing with small pores (10 nm) chosen in order to prevent the protein from binding to the bonded silica. General retentions (k') and selectivity factors (α) were similar to those observed with the covalently bound α_1-acid glycoprotein bonded phase. An interesting exception was promethazine, which was unresolved on the protein column but well resolved (α up to 1.3) on the diol column with the protein in the mobile phase.

The 3-benzyl derivative of diazepam has been resolved in the author's laboratories (45) using an experimental 30-nm pore bonded phase of the Pirkle type (46). The acid 3,5-dinitrobenzoyl leucine was covalently coupled to an aminopropyl large-pore silica (5 μm). The selectivity factor (α) for this diazepam derivative was 5.04 (mobile phase 10% isopropanol in hexane). This value is significantly greater than the usual value (4.60) for the corresponding small-pore bonded phase (Bakerbond Chiral Phase). Data to explain this difference, if it is real, are not yet available.

HIGH MOLECULAR WEIGHT MOLECULES: MONOCLONAL ANTIBODIES

There is a general consensus that molecules of molecular weights greater than 15,000 are best chromatographed on bonded phases that have pore sizes of at least 30 nm (47–50). Hearn and Grego (10) believe the improvement is partially

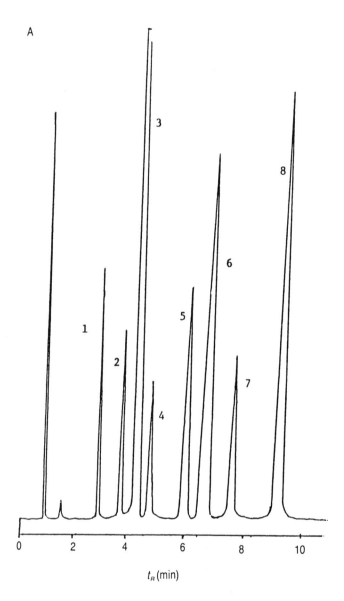

FIGURE 8.   Separation of tricyclic antidepressants. (a) Column: Bakerbond Cyanopropyl (5 μm, 11-nm pore); (b) column: Bakerbond Wide-Pore Cyanopropyl (10 μm, experimental phase, 30-nm pore). Mobile phase: 50:25:25 acetonitrile/methanol/5 mM KH₂PO₄, pH 7.2; flow rate: 2 mL/min. Detection: 254

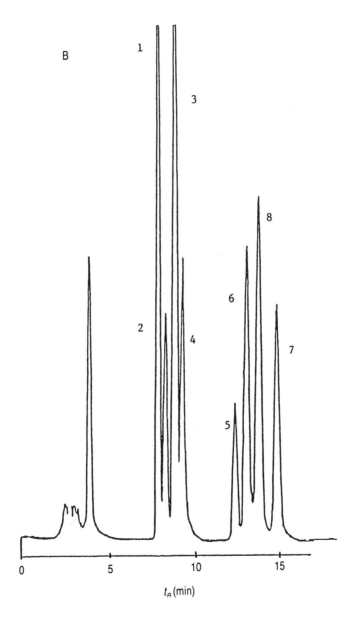

nm, 0.08 AUFS; injection volume: 50 μL. Peaks: 1 = trimipramine, 2 = doxepin, 3 = amitriptylene, 4 = imipramine, 5 = desmethyldoxepin, 6 = nortriptyline, 7 = desipramine, 8 = protriptyline.

achieved in many cases by better selectivity. Wilson and co-workers (9), however, have observed improvements in protein recovery, capacity, and peak sharpness for 30-nm pore supports — especially for proteins whose molecular weight is greater than 15,000.

Because the chromatography of proteins is discussed elsewhere in this book, this section will not review the topic. Nonetheless, note should be made of a significant research program under way in the author's laboratories directed toward the purification of monoclonal antibodies. Because these compounds will be used as pharmaceuticals to an increasing degree in the future (51), it is worthwhile to describe their analysis.

Monoclonal antibodies (immunoglobulins) are globular proteins with molecular weights varying from 140,000 Da (IgG) to 1,000,000 Da (IgM). All high performance bonded phases used to analyze and purify these substances have pore sizes of at least 30 nm (52–55). Peak sharpness is not only a function of stationary phase or mobile phase properties, but also depends on the homogeneity of the immunoglobulin sample (Figure 9). Monoclonal antibodies give much sharper peaks than do even highly purified polyclonal antibodies because of the polydispersity of the latter.

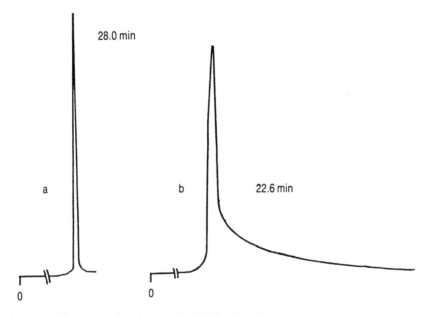

FIGURE 9.   Chromatography of a monoclonal IgG and a polyclonal IgG. Column: 250 mm × 4.6 mm, 5 μm, Bakerbond MAb-F; gradient: 10 mM KH_2PO_4, pH 6.0 to 130 mM KH_2PO_4, pH 6.8 over 60 min; flow rate: 1 mL/min; pressure: 1100 psi; detection: 280 nm, 0.64 AUFS. (a) Monoclonal IgG, from mouse ascites fluid; sample: 20 μL, 4 mg/mL (purified by J.T. Baker Chemical Company); (b) polyclonal human IgG; sample: 250 μL, 2 mg/mL (purified by Miles Laboratories).

The twofold aim of monoclonal analysis and purification is to separate the immunoglobulins from other proteins in ascites fluids or cell culture supernatants and to separate the monoclonal antibody from the host polyclonal antibodies. Previously, immunoglobulins have been separated using a weak anion exchanger (DEAE-type) or by affinity chromatography techniques using protein A. Unfortunately, as a result of the development of monoclonal antibodies, the diversity in chemical and physical properties of immunoglobulins has been clearly revealed. In the past this diversity has been masked by the polyclonal nature of the isolated immunoglobulins.

Monoclonal antibodies can vary greatly in their ability to bind to chromatographic material. No single chromatographic technique can be expected to be used successfully to purify them all; therefore, a multidimensional approach has been used in the author's laboratories to devise a chromatographic system that should be able to analyze and/or purify any monoclonal antibody. The approach uses the

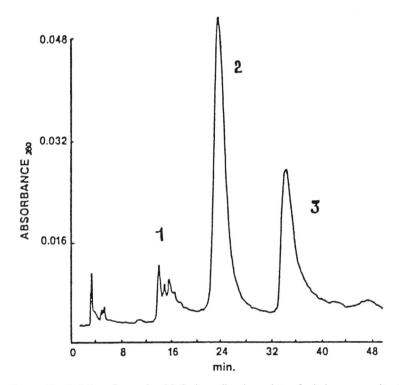

FIGURE 10.   Isolation of monoclonal IgG_1 from albumins and transferrin in mouse ascites fluid. Column: Bakerbond MAb, 250 mm × 4.6 mm, 5 μm; linear gradient from 10 mM KH_2PO_4, pH 6.80 to 130 mM KH_2PO_4, pH 6.4 over 1 h; flow rate: 1 mL/min; pressure: 900 psi; detection: 280 nm, 0.08 AUFS; injection volume: 15 μL of mouse ascites fluid diluted (1:4) with 10 mM KH_2PO_4, pH 6.8, to 75 μL. Peaks: 2 = IgG (greater than 95% purity), 1 = transferrin, 3 = albumin.

techniques of ion-exchange and hydrophobic-interaction chromatography with large-pore (30-nm) silica-based bonded phases. Figures 10, 11, and 12 illustrate three modes of chromatography used in this technology.

The ion exchanger used in Figure 10 (Bakerbond MAb) binds most proteins in mouse ascites fluid, including immunoglobulin, albumins, and transferrin. The monoclonal antibody fraction is eluted while the albumins are still adsorbed to the bonded phase. In Figure 11 an entirely different mode of chromatography is used to purify cell culture supernatant; the full mechanism of separation is unknown at this stage. The bonded phase in this case (Bakerbond Ab-X) allows all albumins and transferrin to pass through the column while monoclonal antibodies of all types and subclasses are bound. In the third mode (Figure 12), the purified antibody can be separated from the host antibodies by hydrophobic-interaction chromatography (Bakerbond Wide-Pore HIC).

All monoclonal antibodies tested so far in the author's laboratories — including polyclonal human immunoglobulin from plasma — can be purified using Bakerbond MAb and/or Bakerbond Ab-X or Bakerbond Wide-Pore HIC.

CONCLUSIONS

In the applications described for monoclonal antibodies, the bonded phases used had pore diameters of 30 nm or larger. This pore-size range has been well established as a requirement for effective chromatography of high molecular weight macromolecules (MW $\geqslant$ 15,000). The steric effect is clearly important in these examples.

For small molecules chromatographed on 30-nm supports, gains in resolution are obtained less often. Those compounds of pharmaceutical interest that give better selectivity and increased efficiency on such packings comprise one group of several that are now being analyzed (56,8,22). The advantage of 30-nm pore bonded phases in these cases is their better selectivity compared to that of small-pore materials ($<$ 15 nm). The reasons for this improved selectivity are generally unknown, and they may not ultimately be related to pore size at all. But as Hearn and Grego (10) and Pearson and co-workers (26) have suggested, in relation to large molecules, the better selectivity may depend on the physicochemical factors previously discussed.

In any case, in view of the unique selectivity shown by wide-pore bonded phases in the applications mentioned in this chapter, researchers should investigate other areas of use for these columns in small molecule separations. Work at the manufacturer's laboratories using Bakerbond Wide-Pore bonded phases for the analysis of vitamins (Figure 7), PTH-amino acids (56), and tricyclic antidepressants (Figure 8) has indicated that worthwhile increases in resolution are sometimes obtained on such large-pore columns.

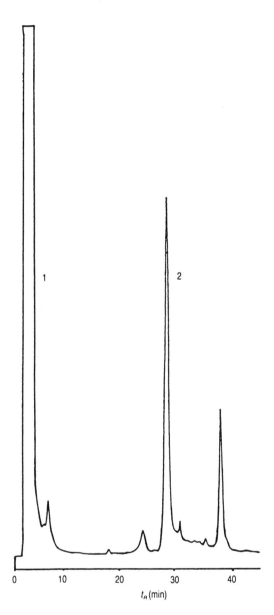

FIGURE 11.   The separation of IgG₁ from other proteins in mouse ascites fluid. Column: Bakerbond Ab-X, 250 mm × 4.6 mm, 5 μm; linear gradient from 10 mM KH₂PO₄, pH 6.0 to 125 mM KH₂PO₄, pH 6.8 over 1 h; flow rate: 1 mL/min; pressure: 1000 psi; detection: 280 nm, 1.28 AUFS; injection volume: 1 mL of diluted (1:3) mouse ascites fluid. Peak 1 = albumins and transferrin, peak 2 = IgG₁ (95% pure by SDS-PAGE).

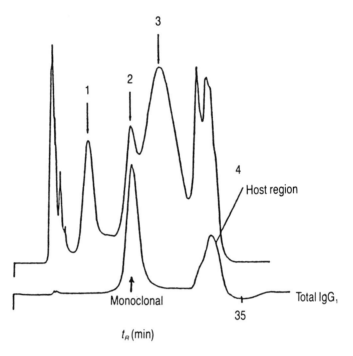

t_R (min)

FIGURE 12.   The separation of proteins in mouse ascites fluid by hydrophobic-interaction chromatography. Top chromatogram is that of ascites diluted with 1 M (NH$_4$)$_2$SO$_4$ (1:4). Lower chromatogram
is that of the total IgG$_1$ fraction purified from the ascites. Column: Bakerbond Wide-Pore HIC, 250 mm
× 4.6 mm, 5 μm; linear gradients from 1 M (NH$_4$)$_2$SO$_4$ + 25 mM KH$_2$PO$_4$, pH 7.0 to 25 mM KH$_2$PO$_4$
pH 7.0 over 45 min; flow rate: 1 mL/min; detection: 280 nm, 1.28 AUFS; injection volume: 0.5 mL.
Peaks: 1 = transferrin, 2 = monoclonal IgG$_1$, 3 = albumins, 4 = host IgG region.

Acknowledgments

    The author gratefully acknowledges discussions with Dr. Laura Crane of J.T. Baker's Research Lab-
oratories and Dr. Michael Flashner of Triton Biosciences in the preparation of this manuscript. The
useful comments and the elegant chromatography of Drs. Steven Berkowitz (Figure 12), David Nau
(Figures 9 and 11), and Michael Flashner (Figures 4 and 10) and Mr. George Wachob (Figure 7) in
their work at J.T. Baker is acknowledged. The assistance of JoAnne Volkert in manuscript preparation
is also acknowledged.

REFERENCES

(1) J. Porath and P. Flodin, *Nature* **183**, 1657 (1959).

(2) J.C. Moore, *J. Polym. Sci.*, Part A, **2**, 835 (1964).

(3) W.W. Yau, J.J. Kirkland, and D.D. Bly, *Modern Size Exclusion Liquid Chromatography* (John
    Wiley & Sons, New York, 1979).

(4) L.R. Snyder and J.J. Kirkland, *Introduction to Modern Liquid Chromatography* (John Wiley & Sons, New York, 1979), pp. 518, 529.

(5) A.P. Graffeo, paper presented at Association of Official Analytical Chemists Meeting, Washington, DC, October, 1977.

(6) R.W. Stout, J.J. DeStefano, and L.R. Snyder, *J. Chromatogr.* **282**, 263–286 (1983).

(7) M.T.W. Hearn and B. Grego, *J. Chromatogr.* **282**, 541–560 (1983).

(8) C. DeJong, G.J. Hughes, E. Van Wieringen, and K.J. Wilson, *J. Chromatogr.* **242**, 345–359 (1982).

(9) K.J. Wilson, E. Van Wieringen, S. Klauser, M.W. Berchtold, and G. J. Hughes, *J. Chromatogr.* **237**, 407–416 (1982).

(10) M.T.W. Hearn and B. Grego, *J. Chromatogr.* **296**, 61–82 (1984).

(11) M.T.W. Hearn, *Meth. Enzymol.* **104**, 190 (1984).

(12) K.K. Unger, *Porous Silica,* Journal of Chromatography Library, Volume 16 (Elsevier, Amsterdam, 1979), pp. 15–42.

(13) S. Brunauer, R.Sh. Mikhail, and E.E. Bodor, *J. Colloid Interface Sci.* **24**, 451 (1967).

(14) M.E. Van Kreveld and N. Van Den Hoed, *J. Chromatogr.* **83**, 11 (1973).

(15) M.J. Cope, paper presented at Ninth International Symposium on Column Liquid Chromatography, Edinburgh, Scotland, July 1–5, 1985.

(16) Y. Kato, T. Kitamura, and T. Hashimoto, paper 105 presented at The Fourth International Symposium on HPLC of Proteins, Peptides and Polynucleotides, Baltimore, Maryland, December 10–12, 1984.

(17) D. Barby, in *Characterization of Powder Surfaces,* G.D. Parkitt and K.S.W. Sing, eds. (Academic Press, New York, 1976), p. 353.

(18) R.K. Iler, *The Chemistry of Silica* (John Wiley & Sons, New York, 1979), pp. 478–492.

(19) J. Thompson (Lord Kelvin), *Philos. Mag.* **47**, 448 (1871).

(20) M.T.W. Hearn and B. Grego, submitted for publication.

(21) L.R. Snyder, M.A. Stadalius, and M.A. Quarry, *Anal. Chem.* **55**, 1413A–1430A (1983).

(22) L.C. Sander and S.A. Wise, *J. Chromatogr.* **316**, 163–181 (1984).

(23) P.P. Wickramanayake and W.A. Aue, *J. Chromatogr.* **195**, 25–34 (1980).

(24) K. Karch, I. Sebastian, and I. Halasz, *J. Chromatogr.* **122**, 3–16 (1976).

(25) G.J. Kennedy and J.H. Knox, *J. Chromatogr. Sci.* **10**, 549 (1972).

(26) J.D. Pearson, N.T. Lin, and F.E. Regnier, *Anal. Biochem.* **124**, 217–230 (1982).

(27) K.H. Harrison, V.I. Millar, and T.L. Yates, "Comprehensive Guide to Reversed-Phase Materials for HPLC," available from The Separations Group, Hesperia, California.

(28) M.T.W. Hearn, in *HPLC — Advances and Perspectives,* Volume 3, Cs. Horváth, ed. (Academic Press, New York, 1983), pp. 87–155.

(29) H.P.J. Bennett, H.M. Hudson, C. McMartin, and G.E. Purdon, *Biochem. J.* **168**, 9 (1977).

(30) J. Rivier, R. McClintock, R. Galyean, and H. Anderson, *J. Chromatogr.* **288**, 303–328 (1984).

(31) J. Rivier, C. Rivier, J. Spiess, and W. Vale, *Anal. Biochem.* **127**, 258 (1983).

(32) J. Rivier, R. McClintock, R. Eksteen, and B.L. Karger, in *Proceedings of the 1982 FDA-USP Workshop on Drug and Reference Standards for Insulins, Somatropins and Thyroid-axis Drugs* (United States Pharmacopeial Convention, Inc., Rockville, Maryland, 1982), pp. 554–564.

(33) D.J. Smith, R.M. Venable, and J. Collins, *J. Chromatogr. Sci.* **23**, 81–88 (1985).

(34) J. Rivier and R. McClintock, *J. Chromatogr.* **268**, 112–119 (1983).

(35) H.W. Smith, Jr., L.M. Atkins, D.A. Binkley, W.G. Richardson, and D.J. Miner, *J. Liq. Chromatogr.* **8**, 419–439 (1985).

(36) K. Hayakawa and H. Tanaka, *J. Chromatogr.* **312**, 476–481 (1984).

(37) P.S.L. Janssen, J.W. Van Nispen, R.L.A.E. Hamelinck, P.A.T.A. Melgers, and B.C. Goverde, *J. Chromatogr. Sci.* **22**, 234–238 (1984).

(38) M.P. Henry and M. Flashner, in *Industry and Technology,* Summer 1983, pp. 84–87.

(39) T. Koskinen and P. Valtonen, *J. Liq. Chromatogr.* **8**, 463–472 (1985).

(40) J. Hermansson, *J. Chromatogr.* **269**, 71–80 (1983).

(41) J. Hermansson, *J. Chromatogr.* **298**, 67 (1984).

(42) J. Hermansson, M. Eriksson, and O. Nyquist, *J. Chromatogr.* **336**, 321 (1984).

(43) J. Hermansson and C. Van Bahr, *J. Chromatogr.* **221**, 109 (1980).

(44) J. Hermansson, *J. Chromatogr.* **316**, 585–604 (1984).

(45) M. Zief and M.P. Henry, unpublished results, J.T. Baker Chemical Co., Phillipsburg, New Jersey, 1983.

(46) W.H. Pirkle, J.M. Finn, B.C. Hamper, J.L. Schreiner, and J.R. Pribish, in *Amer. Chem. Soc. Symposium Series,* No. 185, E. Eliel and S. Otsuka, eds. (1982), Chapter 18, p. 245.

(47) J.D. Pearson, W.C. Mahoney, M.A. Hermodsen, and F.E. Regnier, *J. Chromatogr.* **207**, 325–332 (1981).

(48) R.V. Lewis, A. Fallon, S. Stein, K.D. Gibson, and S. Udenfriend, *Anal. Biochem.* **104**, 153–159 (1980).

(49) E.C. Nice, M.W. Capp, N. Cooke, and M.J. O'Hare, *J. Chromatogr.* **218**, 569–580 (1981).

(50) R.J. Ferris, C.A. Cowgill, and R.R. Traut, *Biochemistry* **23**, 3434–3442 (1984).

(51) For example, R.H. Kennett, Z.L. Jonak, and K.B. Bechtol, in *Monoclonal Antibodies,* R.H. Kennett, T.J. McKearn, and K.B. Bechtol, eds. (Plenum Press, New York, 1980), pp. 155–168.

(52) M. Flashner and M.P. Henry, paper 811 presented at The Fourth International Symposium on HPLC of Proteins, Peptides and Polynucleotides, Baltimore, Maryland, December 10–12, 1984 (silica-based, 30-nm pore).

(53) T.L. Brooks, L.H. Stanker, R. Van Wedel, G.S. Ott, J.-C. Chen, R.P. Walker, and H. Juarez-Salinas, paper 810 presented at The Fourth International Symposium on HPLC of Proteins, Peptides and Polynucleotides, Baltimore, Maryland, December 10–12, 1984 (hydroxylapatite-based, large pore).

(54) P. Clezardin, J.L. McGrego, M. Manach, H. Baukerche, and M. Dechavanns, paper 203 presented at The Fourth International Symposium on HPLC of Proteins, Peptides and Polynucleotides, Baltimore, Maryland, December 10–12, 1984 (hydrophilic-resin based, 100-nm pore).

(55) M. Jude Gemski, M.P. Strickler, M.K. Gentry, and B.P. Doctor, paper 234 presented at The Fourth International Symposium on HPLC of Proteins, Peptides and Polynucleotides, Baltimore, Maryland, December 10–12, 1984 (hydroxylated polyether-based, 100-nm pore).

(56) W. Kruggel and R.V. Lewis, *J. Chromatogr.,* in press.

RESOLUTION OF OPTICAL ISOMERS BY LIQUID CHROMATOGRAPHIC TECHNIQUES

Wolfgang Lindner

Institute of Pharmaceutical Chemistry
University of Graz, Austria

Curt Pettersson

Department of Analytical Pharmaceutical Chemistry
Biomedical Center
University of Uppsala, Sweden

Within the last several years, gas chromatography (GC) and especially liquid chromatography (LC) have become increasingly important tools for the resolution of optically active compounds (enantiomers) on an analytical and semipreparative scale, as well as for solution of various stereochemical problems (1). These selective and sensitive stereochemical analytical techniques have been applied to the understanding of stereochemistry and biological activity of drugs and other bioactive compounds (2), and there is also a growing body of knowledge concerning the enantioselective action of chiral drugs.

Biological systems regulate themselves predominantly by stereoselection, so it is not surprising that the stereoisomers of chiral drugs will act differently in the environment of a biological matrix. Consequently, a *racemic* drug (that is, a mixture of stereoisomers) must be seen as a mixture of two different chemical species, each with its own — and often different — pharmacological spectrum. This property is also true for other chiral compounds, such as pesticides. As a result, one antipode is often responsible for a desired pharmacological activity, and the other stereoisomer can be viewed as an unwanted by-product or impurity. In the description of the biological activity of drugs, unpredictable and unnecessary side effects often are attributable to this type of comedicamentation. This situation can even be more severe when racemic drugs are administered together with other drugs. Thus, it is hoped that, in the future, synthetic chiral drugs will be administered in their optically pure form rather than as racemates (3).

To reach that point within the next few years, a great deal of progress will be necessary in sterochemistry, organic synthesis, and analytical chemistry. Enantioselective high performance liquid chromatography (HPLC) will be an active, integrated stereochemical methodology in this field. Recent successes have indicated the potential of this approach. Interest in the separation of stereoisomers by chromatographic methods has been increasing and has been the subject of reviews by Allenmark (4), Audebert (5), Blaschke (6), Davankov (7,8), Lindner (9), Pirkle (10), and Tamegai (11).

BACKGROUND

The present chapter gives readers an overview of stereoselective liquid chromatographic separation techniques covering such topics as discrimination of enantiomers and diastereoisomers, methods development, and the advantages and disadvantages of the technique. It will also cover applications with an emphasis on bioactive compounds. In this context, *stereoisomers* are defined as structural isomers having identical constitution but different spatial arrangement of their atoms. Stereoisomers are further classified by two distinct and independent criteria, namely, symmetry and energy criteria. *Enantiomers* and *diastereoisomers* are stereoisomers classified by symmetry criteria. Enantiomers are asymmetric and chiral and related to each other as an object and a nonsuperimposable mirror image. They are optically active and are also termed *optical isomers.* To discriminate or characterize enantiomers experimentally, the use of a *chiral handle* (asymmetric handles such as chiral solvents, reagents, enzymes, and cavities) is necessary. Diastereoisomers are stereoisomers that do not display an enantiomeric relationship. They differ in energy content and in physical and chemical properties; they need no external references for their discrimination. Diastereomeric molecules differ intermolecularly, which means that some atoms or groups of atoms have different spatial relationships. These definitions are excerpted from Testa's textbook on organic stereochemistry (12), to which readers are referred for more specific information. Based on the fundamental differences between enantiomers and diastereoisomers it becomes obvious that one must have a chiral handle, source, or environment for the chromatographic discrimination of enantiomers, but such is not necessary for the separation of diastereoisomers.

The chromatographic resolution of optical isomers deals mainly with analytical aspects, such as the determination of the ratio of the optical antipodes of enantiomeric mixtures. This use has a great effect on the field of asymmetric and stereoselective synthesis — including biotechnological processes — in the measurement of the *enantiomeric excess* (EE) or the *optical purity* (OP) of the resultant products. Referring to Testa (12), a medium containing an unequal proportion of the R and S enantiomer $(R > S)$ shows optical rotation, which means that the plane of polarized light is rotated by an angle α. For a given solute, α depends on wavelength λ, type of solvent, temperature T, and concentration c of the solute in the medium. Considering this relationship, the values of optical rotation are expressed as *specific rotation* $[\alpha]$ for a particular mixture of enantiomers at a given set of conditions Thus,

$$[\alpha]_\lambda^T = \frac{100\,\alpha_\lambda^T}{Lc} \qquad [1]$$

where
   L = optical path length (dm)
   c = concentration (g/100 mL of solvent).
The OP of a mixture of enantiomers is defined by the relation

$$\%\ OP = \frac{[\alpha]}{[\alpha]_{max}} \cdot 100 = \%\ EE = \frac{(R-S)}{(R+S)} \cdot 100 = \%\ EP \qquad [2]$$

which is equivalent to % EE and % *enantiomeric purity* EP. When R and S are concentrations of the two enantiomers $(R>S)$, $[\alpha]$ is the specific rotation of the enantiomeric mixture and $[\alpha]_{max}$ of one pure enantiomer. The relation of Equation 2 clearly demonstrates the great advantage of a quantitative separation technique for the enantiomers — and thus of determining the optical purity rather than relying only on measurements of specific rotation as in Equation 1. This means that one must ascertain $[\alpha]_{max}$ values as well as the chemical purity of the enantiomeric mixture in order to derive the correct value for c. Specifying the optical purity of a compound only by $[\alpha]$ values is still common, but can easily be misleading due to incorrect values of $[\alpha]_{max}$ (1).

Using optical rotation as a detection principle in HPLC, however, could be quite helpful, especially if one could use such detectors for on-line chiroptical spectroscopic measurements (ORD and CD spectra), resulting in configurational or conformational information concerning the chromatographically resolved chiral compounds. Initial results on this technique were published recently (13).

Chiral resolutions are also applied to the biological matrices of plants, animals, humans, and other organisms fed or contaminated with chiral or prochiral compounds (drugs) because biological systems most often react stereospecifically to these chemicals. The extent of this stereoselectivity is of high interest in all fields of science. In addition to the analytical aspects, and especially in LC, preparative enantioseparations are also of importance, particularly in the drug industry's research laboratories. Preparative chiral LC can be used to purify new chiral compounds useful in themselves or as a chiral source for other reactions. To some extent one can also obtain information about the absolute configuration of the resolved enantiomers from their elution order as determined on well-defined chiral columns.

There is a fundamental difference between enantiomers and diastereomers that must be considered here. Enantiomers have the same physical and chemical properties except in their different interactions with other optically active media (environment); this factor is indispensable in the discrimination (resolution) of enantiomers in chromatographic systems. Diastereoisomers, by contrast, have unequal physicochemical properties. Moreover, chiral media are not necessary to separate this type of stereoisomers, and conventional nonchiral mobile and stationary phases have proved sufficient.

As a result, the resolution of a mixture of enantiomers by liquid chromatography or other separation techniques can be performed in two modes: the indirect method and the direct method. In order to discriminate enantiomers, they must be forced to interact with another optically active (chiral) component, thus forming molecular complexes that behave diastereomerically. The *indirect technique* is based on the formation of covalent bonds between the enantiomeric solutes (the *selectand*, Sa) and a chiral *selector* (So) resulting in diastereoisomers. In this case, So is a *chiral derivatizing agent* (CDR) of high optical purity. This interaction may be expressed as

$$(R\text{-Sa} + S\text{-Sa}) + R\text{-So} \longrightarrow R\text{-Sa-}R\text{-So} + S\text{-Sa-}R\text{-So} \qquad [3]$$
$$\text{enantiomers} \qquad \text{CDR} \qquad\qquad \text{two diastereoisomers}$$

Here, R and S indicate the conformation of the chiral centers in the Sa and So molecules. This method is in principle restricted to solutes (Sa) that possess functional group(s) accessible for quantitative reaction with the CDR. (Further structural requirements of Sa and So, methodological limitations, and possible advantages will be discussed in more detail in the next section.) The *direct separation* of enantiomers is also based on the formation of diastereomeric molecular associates of Sa and So — but via rapid and reversible diastereomeric interactions rather than the irreversible interactions of the indirect mode. In this technique, the chiral selector So could be seen as a chiral source or chiral environment capable of interacting with the selectands, thus forming spatially fixed Sa/So associates that are diastereomeric. The chiral source can be a chiral resolving agent covalently fixed on the stationary phase, a chiral stationary phase (CSP), a chirally imprinted polymeric stationary phase containing so-called *chiral cavities,* or a chiral additive to the mobile phase (see below). The chiral environment, however, must have a certain stereospecific affinity to the enantiomeric selectands (Sa). On the basis of simple molecular complexes of Sa and So molecules, it must be recognized that the chiral recognition of So and Sa can only be possible when at least three active positions (sites) of So and Sa simultaneously interact with each other; at least one of these interactions must be stereoselective.

The three-point interaction model of Figure 1, proposed by Dalgliesh (14), has been used to guide the synthesis of new chiral phases and also to interpret observed enantioselective retention (7–10). Referring to the figure, one can see that a simultaneous three-point interaction between the chiral stationary phase (So) and at

Enantiomeric solutes

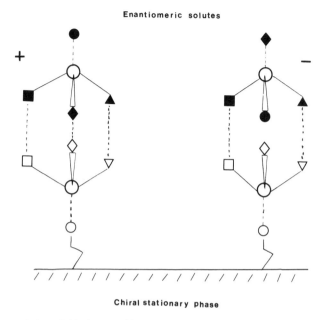

Chiral stationary phase

FIGURE 1.   Schematic binding model between two chiral molecules according to the three-point interaction rule.

least one of the enantiomers (Sa) is necessary in order to distinguish one enantiomer from the other. As a matter of geometric principle, two interactions only are not sufficient to differentiate between two optical isomers. The interactions between the selector and the selectand may be any attractive or repulsive force (hydrogen bonding, dipole-dipole, π-π, electrostatic, hydrophobic, or steric interactions). If one interaction — for instance, a π-π interaction between two aromatic ring systems — defines an axis instead of a point, one additional interaction may be enough. The minimum requirement is to define three points in the three-dimensional molecular arrangement in order to permit discrimination between the two enantiomers. The highest stereoselectivity is, as a rule, obtained when the three interaction groups, of which at least one should be stereochemically dependent, are close to the asymmetrical center. When the groups are situated far from the chiral center, the stereoselectivity generally decreases, probably because of a smaller difference in the average conformations of the diastereoisomeric complexes seen as the distance between interaction groups increases. The rotation around bonds in the diastereomeric complexes may give the enantiomers similar opportunity to adapt a conformation that can interact with the chiral complexing (discriminating) agent. This model of simple Sa/So molecule complexes may become more complicated if one also considers the chromatographic set of conditions in addition to achiral components coming from the mobile and the stationary phase. These normally tend to participate in the formation of the total diaste-reomeric molecule complex via solvation phenomena or nonchiral adsorption caused by the properties of the stationary phase containing the chiral selector groups either covalently or adsorptively bonded. Observations concerning these reflections have recently been published by Pirkle (10,14) and Davankov (15).

The main problem in direct separation of enantiomers is to find a chiral selector that will give a maximum difference in interaction with the two enantiomeric forms. Because the enantiomers differ in three-dimensional arrangements, there will be some difficulty in finding a chiral selector that can give high stereoselectiv-ity and at the same time also have the properties of a general selector that can sepa-rate enantiomers of different classes of compounds. It is sometimes possible to find chiral selectors that are stereoselective for a certain class of compounds (such as amino acids), but even minor changes in the structure of So and/or Sa may pro-duce a complete loss of stereoselectivity.

The same problem of finding a suitable selector — covalently fixed and forming a chiral stationary phase (CSP) — is found in a second direct approach to separa-tion of enantiomers in liquid chromatography using a chiral eluent or mobile-phase additive. The technique is generally based on the formation of labile diaste-reomeric complexes between the enantiomers (Sa) and a chiral complexing agent (So) present in the mobile phase. Generally, chiral recognition with this technique also requires a three-point interaction mechanism between Sa and So. In excep-tional cases, however, a two-point interaction of the diastereomeric Sa/So mole-cule complex (which is by definition insufficient for chiral recognition of the two enantiomeric partners) should be sufficient to achieve enantioseparation in the mobile-phase additive mode — provided, that is, a third and stereochemically de-pendent attraction of the Sa/So molecule complex with the nonchiral stationary phase (reversed-phase) can be performed. This concept was recently discussed by Davankov (15).

The use of chiral eluents has certain advantages in that the technique is easy to operate and can be used with commercially available, highly efficient, nonchiral support material. There are, however, some limitations with the technique, espe-cially for preparative applications, because the mobile phase after separation will be contaminated by the chiral selector, which is usually present in excess. The need for an expensive chiral selector to promote the resolution may also restrict the use of the technique unless it is possible to recover the selector.

INDIRECT SEPARATION OF ENANTIOMERS IN
LIQUID CHROMATOGRAPHY

The indirect mode of separating enantiomers has been used for many years and is still commonly used on both analytical and preparative scales, even when the various limitations of this technique are considered. The preparative aspects are of special interest, because it is often much easier and quicker to set up a nonchiral conventional LC system to resolve a mixture of diastereoisomers in suitable quantities for organic synthesis or other purposes, rather than to find and evaluate a direct enantioselective LC technique, a process which is often quite time-consuming. Before discussing the usefulness of certain chiral reagents, however, one should consider the risks and limitations of the method.

The method is based on the simple chemical reaction of a mixture of enantiomers with an optically active reagent in an appropriate reaction medium. To simplify the notations, the enantiomeric mixture of R/Sa and S/Sa are referred to as R and S, where R and S indicate the concentration of the R and S conformers. The same holds for the chiral reagents S' and (R') [(R') indicates a possible chiral impurity in the chiral reagent].

$$(R+S) \; + \; [S'+(R')] \longrightarrow RS' \; + \; (RR') \; + \; SS' \; + \; (SR') \qquad [4]$$
$$\text{selectand} \qquad \text{reagent} \qquad\qquad \text{reaction products}$$

If the chiral reagent contains no (R'), one should observe two diastereoisomers RS' and SS', that can be resolved as peak A and peak B. If the reagent contains (R'), one will also observe just two peaks, although four reaction products were formed. This results from the fact that (RR') and SS' as well as RS' and (SR') are enantomeric pairs. Thus, only on a chiral LC system could four peaks theoretically be observed. The RS' peak is contaminated with (SR'), and, if the ratio of (R') in S' is high, one derives an incorrect R value. Reagent S' therefore must be optically pure, and the accuracy of the quantified peak ratio of R to S can never be better than the optical purity of the reagent S'. The quantitative determination is particularly subject to errors when one tries to quantify traces of S beside R with a reagent containing (R') in addition to S'.

This systematical error is based on the assumption of a quantitative reaction; one is also forced to consider kinetic resolution phenomena (1). To avoid the latter, the chiral reagent must be added in excess. But in this case another kinetic effect determined by the reagent composition $[S' + (R')]$ cannot be excluded because S' and (R') will have different reaction constants with R and S; consequently, it could lead again to error in the measured R/S ratio caused by the (R') impurity in S'. Third, the reaction conditions must be such that racemization of the chiral centers is negligible either in the solute or in the chiral reagent. A very critical point that must be considered is the fact that various functional groups tend to racemize differently.

When using the direct enantioseparation mode involving a nonchiral derivatization step to make the selectands more suitable to the stereoselective molecule complex binding model, the possible racemization of the enantiomeric Sa molecules must also be seriously considered. Theoretically, partial racemization could also occur during the chromatographic separation step, which is unlikely in the indirect mode but observable in certain cases in the direct mode; the phenomenon is termed *peak coalescence* (16).

As a consequence, when performing the indirect enantioseparation technique the actual optical purity of the chiral reagent must always be determined or known — for example, by reaction with optically pure selectand or with a selectand of known enantiomeric composition. If this is done routinely and with due consideration of the possible risk of erratic analytical results of the determined R/S ratio, the indirect mode of enantioseparation can still be a valuable tool. Trace analysis of one enantiomer besides large amounts of the other will, however, have serious limitations.

Many applications of enantioselective drug monitoring via the indirect mode have been performed because the enantiomers and diastereoisomers (respectively) often must be separated from many other matrix components. Such separation is easier on conventional LC systems rather than on chiral columns with restricted chromatographic conditions — especially of mobile phase composition.

Table I lists some chiral reagents cited in the literature in combination with the classes of substrates with which they were reacted. They are not evaluated in terms of their diastereoselectivity because this strongly depends on the structure of the selectands and the CDR (11). Table I could be expanded significantly if one also considered the many chiral reagents previously used in GC. The indirect GC technique is, however, less important today because direct enantioseparation in

TABLE I

Chiral Reagents for the Separation of Enantiomers via the Indirect Mode

Chiral reagent	Selectand	Diastereomeric derivatives	References
Chiral amines			
C_6H_5-*CH-NH$_2$	benoxaprofen (α-aryl- propionic acids)	amides	17–19
CH$_3$	acids, lactones	amides	20
	pyrethroids	amides	21
	ibuprofen	amide	22
	carprofen	amide	23
	aldoses	alditoles	24
O_2N-C_6H_5-*CH-NH$_2$	isoprenoid acids	amides	25
CH$_3$	citronellic acids	amides	26
$C_{10}H_7$-*CH-NH$_2$	carboxylic acids	amides	27
CH$_3$			
R-*CH-NH$_2$	carboxylic acids	amides	28
COOMe	amino acids	peptides	

TABLE I *(Continued)*

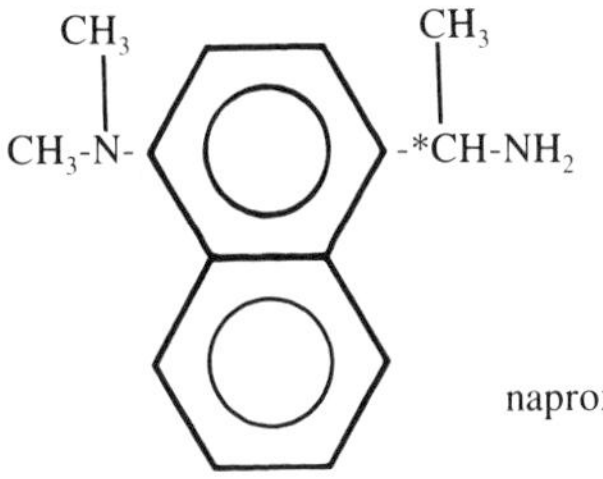	naproxen	amide	29
various optically active amines	chlorphenoxyphenyl-propionic acid	amide	11
Chiral acids (activated)			
C_6H_5-*CH(O-X)-COOH X=H	alcohols	esters	30
X=CH_3	amines	amides	31
$C_{10}H_7$-*C(CH_3)(OCH_3)-COOH	amino acid esters	amides	32
	amphetamine analogues	amides	33
C_6H_5-*C(CF_3)(OCH_3)-COOH (MTPA)	bromhydrines	esters	34
	oxfenicine	amide	35
	amine	amide	
	alcohol	ester	36
	terpenoid alcohols	ester	
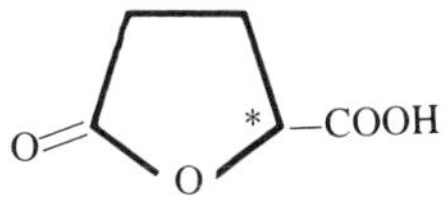 (TOF)	alcohol	ester	37
Chiral acids			
C_6H_5-O-CO-NH-*CH(R)-COOH	warfarin	ester	38
	amino acid ester	amides	39
(X)HN-*CH(R)-COOH (X) = BOC	propranolol	amide	40
	amino acids	peptides	41
	thyronine hormones	peptides	42
	penicillamine	peptide	43

(CONTINUED)

TABLE I *(Continued)*

CF$_3$-CO-N⎯⎯*⎯COOH

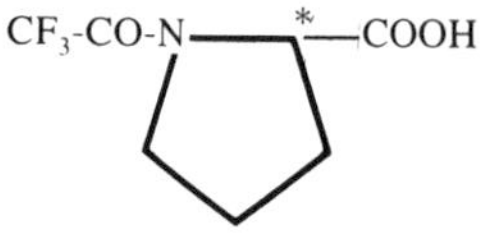	propranolol	amide	44,45

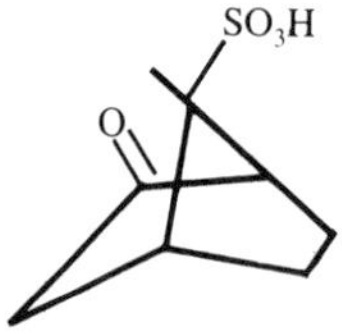

SO$_3$H

proxyphylline	camphanate	46
amines	camphanate	47
amino acid esters	camphanate	48

-*CH-COOH
CH$_3$

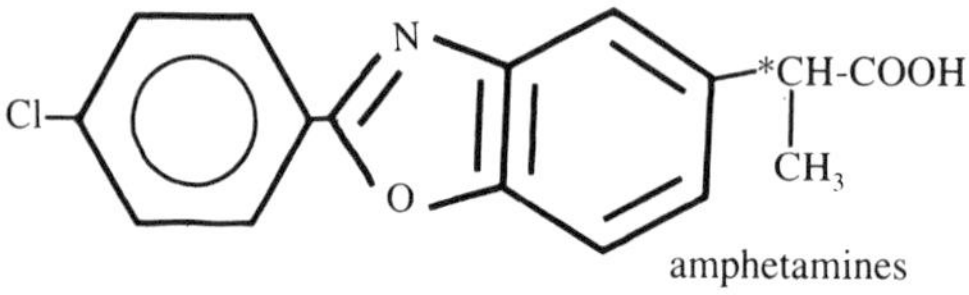

amphetamines	amides	17

CH$_3$
|
Ar-*CH-COOH

amines	amides	17

COOH

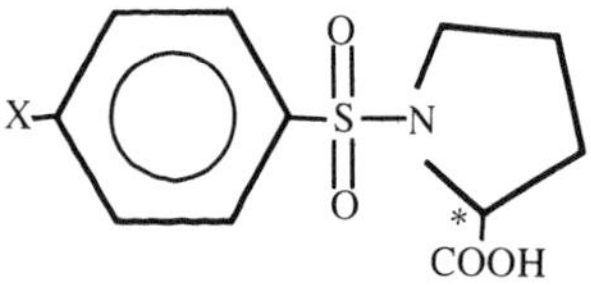

amine	amides	49

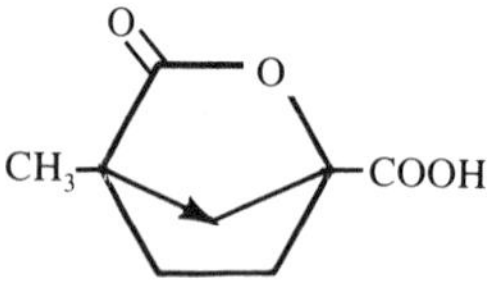

CH$_3$⎯ ⎯COOH

amines	amides	50

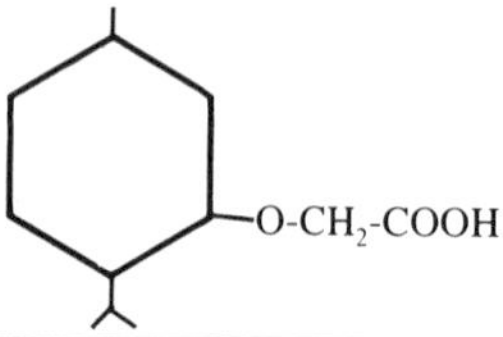

-O-CH$_2$-COOH

benzo[a]pyrene diols	esters	51
warfarin	ester	52

C$_6$H$_5$-*CH-COOH
|
C$_2$H$_5$

cyclophosphamide	ester	53

TABLE I *(Continued)*

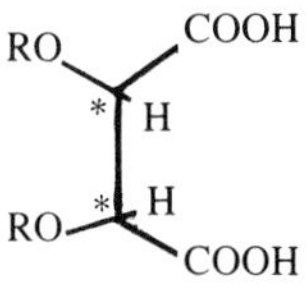	beta-blockers	esters	54

Isocyanates, thioisocyanates

	atropin	ureas	55
	amphetamines	ureas	56
	oxprenolol	ureas	57
	propranolol	ureas	58

	beta-blockers	ureas	59
	amines	ureas	10
	lactomes	ureas	10
	alcohols	carbamates	60

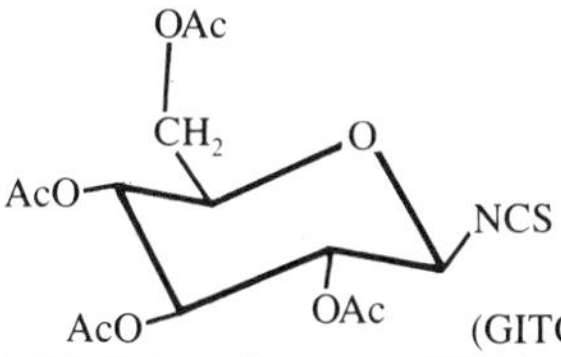	amphetamines	thioureas	56
	ephedrines	thioureas	61,62

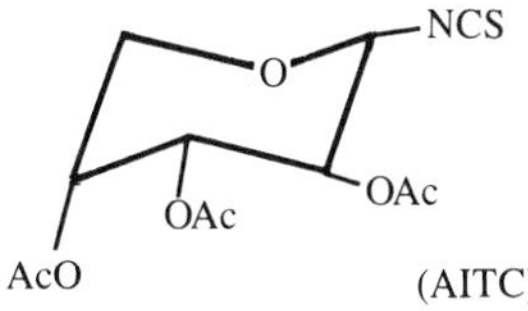	amino acids	thioureas	63–65
	beta-blockers		

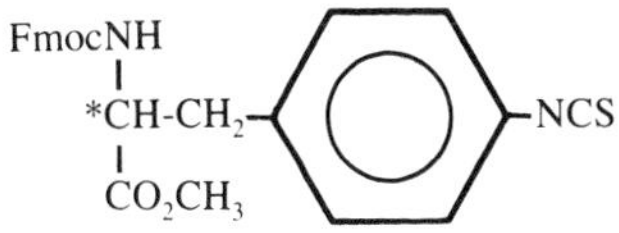 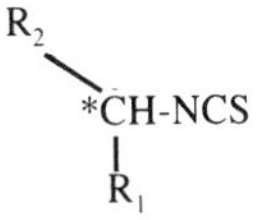	sequence analysis of amino acids	thiohydantoins	66,67

(CONTINUED)

TABLE I *(Continued)*

Miscellaneous

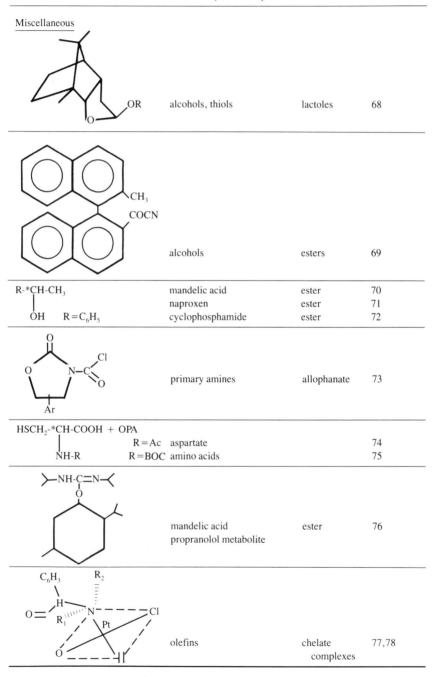

	alcohols, thiols	lactoles	68
	alcohols	esters	69
	mandelic acid	ester	70
	naproxen	ester	71
	cyclophosphamide	ester	72
	primary amines	allophanate	73
R=Ac	aspartate		74
R=BOC	amino acids		75
	mandelic acid	ester	76
	propranolol metabolite		
	olefins	chelate complexes	77,78

GC has become highly successful (16). Even enantioselective GC has its limitations, especially when it comes to enantioseparation of polar or large chiral compounds or when relatively large quantities of racemic mixtures must be resolved. In this case, preparative HPLC techniques seem to be the methods of choice.

When discussing *diastereoselectivity* as a term for the chromatographic resolvability of diastereomeric derivatives, the configurational and conformational aspects of the isomeric molecules also must be considered. As a general rule, to obtain good and sufficient separation factors of the peaks of the diastereoisomers, the optically active centers contributed by the solute (Sa) and the selector reagent (So) in the diastereomeric molecules should be close to each other. Rigid and relatively large substituents (for instance, phenyl or naphthyl groups) on both the Sa and So molecule will support this trend.

A primary goal of the indirect enantioseparation technique is to design diastereomeric molecules that significantly differ in their physicochemical properties and thus to achieve easy separation on nonchiral chromatography systems. All types of intramolecular steric interactions and binding forces (hydrophobic, hydrogen bonding, dipole, and electrostatic interactions) should be taken into consideration to fix certain molecular structures. It is not a necessity in this case as it is in the direct enantioseparation mode, but incorporating additional intramolecular binding increments via the derivatization step can certainly be advantageous. Its significance can be seen, for example, in the separation of racemic beta blockers via derivatization with tartaric acid derivatives, thus forming tartaric acid monoesters (54). Because of the free carboxylic and the amine function (Figure 2), this type of zwitterion tends to form an intramolecular ring structure depending on the pH of the mobile phase. The ring formation will be likely only if the zwitterionic form is dominant. Without the additional binding increment, the diastereoselectivy expressed as α values (relative retention) of the pair of diastereomeric derivatives would be much lower. As an example, α drops from 2.7 to 1.1 in Figure 3.

To support the above statement — a rigid and even bulky structure of the diastereoisomer is the main factor for good diastereoselectivity — recent reports made by Buck (75) should be cited. Figure 4 shows a separation of α-amino acids derivatized with chiral t-BOC-L-cysteine and nonchiral o-phthalaldehyde (OPA) reagent, resulting in a diastereomeric adduct in which the chiral centers are not close to each other. As a rule, this situation should lead to a very small resolution factor, but here the amino acid becomes part of a rigid heterocycle that can participate to some extent in an intramolecular ring formation with the carboxyl function of the chiral reagent. Thus, the α values become sufficient to resolve almost all common amino acids by this indirect enantioseparation technique. The authors claim that their derivatization procedure shows no racemization and that it is suitable for the quantification of amino acid enantiomers in peptide hydrolysates.

Derivatives of α-amino acids were commonly used as chiral reagents for enantioseparation of α-amino acids that form diastereomeric dipeptides (42). It was al-

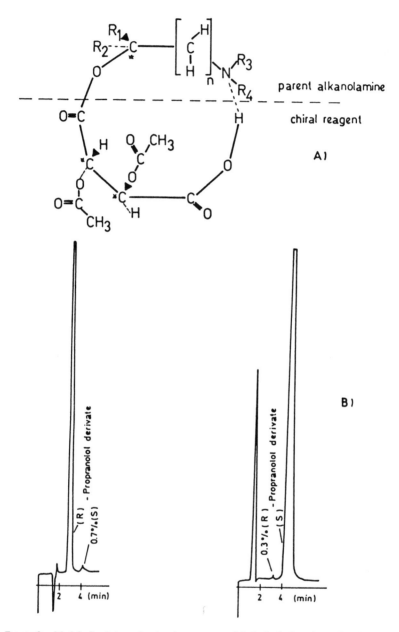

FIGURE 2.   Model of an intramolecular ring structure of O,O-disubstituted tartaric acid monoester of alkanolamines (a); HPLC separation of (R,S)-propranolol, derivatized with (R,R)-O,O-diacetyltartaric acid anhydride (b). Column: 250 mm × 4.6 mm, 5-μm Spherisorb ODS; mobile phase: 2% acetic acid in water (adjusted with ammonium to pH 3.7)/methanol (50:50). (Reprinted, with permission, from Reference 54.)

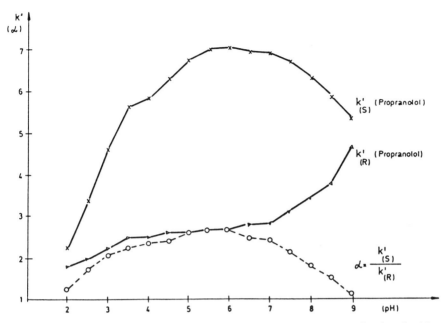

FIGURE 3.   Relative retention values of the zwitterionic molecule (Figure 2a) as a function of mobile phase pH. (For experimental conditions consult Figure 2a and Reference 54.)

so observed, however, that the existence of a protecting group on the chiral amino function of the reagent tended to diminish the diastereoselectivity; this reduction is claimed to stem from the bulkiness of these groups, which equalizes the overall physicochemical properties of the diastereoisomers. As a consequence, the protecting groups (t-BOC groups) must be cleaved to get sufficient resolution factors. These results fit into the above mentioned speculation that an important point for creating good diastereoselectivity is to get additional intramolecular binding points activated; if they happen to be close to the chiral centers, it is even more advantageous. This is illustrated by the following example:

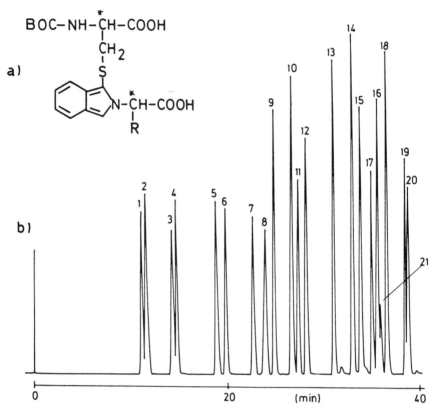

FIGURE 4.   Proposed structure of the amino acid derivatives formed with OPA and *t*-BOC-L-cysteine (a). Chromatogram of the diastereomeric amino acid derivatives (b). Column 220 mm × 4.6 mm, packed with 5-μm RP$_8$; mobile phase: A = 0.05 *M* sodium phosphate buffer (pH 7.0) with 1% tetrahydrofuran; B = methanol. Peaks: 1, 2 = L -,D-Asp; 3, 4 = L-,D-Glu; 5, 6 = L-,D-Asn; 7, 8 = L-,D-Thr; 10 = L-,D-Ala; 11, 12 = L-,D -Tyr; 13, 14 = L-,D-Val; 15, 16 = L-,D-Ile; 17,18 = L-,D-Leu; 19, 20 = L-,D-Lys; 21 = impurity. (Reprinted, with permission, from Reference 75.)

In this case, intramolecular electrostatic interactions could also be postulated.

Chiral derivatization is not restricted to solutes that contain functional groups such as amines or hydroxyl groups, but can also be used for the indirect enantioseparation of chiral olefins by LC (77,78). As may be seen in Figure 5, the chiral platinum complex (I) can reversibly react with chiral olefins, forming two pairs of diastereomeric mixed complexes by exchanging the olefin ligands. These complexes are stable and easily resolvable under normal LC conditions. Besides the analytical aspects of this approach, it should be noted that semipreparative-scale separations of chiral olefins are possible, as seen in Figure 6. Because of the reversability of the ligand exchange mechanism under mild conditions, preparative-scale enantioseparation is also possible.

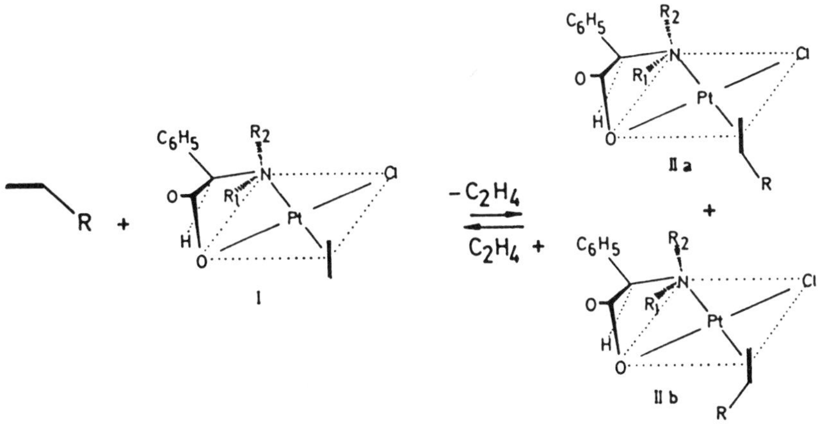

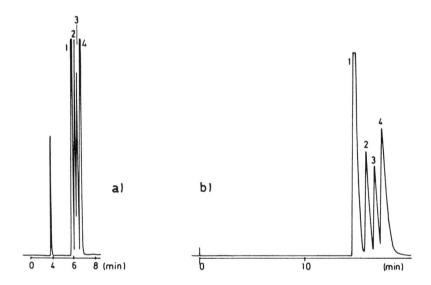

FIGURE 5.   Proposed structure of the diastereomeric platinum complexes resolvable by chromato-graphic techniques. (Reprinted, with permission, from Reference 78.)

FIGURE 6.   Comparative analytical-scale (a) and preparative-scale (b) HPLC separation of diaste-reoisomers resulting from reaction of vinylnorbornane with complex I in Figure 5. Conditions: (a) col-umn: 300 mm × 4.6 mm, 5-μm LiChrosorb Si 60; mobile phase: n-heptane/dichlormethane/isopropanol (60:37:3); (b) column: 250 mm × 21.2 mm, 8-μm Zorbax; mobile phase: n-heptane/dichlormethane/isopropanol (60:38.5:1.5). Peaks: 1 and 4 = diastereomers belonging to (+)-vinylnorbornane, 2 and 3 = diastereomers belonging to (−)-vinylnorbornane. (Reprinted, with permission, from Reference 75.)

Conclusion

Considering its risks and limitations, the indirect enantioseparation technique could be a valuable analytical technique, especially for HPLC. Total freedom is given for the choice of the type of chiral groups, chromophores, spacers, and additional (nonchiral) binding groups incorporated via derivatization into the final diastereomeric reaction product, thus creating exceptional separation factors. Analytically, this technique is important in enantioselective drug monitoring in biological samples, although trace analysis of one enantiomer beside another is problematic and should be performed by direct resolution techniques. The greatest potential of this HPLC technique may be the semipreparative separation and isolation of diastereoisomers in all fields of organic stereochemistry, such as organic synthesis or pharmaceutical chemistry, especially if one considers diastereomeric derivatives that can smoothly be cleaved to their parent enantiomers.

DIRECT SEPARATION OF ENANTIOMERS IN LIQUID CHROMATOGRAPHY: SEPARATION ON HYDROGEN-BONDING CHIRAL STATIONARY PHASES

Compared to the *indirect separation* technique, the *direct separation* of enantiomers is based on a reversible formation of diastereomeric molecule associates of Sa and So partners. The forces to form these spatially fixed, mixed molecule complexes can be hydrogen bonding, dipole-dipole, π-π, electrostatic, and hydrophobic interactions, whereby at least one binding point must be sterically controlled (attraction or repulsion) by chiral centers of the Sa and So molecules.

Derivatives of dipeptides have been used successfully as chiral stationary phases for separation of racemates in gas chromatography (79,80). Differences in multipoint hydrogen bonding between the enantiomers and the amide function(s) of the dipeptide and steric factors are regarded to be the basis of the observed resolutions (80 and references therein).

Hara and co-workers transferred the principle of using a hydrogen-bonding chiral stationary phase into normal-phase LC (81,82). Studies have been performed with *N*-formyl-L-valine (FVA) (81,82) and similar *N*-acyl-L-valine homologues (Figure 7) as well as with other *N*-blocked amino acids (83) chemically bonded to modified silica supports. The synthesis generally starts with the grafting of an aminopropyl spacer group to the silica surface. The second amide function is formed when the amine reacts with the carboxylic group in the valine molecule. A similar solid phase with the amine part of the valine molecule bonded to the matrix has been made using *tert*-butylvalineamide as the chiral moiety (84).

When applying these chiral phases, the commonly used mobile phase is *n*-hexane with a small amount of a polar modifier — for example, 2-propanol or ethylacetate — added to regulate the retention of the solutes. The content of polar compounds

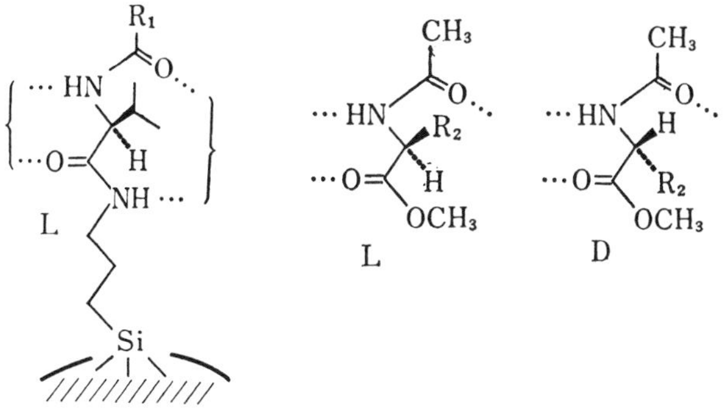

FIGURE 7. Structure of a hydrogen-bonding chiral stationary phase. The two amide groups of the grafted moiety can interact with a solute having two active functions, for example, amide and ester groups of an amino acid derivative. (Reprinted, with permission, from Reference 82.)

should be kept at a minimum in order not to interfere with the stereoselective interaction between the enantiomers and the stationary phase. Enantiomers of amino acids and dipeptides that must be derivatized on both the amine and the carboxylic groups have been separated on these diamide phases. The separation factor, α (that is, the ratio of the capacity factor k' for the second eluted enantiomer to k' for the first eluted enantiomer) with the FVA-column was between 1.05 and 1.39, depending on the amino acid and on the nature of the protecting groups (Tables II and III) (85). The N-acyl butylester derivatives in general seem to give the highest separation factors. Stationary phases based on urea derivatives of L-valine were found to give slightly

TABLE II

Resolution of D- and L-Leucine Derivatives of $(CH_3)_2CHCH_2CH(COOR^1)NHCOR^2$ on N-formyl-L-valylaminopropyl Silica Gel Column*

Derivative		Mobile phase (% v/v of 2-PrOH in n-hexane)	Capacity factor (k')		Separation factor (α)
R^1	R^2		D	L	
Me	H	6	3.25		1.00
Me	Me	6	3.35	3.66	1.09
Me	Et	4	2.36	2.63	1.11
Me	i-Pr	2	3.00	3.39	1.13
Me	t-Bu	0.5	2.65	2.93	1.11
Me	Me	4	5.50	6.05	1.10
Et	Me	3	4.98	5.74	1.15
i-Pr	Me	2.5	4.71	5.65	1.20
t-Bu	Me	4	2.03	2.54	1.25
t-Bu	Me	2	4.84	6.70	1.38

* Reprinted, with permission, from Reference 85.

higher stereoselectivity with values as high as 2.36 for *N*-acyl-DL-leucine-*tert*-butyl ester (86). (Chiral selectors with diamide structure have lately been used as chiral additives in the mobile phase and will be discussed below.)

ENANTIOSELECTIVE CHROMATOGRAPHIC SYSTEMS BASED ON CHARGE-TRANSFER INTERACTIONS

Techniques based on electron donor-acceptor (π-π) interactions have been applied to enantiomeric separations in liquid chromatography. The chiral selector possessing either electron-donating or electron-accepting properties can be present in the eluent or may either be adsorbed (coated) onto, or chemically bonded to, the solid support.

Fluoroalcohols

From nuclear magnetic resonance (NMR) studies, Pirkle found that optically active fluoroalcohols gave two-point, chelate-like adducts by hydrogen bonding with different types of solutes, such as sulfoxides (87). The principle of diaste-

TABLE III

Resolution of Enantiomers of *N*-Acetyl Amino Acid *tert*-Butyl Esters on *N*-Formyl-L-valylaminopropyl Silica Gel Column*

Separation	Amino acid	Strong solvent in *n*-hexane (% v/v)	k' D	k' L	Resolution α	Resolution R_s
1	Leu	Et$_2$O (80)	3.12	4.33	1.39	4.21
2	Val	Et$_2$O (80)	3.41	4.69	1.38	4.17
3	Nle	Et$_2$O (80)	3.13	4.28	1.37	3.99
4	Nva	Et$_2$O (80)	3.63	4.89	1.35	3.88
5	Abu	Et$_2$O (80)	4.12	5.37	1.30	3.39
6	Ala	Et$_2$O (80)	4.83	5.95	1.23	2.66
7	Ile	Et$_2$O (80)	3.16	4.32	1.37	4.00
8	O-*t*-BuSer	Et$_2$O (80)	2.13	2.82	1.32	3.12
9	O-AcTyr	Et$_2$O (80)	7.67	9.33	1.22	2.70
10	O-*t*-BuAsp	Et$_2$O (80)	2.57	3.10	1.21	2.13
11	O-*t*-BuGlu	Et$_2$O (80)	3.57	4.33	1.21	2.30
12	S-BzlCys	CH$_2$Cl$_2$ (30)	1.77	2.31	1.31	2.85
13	*N*-*t*-BuTrp	CH$_2$Cl$_2$ (30)	1.88	2.58	1.37	3.50
14	PheGly	CHCl$_3$ (30)	2.28	3.02	1.32	3.18
15	Phe	CHCl$_3$ (30)	1.96	2.71	1.38	3.64
16	*N*-AcLys	2-PrOH (12)	6.83	7.17	1.05	0.60
17	Gln	2-PrOH (8)	17.63		>1.00	>0.30
18	Pro	2-PrOH (4)	2.77		1.00	

*Reprinted, with permission, from Reference 85.

reomeric adduct formation was later used in LC with the fluoroalcohol (selector) added to the mobile phase (88). The main work, however, has been devoted to systems with chiral selectors grafted to the support.

Chiral stationary phases (CSPs) containing 2,2,2-trifluoro-1-(9-anthryl)ethanol chemically bonded to silica have the advantage of promoting enantioselective retention for a variety of different solutes, such as sulfoxides, lactams, and derivatives of alcohols, amines, amino acids, hydroxy acids, and mercaptans (89,90). A model based on the three-point rule offered above was proposed to explain the stereoselectivity obtained with the chiral fluoroalcohol (Figure 8). Two diastereomeric complexes arise by formation of two hydrogen bonds between the enantiomers and the hydrogen of the hydroxyl and carbonyl functions of the anthryl alcohol. Furthermore, one of the enantiomers should have the possibility of an additional interaction between the electron-donating anthryl ring and a π-acidic function in the solute (90). For solutes lacking electron-accepting properties (π acids), derivatization with, for example, 2,4-dinitrofluorobenzene or 3,5-dinitrobenzoyl chloride — that is, with substances known to be strong π acids — will introduce the necessary properties.

(*R*)-*N*-(3,5-Dinitrobenzoyl)phenylglycine

The roles of the selector and selectands in offering and/or accepting stereoselective binding points can in principle also be reversed; that is, racemic mixtures of anthryl alcohols can be resolved using a π-accepting chiral stationary phase (10 and references therein). Extensive research has been carried out with (*R*)-*N*-(3,5-dinitrobenzoyl)phenylglycine [(*R*)-*N*-(3,5-DNB)phenylglycine; Figure 9] as the chiral stationary phase (10,91 and references therein). The phenylglycine derivative may either be ionically or covalently bonded to aminopropyl silica.

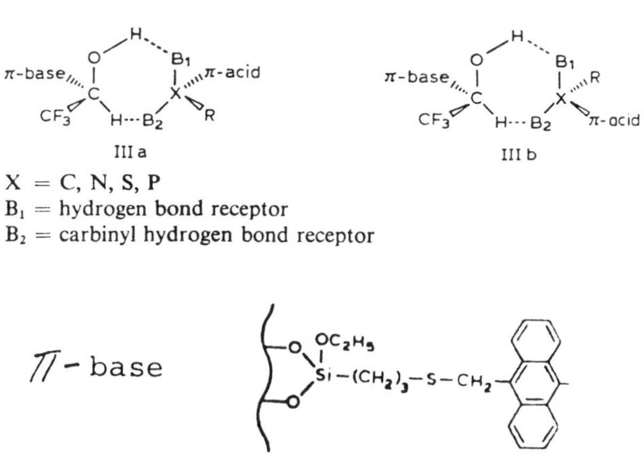

$X = C, N, S, P$
B_1 = hydrogen bond receptor
B_2 = carbinyl hydrogen bond receptor

FIGURE 8. Model for stereoselective interaction between anthrylfluoroalcohol (selector) and solute (selectand). (Reprinted, with permission, from Reference 90.)

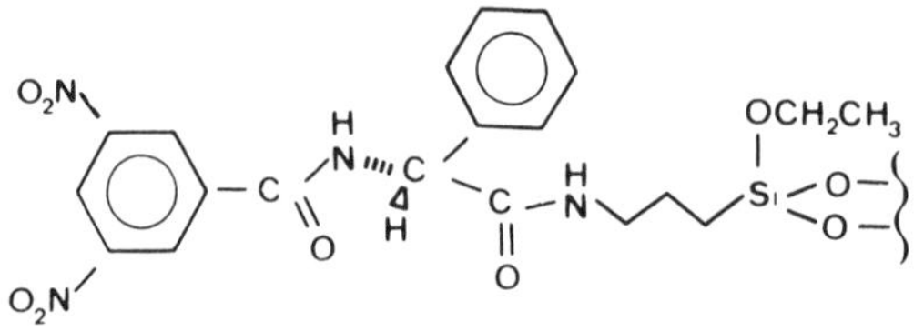

FIGURE 9.   Structure of (R)-N-(3,5-dinitrobenzolyl)phenylglycine. CSP I: covalently bound; CSP II: ionically bound.

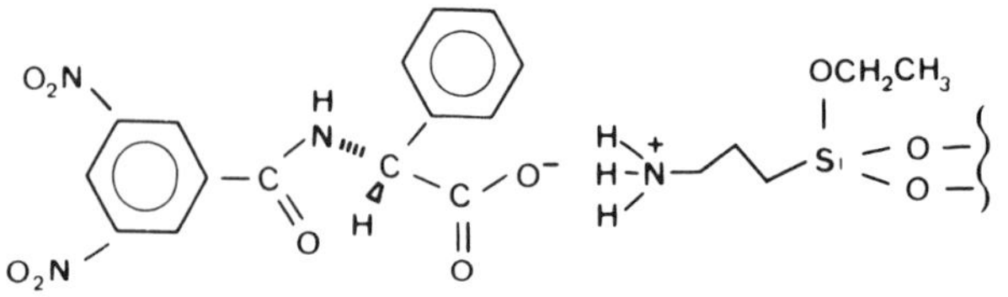

Results of several applications have shown that the (R)-N-(3,5-DNB)-phenylglycine support also is a resolving phase with a broad spectrum, as is indicated in Table IV. This table gives examples of substances that have been resolved, including some drugs. Enantiomers of norephedrine (96) and ephedrine (97) have been separated as 2-oxazolidine derivatives on the (R)-N-(3,5 DNB)phenylglycine phase. The same chiral stationary phase was used to resolve enantiomers of some antiinflammatory agents — fenoprofen, benoxaprofen, ibuprofen, and naproxen (102) — as well as enantiomers of some acetylcholinesterase and carboxylesterase inhibitors (105). Recently, studies on the bioavailability of propranolol have been performed on the (R)-N-(3,5-DNB)phenylglycine phase. A chromatogram illustrating the separation of enantiomers of the 2-oxazolidine·derivative of (R,S)-propranolol is given in Figure 10 (95).

Chiral rational recognition based upon molecular models as a mean of predicting stereoselectivity and elution order for the enantiomers has been discussed (98). The models are used to illustrate the differences in interactions (dipole-dipole, hydrogen-bonding, charge-transfer, hydrophobic, and steric interactions) between the enantiomers and the (R)-N-(3,5 DNB)phenylglycine phase. The absolute configuration may in principle be determined from the retention order for the enantiomers, the main problem being accurate prediction of the relative stabilities between the diastereomeric complexes. Certain precautions should, however, be taken with regard to the possibility of assigning *absolute* configuration from *relative* retentions for the enantiomers. Within one class of compounds (carbinols) the

TABLE IV

Enantiomeric Separation on (R)-N-(3,5-dinitrobenzoyl)-
phenylglycine Phases

Solutes	References
N-Acyl-amines	10, 92, 93
amphetamine derivative	92
N-Acyl-heterocyclic amines	94
Amides	10
Aminoalcohol derivatives	10
propranolol	10, 95
norephedrine	96
ephedrine	97
Arylalkyl alcohols	10, 98
Benzocycloalkenols	98, 99
Bi-β-naphthols and analogues	10, 100
Carboxylic acids (pyrethroid insecticides)	101
Dihydrodiols and tetrahydrodiols	104
of benzo[a]pyrene and benz[a]anthracene	
Fluoroalcohols	10, 90, 100
Hydantoins	10, 100
Organophosphinates	105
Sulfoxides	10, 100

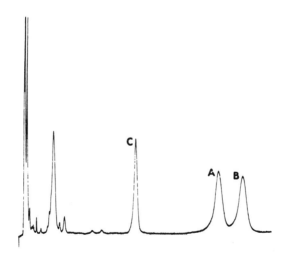

FIGURE 10.   Chromatogram from extraction of whole blood sample spiked with 50 ng/mL racemic propranolol. Column: 250 mm × 4.6 mm Pirkle Type 1-A (Regis Chemical, Morton Grove, Illinois) with an aminopropyl packing of 5-μm spherical particles modified with (R)-N-(3,5-DNB)phenylglycine. Peaks: A = oxazolidone corresponding to L-propranolol; B = oxazolidone corresponding to D-propranolol; C = oxazolidone corresponding to internal standard. (Reprinted, with permission, from Reference 95.)

retention order for the enantiomers was reversed, although the absolute configuration was the same (99). It has also been observed that stereoselectivity and retention order for enantiomers of the same compound can differ for the ionically and covalently bonded (R)-N-(3,5-DNB)phenylglycine phases (92).

Weems and Yang (104) found that certain racemic dihydrodiols and tertrahydrodiols of benzo[a]pyrene and benzo[a]anthracene were resolved on the (R)-N-(3,5-DNB)phenylglycine stationary phase, although other members of the same class of compounds did not show any stereoselective retention.

Substituents may either increase the difference in relative stabilities of the diastereomeric complexation by facilitating a discriminating interaction for one of the enantiomers (π-π interaction or decrease of the stereoselectivity) by preventing the necessary stereoselective interaction by, for example, steric repulsion. Often the solutes must be derivatized before being chromatographed in systems with these chiral phases, because they generally are used with nonpolar mobile phases. The nature of the reagent may also affect stereoselectivity; for example, the amide derivatives of acids usually are easier to resolve than esters of the same acid (102). In the case of tropic acid, no difference in retention at all was observed for ester derivatives of the enantiomers (103).

The variety of substances that can be separated on the (R)-N-(3,5-DNB)-phenylglycine phase has had a strong effect on the research and development of new chiral phases. Oi and coworkers (106–109) have synthesized and studied phases similar to (R)-N-(3,5-DNB)phenylglycine (Table V). Enantiomers of derivatized amines, acids, amino acids, and alcohols including a fungicide were separated using this support.

TABLE V

Chiral Stationary Phases Based on Modified Aminopropyl Silica

Chiral stationary phase		Solutes	Reference
A		Amines and amino acids as N-3,5-DNB*	107
		Carboxylic acids as 3,5-DNA[†]	107
B		3,5-dinitrophenylurethane derivatives of alcohols	108
		Amines and amino acids as N-3,5-DNB	106
C		Carboxylic acids as 3,5-DNA	107
		Amines and amino acids as N-3,5-DNB	109
D		Carboxylic acids as 3,5-DNA	109
		3,5-dinitrophenylurethane derivatives of alcohols	108

* N-3,5-DNB = N-3,5-dinitrobenzoyl derivatives
[†] 3,5-DNA = 3,5-dinitroanilide derivatives

Chiral stationary phases based on (*R*)-*N*-(3,5-DNB)phenylglycine have also been applied to direct separation of enantiomers in thin-layer chromatography (TLC) (110), as well as for preparative resolution of racemates in column LC (100). Figure 11 shows a chromatogram in which 1.6 g of a hydantoin has been injected (100). This gives an indication of the potential capacity of these phases to produce optically pure substances on a semipreparative or preparative scale in column LC. Columns packed with the (*R*)-*N*-(3,5-DNB)phenylglycine phase ionically or covalently bonded are now commercially available (Regis Chemical Co., Morton Grove, Illinois, and J.T. Baker Chemical Co., Phillipsburg, New Jersey) for both analytical and preparative applications.

As was previously indicated, few differences result whether the π acid group is grafted on the stationary phase or is part of the chiral solute; the group only needs to come close enough to undergo π-π interactions. Consequently, chiral phases have been synthesized containing a weak π base, as was recently shown by Arm and co-workers (111). The chiral amide phase, based on phenylethyl amide, exhibits, for example, enantioselectivity for amide-type solutes substituted with a π acid functionality. It is interesting to note that the enantioselectivity can be observed using nonaqueous (but to some extent also with aqueous) mobile phase conditions, which is further indication of the enantioselective dipole stacking mechanism (14) that competes with the hydrogen-bonding mechanism. This concept

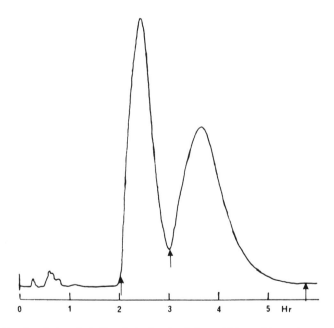

FIGURE 11.   Resolution of a 1.60 g of a hydantoin. Column: 30 in. × 2 in. aminopropyl-silanized 40-μm silica particles modified with (*R*)-*N*-(3,5-DNB)phenylglycine. (Reprinted, with permission, from Reference 100.)

may also be valid for chiral separations performed on those chiral phases based on tartaric acid amide derivatives (112).

Other Chiral π Selectors

Enantiomers of helicenes (chiral arenes) have been separated in liquid chromatographic systems with π-accepting chiral selectors. The chiral resolving agents, listed in Table VI, are coated or are ionically or covalently attached to the matrix, usually silica gel. Gil-Av and co-workers have used 2-(2,4,5,7-tetranitro-9-fluorenylidoneaminoxy)propionic acid (TAPA) (I in Table VI) and homologues thereof for separation of the two enantiomeric forms of several helicenes (113) that are chiral as a result of their cylindrical helix structure. The effect of the composition of the mobile phase on chromatographic performance — and the effect of changing the structure of the π base (solute) or the π acid (chiral stationary phase) on stereoselectivity — led the authors to propose a model based on complementary chiral concavity and convexity for the stereoselective interaction (Figure 12). The P-(+)-[14]helicene is believed to fit less suitably than does the corresponding enantiomer over the cone formed by the H and CH_3 groups at the asymmetrical carbon atom in the TAPA molecule and therefore is less retained (Figure 12).

TABLE VI

Chiral Stationary Phases with π-π Selectors

		Selector	References
I		2-(2,4,5,7-tetranitro-9-fluorenylidone aminoxy)propionic acid (TAPA)	113,114,115
II		Riboflavin	116
III		Binaphthylphosphoric acid	114
IV		N-(2,4-dinitrophenyl)-L-alanine	117
V		P-(+)-hexahelicene-7,7'-dicarboxylic acid	118

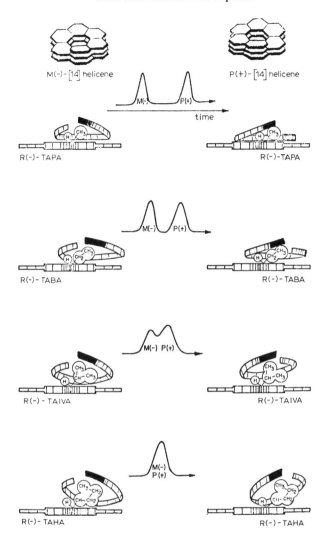

FIGURE 12.   Suggested explanation for the gradual decrease in resolution with increasing size of the alkyl-group at the asymmetric carbon of TAPA homologues. The chromatograms show the resolution of [14]helicene. (Reprinted, with permission, from Reference 113.)

The effect of enantiomeric discrimination decreases as the bulkiness of the groups attached to the chiral center increases. Enantiomers of derivatives of the carcinogenic and mutagenic polyaromatic hydrocarbons, such as *trans*-4,5-dihydroxy-4,5-dihydrobenzo[a]pyrene, have been separated on the TAPA phase with dichloromethane/methanol as the mobile phase. A separation obtained for the enantiomers of the benzo[a]pyrene derivative is shown in Figure 13 (115).

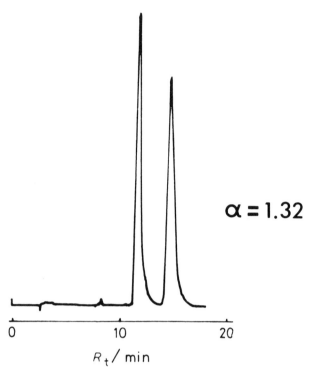

$\alpha = 1.32$

FIGURE 13.   Resolution of (±)-*trans*-4,5-dihydroxy-4,5-dihydrobenzo[a]pyrene. Column: 150 mm × 4.6 mm packed with TAPA covalently linked to silica; mobile phase: MeOH/CH$_2$Cl$_2$ (4:11). (Reprinted, with permission, from Reference 115.)

Chiral helicenes also interact stereospecifically with riboflavin (II in Table VI) (116) and binaphthylphosphoric acid (III in Table VI) (114). The phosphoric acid derivative is especially well adapted to helicenes containing heteroatoms or electron-donating substituents (114).

The use of a much simpler molecule, *N*-(2,4-dinitrophenyl)-L-alanine (IV in Table VI) chemically bonded to silica, indicates that complementary topographic interactions might not be necessary in order to obtain stereoselective retention of substances with helical structures (117). The chiral selectors I–IV (Table VI) have a limited range of applications mainly restricted to substances with helical structures; the dipotassium salt of the helicene, *P*-(+)-hexahelicene-7-7 '-dicarboxylic acid (V in Table VI) coated silica, promotes different retention for the D and L forms of amino acid derivatives (118). The isopropyl esters of amino acids gave higher stereoselectivity than the methylesters. An exception is leucine, which could not be resolved at all on this chiral phase.

SEPARATION OF ENANTIOMERS USING CHIRAL CROWN ETHERS

Crown ethers can produce very stable complexes, a characteristic that has long been used in organic synthesis (phase-transfer catalysis) (119). The selectivity obtained in the complexation has also led to construction of efficient separation systems based on these macrocyclic polyethers. The crown ethers can form complexes with several different solutes, including alkali metal and ammonium salts.

Introducing chiral elements such as binaphthyl rings into the macrocyclic polyether results in complexing agents that may interact differently with the two enantiomeric forms of a compound. The binaphthalene ring attached to the cyclic ether introduces a chiral barrier and gives the crown ether a rigid structure that also restricts the possibility of reorganization upon complexation. Cram and co-workers (120 and references therein) have synthesized different binaphthyl crown eithers, mainly derivatives of 22-crown-6 (Figure 14) and studied their stereoselective complexation with ammonium ions (amino acids and derivatives thereof).

Two diastereomeric complexes with different stabilities are formed when the crown ether (host) interacts with the two enantiomers (guests) as described in equations 5 and 6,

$$(R,R)\text{-Host} + (R)\text{-Guest} \quad \overset{K_R}{\rightleftharpoons} \quad (R,R)\text{-Host-}(R)\text{-Guest} \qquad [5]$$

$$(R,R)\text{-Host} + (S)\text{-Guest} \quad \overset{K_S}{\rightleftharpoons} \quad (R,R)\text{-Host-}(S)\text{-Guest} \qquad [6]$$

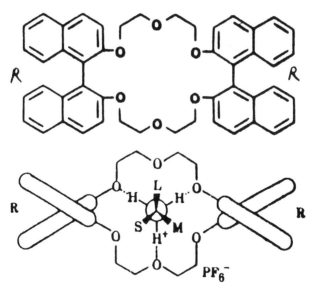

FIGURE 14. Suggested model for stereoselective interaction with the chiral crown ether (R,R)-dibinaphthalene-22-crown-6. Large, Medium and Small groups at the asymmetrical center.

Differences in distribution for enantiomers between water and chloroform containing crown ethers as complexing agents were studied. The crown ethers were synthesized to give a complementary structure to the enantiomers to be separated and contained binding sites for the amine and carboxylic functions in the amino acid or in the amino acid derivatives. The NH_3 group is fixed by hydrogen bonds in a tripole arrangement to the ether oxygens in the cyclic ring (Figure 14). The binding of the binaphthalene rings to the cyclic polyether results in a chiral cavity into which the other groups of the solute can penetrate. Additional interactions, such as steric effects, dipole-dipole and π-π interactions, and hydrogen bonding between the host and the guest, promote a high degree of stereoselectivity in the complexation process. The proposed models for stereoselective complexation have been supported by results from NMR studies and by observed changes in stereoselectivity from a systematical variation of the structure of the crown ether and the solute, respectively. The six-membered ether ring was found to give the highest stereoselectivity; an increase or decrease of the ring size resulted in a lower stereoselectivity.

Enantiomers of alkyl and ammonium ions have been separated using the principle of host/guest complexation in LC. First liquid-liquid chromatographic systems with H_2O-$NaPF_6$ or H_2O-$LiPF_6$ as the stationary phase adsorbed on Celite or silica gel as support material were studied. The optically active crown ether was added to the mobile phase, which was usually chloroform and different retention for the enantiomers of α-phenylethylammonium PF_6, methylphenylglycinate PF_6, and methyl-p-hydroxyphenylglycinate PF_6 was observed. A good correlation between separation factor and the enantiomeric distribution constant (EDC) — that is, the ratio of the distribution constant between the aqueous and organic phase for the more complexed enantiomer to the less complexed enantiomer — was obtained, indicating that the support material has an insignificant effect on stereoselective complexation. Later, chiral stationary phases were made by Cram and colleagues by covalently binding a crown ether to solid matrices (120). Impressively high separation factors could be obtained, for example, $\alpha = 26$ for the ClO_4 salt of p-hydroxyphenylglycine ester. Racemic solutions of underivatized amino acids and amino acid esters were resolved on the phase having a crown ether grafted to the resin. It is also possible to synthesize chiral phases with the macrocyclic crown ether fixed to silica via –Si-$(CH_2)_3$-O-Si bonds to position 6 in the naphthalene rings (Figure 15) (120). An example of the separation of the enantiomers of a primary amine on the silica-bonded crown ether phase is illustrated in Figure 15. The degree of differentiation between the enantiomers, that is, the stereoselectivity, depends upon the configuration of selector (crown ether) and selectands (solute) and also on temperature, organic solvent, and counter ion.

Optically active crown ethers have also been used in liquid membrane electrodes for the determination of enantiomeric excess of chiral ammonium ions (121).

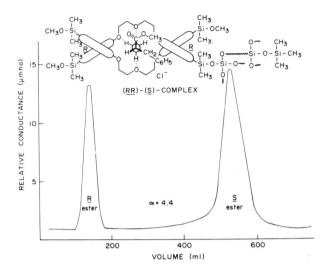

FIGURE 15.   Chromatographic optical resolution of methyl phenylalaninate hydrochloride salt on crown-ether bonded silica phase. (Reprinted, with permission, from Reference 120.)

SEPARATION OF ENANTIOMERS USING POLYMERIC CHIRAL STATIONARY PHASES

Naturally occurring biopolymers, such as cellulose and starch, were some of the earliest phases used for separation of enantiomers in liquid chromatography. Recently, much interest has been focused on the resolving capacity of synthetic polymers (6). Optically active polymers can also be synthesized by binding (in copolymerization) an optically active solute to an inactive backbone (122). Formation of polymers with helical structures from optically inactive substances in the presence of chiral catalysts is also possible (123). The resolution mechanism for most polymers is considered to be an unequal distribution of the enantiomers into chiral cavities of the polymer. This means that the antipodes, as they pass through the column, are retained to different degrees by the stationary polymeric phase. The term *inclusion chromatography* (124) has been applied to this technique, in which the separation is attributable to different fitting of the enantiomers to chiral cavities in a complexing agent.

Microcrystalline Triacetyl-Cellulose

Microcrystalline triacetyl-cellulose, first introduced by Hesse and Hagel in 1973 (124), has been successfully used for separation of the enantiomers of a variety of compounds, including barbiturates (125). Chromatograms from the resolution of some barbiturates with triacetyl-cellulose as the chiral stationary phase are shown in Figure 16.

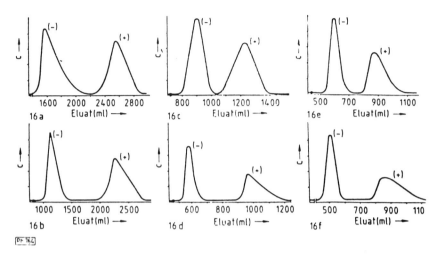

FIGURE 16. Chromatographic resolution of racemates on microcrystalline cellulose-triacetate. Column: 85 cm × 2.5 cm; mobile phase: 95% ethanol; pressure: 1.2 bar; flow rate: 50 mL/h. Samples: (a) 160 mg methyleudan; (b) 201 mg methylphenobarbital; (c) 161 mg hexobarbital; (d) 209 mg methylcyclobarbital; (e) 200 mg dimethylcyclohexylbarbituric acid; (f) 205 mg cyclohexylethylmethylbarbituric acid. (Reprinted, with permission, from Reference 125.)

The enantiomers are believed to penetrate differently into the cavities that are formed between the d-α-glucose units, and it has generally been found that enantiomers of compounds having a ring structure (126) or with other bulky rigid groups (127,128) show different retention on the triacetyl-cellulose phase. The helical structure of cellulose is largely preserved during the heterogenous acylation and is vital for the resolution, whereas recrystallized cellulose did not possess any resolving capacity for enantiomers (124).

The triacetyl-cellulose phase is used in a swollen state, which restricts the pressure that can be applied on these phases. Limited pressure stability and the relatively low column efficiency obtained with triacetyl-cellulose columns have mainly led to semipreparative or preparative applications.

Chiral stationary phases based on homogenous acetylated cellulose or other cellulose derivatives and coated on silinazed macroporous silica gel have recently been introduced by Okamoto and co-workers (129). These phases are believed to have different and/or higher stereoselectivity than the heterogenous cellulose phase and also to possess higher compressive strength and durability.

β-Cyclodextrin

Enantiomeric separation by inclusion complexation has also been achieved with β-cyclodextrin. The chiral cavity created when the glucose units are polymerized via β-(1,4) glycosidic linkage to form the β-cyclodextrin molecule can discriminate between optical isomers. Different approaches have been used to synthesize β-cyclodextrin chiral stationary phases for direct separation of enantiomers (130 and references therein).

Small spherules (fractions of 63–90 μm and 90–125 μm) of a cross-linked product containing 58% of β-cyclodextrin were used for separation of enantiomers of indole alkaloids (Figure 17) (131). The amount of ($\pm$)-vincadifformine injected in this experiment was 0.5 g. Enantiomers with optical purity of 92.5% ($-$ form) and 98.2% ($+$ form), respectively, could be obtained after recrystallization of the collected fractions. Retention and stereoselectivity of the alkaloids was regulated by chemical composition of the buffer and by pH in the mobile phase. Recently, supports with high separation efficiency were produced with β-cyclodextrin chemically bonded to silica particles (130, 132). The columns still had somewhat lower efficiency than that generally obtained in reversed-phase chromatography, which might be attributable to slow kinetics of the mass transfer between the mobile and the stationary phase that is accompanied by the strong complexation with β-cyclodextrin (131). Stereoselective retention has been obtained for propranolol, barbiturates, dansyl and naphthyl derivatives of amino acids (131), and carboxylic acids (mandelic, cyclohexylphenylglycolic, and dicyclohexylphenylacetic acids) (132) using aqueous mobile phases and β-cyclodextrin/silica as the stationary phase. One type of the silica-based cyclodextrin column is now commercially available (Advanced Separation Technology, Whippany, New Jersey).

Polyacryl- or Polymethacryl-Based CSP

The idea of making optically active stationary phases for liquid chromatography by polymerization of chiral monomers has been realized by Blaschke and co-workers (6 and references therein). The polymers consist of polyacrylamide or

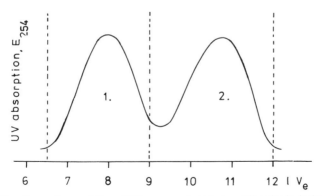

FIGURE 17.   Preparative resolution of 500-mg racemic ($\pm$)-vincadifformine by inclusion chromatography. Column: 90 cm × 5 cm CDP-25 β-cyclodextrin bead-polymer, particle size 90–125 μm; mobile phase: phosphate buffer, pH 9.5. Peaks: 1 = (+)-vincadifformine (11 β); 2 = ($-$)-vincadifformine (11 α). (Reprinted, with permission, from Reference 131.)

TABLE VII

Enantiomeric Separation of
Pharmaceuticals on Optically
Active Polyacryl and Polymethacryl
Chiral Stationary Phases

Pharmaceutical	Reference
Biglumide	135
Chloroquine	136
Chlorothalidone	137
Hexobarbital	133
Mephenytoin	133
Methyprylone	133
Oxazepam	136
Phenprocoumon	133
Thalidomide	135

polymethacrylamid of phenylethyl or ephedrine derivatives. The structure of the monomeric unit and the cross-linking agent used have a strong influence on stereoselectivity. The enantioselectivity obtained on these phases are also strongly dependent on the polymerization procedure — for example, solvent, initiating agent, and degree of cross-linking.

The mechanism for resolution is not known in detail, but it has been suggested that the enantiomers fit differently into asymmetrical microcavities in the polymers (133). Hydrogen-bonding and π-π interactions may also play a role in the retention behavior between the enantiomeric solutes and these polymeric phases.

Besides enantiomers of mandelic acid and mandelic acid derivatives, the racemates of N-acyl amino esters, benzoin derivatives, N-substituted acetamides, and N-substituted phenylacetamides have also been resolved (134). It has been difficult so far to predict the retention order for the enantiomers and the structural requirements necessary for resolution on these polyacryl- or polymethacryl-based phases, but they seem to be especially well-suited to separate enantiomers with amide and imide structures. Table VII gives a summary of some pharmaceuticals that have shown stereoselective retention on these chiral supports. A chromatogram to demonstrate the resolution capacity of these chiral phases is given in Figure 18, showing the complete resolution of ($\pm$)-thalidomide (135).

The acryl polymers are soft gels that are used at low pressures and with nonpolar mobile phases (benzene, benzene/dioxane, or benzene/cyclohexane). The organic mobile phases and the relatively high capacity of the gels have made them well-suited for semipreparative and preparative resolutions of racemic mixtures.

(+)-Poly(triphenylmethyl methacrylate)

Okamoto and co-workers recently introduced (+)-poly(triphenylmethyl methacrylate) [(+)-PTrMA] (Figure 19a) as a chiral LC stationary phase (123,138).

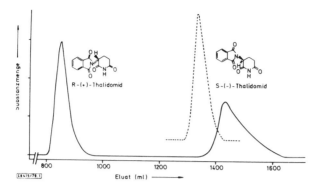

FIGURE 18. Chromatogram of 51.5 mg of thalidomide on 65 g of poly[(S)-N-(1-cyclo-hexylethyl)]methacrylate phase. Column: 80 cm × 2.3 cm; mobile phase: benzol/dioxane (4:1). ----- = Chromatogram of (−)-enantiomer increasing the temperature to 50 °C after elution of 1260 mL. (Reprinted, with permission, from Reference 135.)

The chirality of the polymer is caused by helicity formed in the presence of chiral anionic catalysts. After grinding to make 20–44 μm particles, the polymer may be used as support material packed in the column (123) or it may be coated on the surface of silianzed silica-gel particles (139).

The silica-(+)-PTrMA impregenated phase has some advantage over the polymeric phase: first, the retention of the solute is lower, which produces a shorter analysis time; second, the effciency is higher on the impregnated phase; and, third, the coated phase is said to have better durability.

The resolution principle with the (+)-PTrMA phases is not fully understood, but it has been observed that a polar mobile phase such as methanol gives higher stereoselectivity than does a nonpolar phase such as hexane. According to the authors, this behavior indicates that nonpolar interactions are the main cause for the resolution of enantiomers (139). Different types of optically active compounds (alcohols, amines, and esters) and especially those compounds containing an aromatic ring structure can be resolved on the (+)-PTrMA phases. Derivatives of some 2,2′-disubstitued-1,1′-binaphthyls show low or no stereoselectivity on the polymeric support, although they can be resolved when chromatographed on the silica-gel coated phase (139).

The separation of the four stereoisomers of phenothrin (3-phenoxybenzyl chrysanthemate, a pyrethroid insecticide) is shown in Figure 19b. The *cis* and *trans* forms as well as their enantiomeric forms were separated on (+)-PTrMA impregnated macroporous silica gel (139). Further applications with the coated (+)-PTrMA phase are to be expected, as it has now become commercially available (Chiralpack-OT; Jasco Co., Easton, Maryland).

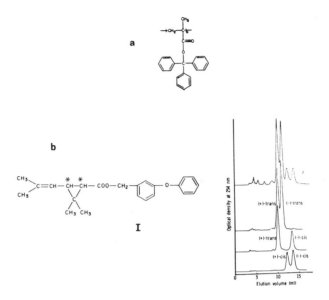

FIGURE 19.   Structure of (+)-poly(triphenylmethyl methacrylate) [(+)-PTrMA] (a); separation of phenothrin isomers with the (+)-PTrMA-coated silica gel column (b). Column: 25 cm × 0.46 cm; mobile phase: methanol; flow rate: 0.50 mL/min; temperature: 20 °C. The chromatogram is of a mixture of (±)-*trans* I and (±)-*cis* I. (Reprinted, with permission, from Reference 139.)

Imprinted Phases

The idea of forming selective adsorbents by molding them with a solute was originally suggested by Pauling (140). The principle of imprinted silica gel involves coagulation of siliac acid in the presence of a chiral solute and then removal of the solute to the greatest extent possible by washing the silica gel with different solvents (141). A footprint of the enantiomer present during the coalugation is formed, to which the enantiomer selectively fits.

A similar technique to make imprints in a polymer was published by Wulff and Vesper (142). A 4-dinitrophenyl-α-D-mannopyranoside derivative of *p*-vinylphenylboronate was copolymerized to form a macroporous polymer; then the 4-dinitrophenyl-α-D-mannopyranoside was removed by hydrolysis, leaving a chiral imprint in the polymer. Enantioselectivity as high as α=2.32 was obtained for 4-dinitrophenyl-α-D-mannopyranoside, but the stereoselectivity was found to depend strongly on the degree of cross-linking. The fact that these imprinted phases are difficult to reproduce and that they have a limited pressure stability and low column efficiency have restricted their use as chiral stationary phases in LC; its spectrum on enantioselectivity also seems rather limited.

Bonded Proteins

Proteins and enzymes are macromolecules that can interact stereoselectively with small ligands. It is well known that enantiomers of drugs may bind differently to albumin in plasma (143); chromatographic systems for separation of enantiomers have been designed with albumin either covalently bound to a surface or added to the mobile phase as a complexing agent.

In 1973 Stewart and Doherty indicated that they had succeeded in separating the enantiomers of tryptophan on an albumin-agarose column (144). Bovine serum albumin (BSA) was coupled to Sepharose via a succinoylaminoethyl spacer. Further studies showed that albumin phases could promote stereoselective retention of other substances as well, such as racemic oxazepam esters (145) and warfarin (146,147). The possibility of evaluating binding constants to albumin with these chromatographic systems have also been studied (146). Albumin from different species was bound to Sepharose-4B and was found to give quite different chiral discrimination. This resultion was probably a result of differences in amino acid sequence at the binding site(s) of albumin (147).

Studies with albumin chemically attached to silica have been published (148–150). This support has recently become commercially available (Resolvosil; Macherey-Nagel & Co., Düren, West Germany). The use of small, rigid silica particles increases column efficiency and permits the use of higher column pressures. The capacity of the columns is, however, relatively low, and the efficiency is still moderate ($h = 70$), a factor that has restricted the use of the column to analytical applications for compounds with good detection properties. The retention of the enantiomers is regulated by pH, ionic strength, and content of organic modifier in the mobile phase. The effect of changes in pH are complicated because they may also affect the binding properties of albumin (151) and thus the stereoselectivity.

It is difficult to predict any rational chiral recognition because neither the structure of the stereoselective binding sites nor the effect on the tertiary structure of the protein when attached to a solid matrix are known. Hydrophobic interactions, electrostatic attractions, and π-π interactions have been suggested to be responsible for the chiral resolution of the enantiomers. Enantiomers of, for instance, amino acid derivatives and sulfoxides have been separated on albumin-silica gel phase. A chromatogram illustrating the separation of (R,S)-omeprazole, a potential drug for gastric acid inhibition, is shown in Figure 20 (150).

Racemic drugs that are basic generally do not show different retention on albumin phases but may be separated on an α-glycoprotein-silica phase (152,153) as demonstrated in Figure 21. The glycoprotein is coupled to epoxide-activated silica gel and forms a chiral stationary phase that is said to be stable for months (153). The parameters that control retention are similar to those controlling retention on the albumin-silica phase (that is, pH, ionic strength, and organic modifier). Addition to the mobile phase of an amine that can interact with the glycoprotein will also decrease the retention of the enantiomers (153).

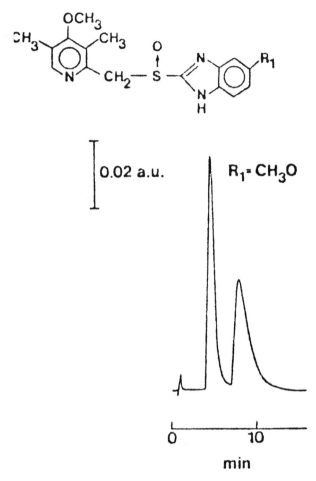

FIGURE 20.   Separation of (*R,S*)-omeprazole on immobilized BSA-silica gel column. Mobile phase: 0.08 *M* phosphate, pH 5.80; flow rate: 2.0 mL/min. (Reprinted, with permission, from Reference 150.)

Stereospecific antibodies for (+)- and (−)-abscisic acid were used for separation of enantiomers in a metabolic study (154). From antisera raised in rabbits, IgG fractions were prepared and then grafted to Sepharose particles. The modified Sepharose particles were used as a chiral stationary phase.

CHIRAL ION-EXCHANGERS

Ion-exchange chromatography and ion-pair chromatography are the primary techniques for LC separation of ionic substances. Asymmetrical ion exchangers

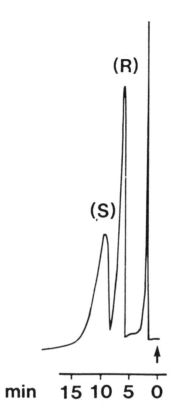

FIGURE 21.   Separation of (*R*)- and (*S*)-disopyramide on an α-acid glycoprotein silica column (150 mm × 3.2 mm) using mobile phase of phosphate buffer (pH 7.06; μ = 0.05) containing 1% L-propanol. (Reprinted, with permission, from Reference 152.)

have also been used in attempts to resolve racemates (155 and references therein).

The separation of enantiomers of mandelic acid has been studied on different anionic ion exchangers containing alkaloids or amino acids as chiral selectors. Quinine attached to Amberlite XE-64 gave an optically active anionic resin that retained the enantiomers of mandelic acid differently when chloroform was the mobile phase (156). Other amines that have been used to form chiral ion exchangers are brucine (157) and (−)-*N,N*-dimethyl-α-phenylethylamine (158). In both cases, chloromethylated polystyrene resins were used to attach the amines.

The zwitterionic character of the amino acid is preserved when L-arginine is attached to Sephadex (159). The ion-exchanger gave complete resolution of racemic β-3,4-dihydroxyphenylalanine (DOPA) with a separation factor (α) as high as 1.6 but gave only partial resolution of tyrosine. The authors claim that the resolution is obtained because the enantiomers have different possibilities of interacting with the arginine residue of the stationary phase. They suggest that a three-point interaction can be established between the resin and the D-enantiomer: an electrostatic quadrupole attraction between amino and carboxylic groups in DOPA and the arginine molecule is formed. Furthermore, the phenolic function(s) in D-DOPA can interact with the basic ε-amino group of arginine. This third electrostatic interaction is not pronounced for the other enantiomer (that is, the L form). Without the possibility of a third interaction, phenylalanine did not show any stereoselective interaction with the L-arginine ion exchanger.

Recently a Japanese group reported on the separation of β-hydroxy-D,L-aspartic acid on resins containing L-lysine or L-ornitine as ion-exchanger (160). The resolution of hydroxyaspartic acid was found to depend on pH and the ammonium-ion derivative in the mobile phase. No enantiomeric separation was observed with pyridinium acetate, but with ammonium acetate the enantiomers showed different retention times. A drawback with the resins is the relatively low column efficiency and the spontaneous degradation of the support. The optically active amino acid bleeds the styrene-divinylbenzene copolymers.

LIGAND-EXCHANGE CHROMATOGRAPHIC RESOLUTION OF ENANTIOMERS WITH COVALENTLY BONDED LIGANDS

Ligand-exchange chromatography (LEC) has become a popular technique for separation of enantiomers, especially of amino acids and amino acid derivatives. A huge amount of data on ligand-exchange chromatographic systems has been accumulated during the years, and the technique has been reviewed several times. The most exhaustive treatments are given in articles by Davankov (7,8 and references therein), Audebert (5), and Lindner (9).

The principle of chiral ligand-exchange chromatography is formation of diastereomeric ternary complexes between the selector ligand (L -So), a transition metal (Me), and the selectand ligands (D, L-Sa), as written in Equation 7:

$$2 \text{ (L)-So-Me-(L)-So} + \text{(D,L)-Sa} \rightleftharpoons$$
$$\text{(L)-So-Me-(L)-Sa} + \text{(L)-So-Me-(D)-Sa} + 2 \text{ (L)-So} \quad [7]$$

When the column is packed with a chemically bonded ligand phase prepared dominantly from α-amino acids as the chiral So and loaded with a metal ion such as Cu(II), the injected ligands, chiral organic, and coordinating compounds can form mixed-chelate complexes with different stabilities; the enantiomer that gives the less stable complex is eluted first from the column. Various intramolecular interactions between the ligands in the ternary complex molecules are one of the causes for different stabilities of the diastereomeric complexes. The solid phase and spacer groups used as the matrix for grafting of the chiral selector ligand may also affect the stereoselectivity by participating in the coordination complex or by interaction — by, for example, steric repulsion — with the ligands in the metal complexes. Except for the structure of selector ligands, the kind of transition metals used also can influence the complexation. Systems with copper, cobalt, nickel, and other metal ions have been studied, but in general copper seems to be superior for separation of enantiomers. Other mobile phase constituents that can coordinate with the transition metal may also affect the resolution (7–9).

A prerequisite for resolution in LEC is, of course, that besides the selector molecules the selectand solutes can coordinate (complex) with the metal ion used. Although most principal studies with ligand-exchange chromatography have been made with enantiomers of α-amino acids and derivatives thereof, it is also possible to separate enantiomers of other classes, including carboxylic acids (161), amino alcohols as Schiff bases (162), barbiturates, hydantoins, and succinimide derivatives (9). Table VIII gives a survey of some ligand-exchange systems with chemically bonded selector ligands. The table includes the chiral ligand attached to the support, type of matrix, the metal ions used, and the type of racemate that was separated.

Davankov and co-workers have made systematic studies with several different ligands — for example, α-amino acids — bonded to poly(styrene-divinylbenzene) (7,8). The cyclic amino acids showed the highest stereoselectivity, and this has generally proved true for other matrices as well. A model to explain the difference in retention order for enantiomers with bidentate and tridentate α-amino acids is given in Figure 22 (8). Features of this model include the conformation and the participation of the N-benzyl moiety from the polystyrene backbone and the possible coordination of a water or ammonia molecule in the second axial position. That molecule will destabilize the LL complex of a bidentate amino acid, whereas fixation of the water molecule will not affect the stability of the DL complex (upper part of Figure 22). Consequently, the D form of the bidentate amino acid will be more highly retained. The lower part of Figure 22 illustrates the situation for tridentate amino acids such as hydroxyproline. The LL complex is stabilized by the additional coordination of the hydroxyl group to the metal ion. For the DL complex, however, there is no chance of coordination. The nature of the supports and mobile-phase components can affect the complexation and may give an elution order that is contradictory to what is predicted from the models above. The retention order of proline, a bidentate amino acid, is the reverse of what is expected.

Polyacrylamide gels have also been used as a matrix to bind optically active chelating ligands such as L-proline (165). In this case, the D form of α-amino acids is less highly retained than the L form — in contrast to what is found when using an L-proline-polystyrene phase. A possible explanation for the deviating behavior is given in Figure 23. The amide function in the gel is assumed to participate in the complexation process by coordinating to the metal ion, and this might affect the conformation and the relative stabilities of the ternary diastereomeric complexes (DL and LL) (165).

The high probability of forming and deforming complexes during elution through the column when high concentrations of copper are used gives high reten-

TABLE VIII

Chiral Stationary Phases Used in Ligand-Exchange Chromatography

Chemical bonded ligand	Matrix	Metal ion	Racemic analyte	References
More than 20 different L (or D) amino acids	Polystyrene	Cu^{2+}, Ni^{2+}, Zn^{2+}	Various D,L-amino acids, mandelic acid, 2-amino alcohols(1), labeled D,L-amino acids	7,8 and references therein
N-carboxymethyl-L-Val	Polystyrene	Cu^{2+}	D,L-amino acids	163
(R)-N,N'-dibenzyl-1,2-propanediamine	Polystyrene	Cu^{2+}	D,L-amino acids	164
L-Pro,L-amino acids	Polyacrylamide	Cu^{2+}	D,L-amino acids	5,165
L-Pro	Silica gel	Cu^{2+}	D,L-Tyr; D,L-Trp; D,L-Pro	166,167
L(or D)-Pro, L-OHPro L-Val,L-His L-pipecolinic acid	Silica gel	Cu^{2+}, Ni^{2+}, Cu^{2+}, Zn^{2+}	D,L-amino acids; D,L-DOPA	168–170
L-Pro-amide	Silica gel	Cu^{2+}	D,L-Trp, D,L-Tyr, D,L-Phe	171
L-Pro-amide	Silica gel	Cd^{2+}	DNS-D,L-amino acids, barbiturates, hydantoins	9,172–174
t-BOC-L-Pro- or L-Val-amide	Silica gel	Cu^{2+}	DNS-D,L-amino acids; D,L-amino acids	175
Linear polyacrylamide-L-Pro-amide	Adsorbed on silica gel	Cu^{2+}	D,L-amino acids	176
L-Phe-Ala, L-OH-Pro	Polyacrylamide	Cu^{2+}	D,L-amino acids	177
L-Methionine-D-sulfoxide, L-methionine-L-sulfoxide	Polystyrene	Cu^{2+},Ni^{2+}	D,L-amino acids	178
L-His	Silica gel	Cu^{2+}	D,L-amino acids, D,L-DOPA, D,L-mandelic acid	179
L-Pro; L-HOPro	Silica gel	Cu^{2+}	D,L-amino acids	180–182
(R,R)-Tartaric acid amide	Silica gel	Cu^{2+}	D,L-amino acids, α-hydroxy acids, norepinephrine	183

TABLE VIII *(Continued)*

N-ω-(dimeth-yl-siloxyl)-undecanoyl-L-valine	Silica gel	Cu^{2+}	DNS-DL-amino acids	184

tion times and eventually, but not necessarily, high separation factors for the enantiomers. Increasing the concentration of ammonia, which can coordinate with copper and thus act as a competitor for the Sa molecules, will decrease retention and give lower stereoselectivity as a result of a reduction in the frequency of enantioselective coordination by the Sa ligands. Temperature and flow rate of the mobile phase affect column efficiency and thus the ability to resolve enantiomers because resolution (R_s) is a function of both stereoselectivity (α) and column efficiency (N).

In order to increase column efficiency and to improve resolution as well as detection sensitivity, several research groups have begun to use microparticle silica gel as a matrix to which the selector ligand is bonded via a spacer. These columns are used at high temperatures (35°–60°C) to improve mass transfer and column efficiency. Supports I, IV, and V in Figure 24, each of which has free carboxyl

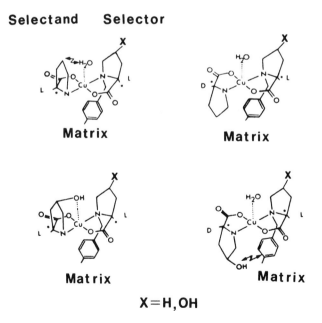

Selectand   Selector

Matrix   Matrix

Matrix   Matrix

X=H,OH

FIGURE 22.   Proposed models for stereoselective interaction for bidentate and tridentate amino acids and a copper-L-proline-containing stationary phase. (Reprinted, with permission, from Reference 8.)

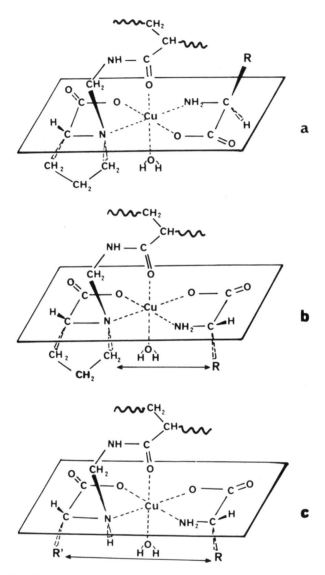

FIGURE 23.   Possible structure of ternary metal complex with polyamide supports. (a) and (b): *trans* and *cis* form with L-proline as grafted ligand; (c): complex with noncyclic amino acid bounded to the polyamide matrix. (Reprinted, with permission, from Reference 165.)

functions, have under optimal conditions resolved a majority of the free α-amino acids. The hydroxyl groups and amide groups of the spacer and the selector are believed to participate in complexation. Dansyl derivatives (DNS) of all common α-amino acids with primary amine function have been resolved on ligand II (Figure

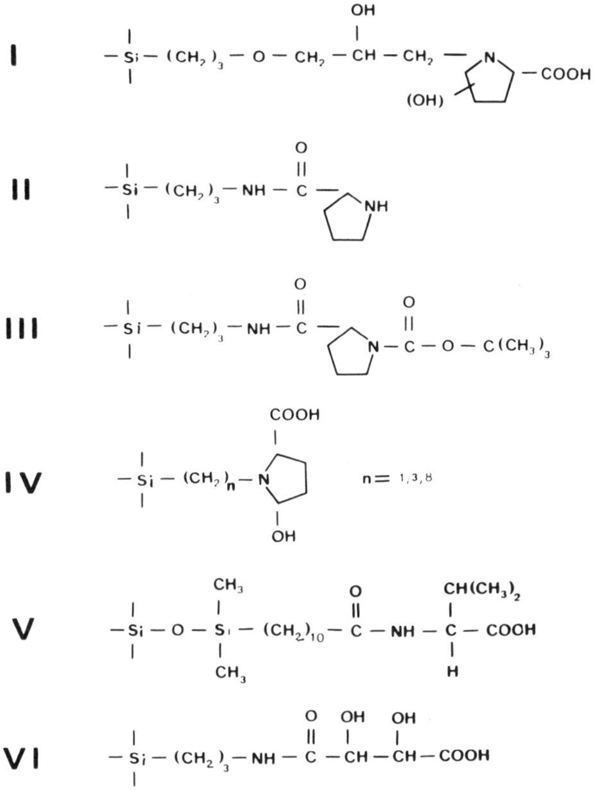

FIGURE 24. Structure of ligands covalently fixed to silica. I. Gübitz et al. (168); II. Lindner et al. (9,173); III. Engelhardt et al. (175); IV. Roumeliotis et al. (181–182); V. Karger et al. (184); VI. Lindner et al. (183).

24) chemically bonded to silica (9). For the DNS-amino acids the stereoselectivity was in the range of 1.04 to 3.16 except for proline, which could not be resolved under these conditions. Several underivatized α-amino acids can also be resolved on these columns. Applying DNS derivatives in chromatography has the advantage of giving high detection sensitivity with UV and fluorescence detectors. An application showing the separation and detection of an impurity of 0.2% of DNS-D-methionine present in the L form is given in Figure 25 (173). Further applications are presented in Figure 26. The enantiomers of methyleudan and hexobarbital are separated in a system with L-prolinamide as support, and 0.1 M ammonium acetate and 1 mM cadmium acetate (pH 9.0)/methanol (7:3) as mobile phase (9). Retention and stereoselectivity may be controlled by pH, organic modifier, and metal ion. If copper is exchanged for cadmium, the elution order between the enantiomers is reversed.

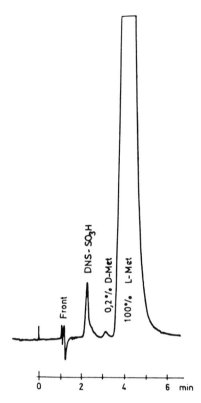

FIGURE 25.   Detection of D-methionine as an impurity in L-methionine, chromatographed as dansyl derivatives on a silica-gel L-proline-amide phase. (Reprinted, with permission, from Reference 173.)

The *N-t*-BOC-blocked L-prolinamide (III) and its valine analogue can also be used for enantiomeric separation of DNS-amino acids as well as for separating enantiomers of unmodified tryptophan and proline (175). The influence of the length of the spacer between the ligand and the silica gel surface was studied with different analogues (n = 1,3,8) of adsorbent IV (Figure 24) (181,182). The experimental results indicate, according to the authors, that the stereoselectivity can be a function of several multiple solute-surface interactions, including complexation, ion-exchange, and hydrophobic interactions. It was also found that silica gel with different surface areas (501 m^2/g and 316 m^2/g) displays different stereoselectivity for amino acids under otherwise equal experimental conditions (181,182).

Karger and co-workers (184) used *N*-ω-(dimethylsiloxyl)undecanoyl-L-valine as a chemically fixed ligand and copper as the metal ion in their studies of diluted phases. In order to minimize possible interference from neighboring ligands on the stereoselective complexation, the surface was diluted with nonchelating groups. These diluted phases promote a higher stereoselectivity than the concen-

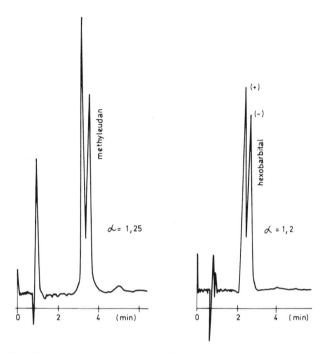

FIGURE 26.   Direct enantioseparation of racemic barbiturates on a L-Pro-amide-CLEC system. Column: 200 mm × 4 mm, 7-μm Lichrosorb Si 100-L-Pro-amide; mobile phase: 0.1 M NH$_4$Ac + 1 mM Cd(Ac)$_2$, pH 9.0/MeOH(700:300); flow rate: 3 mL/min; temperature: 30 °C; detection: UV 254 nm. (Reprinted, with permission, from Reference 9.)

trated ones and improve chromatographic performance by increasing column efficiency. Enantiomers of aminoalcohols that have been derivatized with salicylaldehyde to form Schiff bases have been separated on similar phases (162).

In addition to the chiral ligand selectors based on α-amino acid derivatives, a new chiral ligand grafted onto silica gel has recently been developed (183). This hydroxy acid phase shows enantioselectivity for norepinephrine and some methylated α-amino acids. The same fundamental limitations will be obvious as with all amino acid ligands, namely, that their usefulness as chiral selectors depends strongly on the necessity that the enantiomers to be resolved must be capable of chelating with transition metal ions. This requirement is a major restriction to the molecular structure of the solutes; as seen from the references, the enantioseparation of five-membered chelate ring structures, such as possessed by α-amino acids, is the most favorable.

In summary, the ligand-exchange chromatographic systems based on covalently bonded ligands are usually used at elevated temperatures in order to increase column efficiency. Retention and stereoselectivity are optimized by pH, the nature and concentration of organic modifier, and by choice of metal ion. As has been pointed out, the influence of several components of the mobile phase and of the support material itself on the complexation process may be difficult to foresee, and thus impair the possibility of predicting stereoselectivity and retention order for the enantiomers. Preparative resolutions of amino acids such as proline and threonine have been obtained in ligand-exchange chromatography with grafted ligands (185).

DIRECT RESOLUTION OF ENANTIOMERS USING CHIRAL ADDITIVES IN THE MOBILE PHASE

Ligand-Exchange Chromatography in Systems with Chiral Additives

Ligand-exchange chromatography for separation of enantiomers can also be performed by adding the selector ligand along with the transition metal to the mobile phase (chiral eluent). When injecting a racemic mixture of a compound that can coordinate with the central metal ion, two diastereomeric complexes may be formed between the enantiomer selectands and the selector-metal ion complex (So-M). Several different retention and resolution mechanisms are possible depending on the nature of the ligand added to the mobile phase (So) and on the ligands to be separated (Sa). Mobile phase components such as methanol, bases, ammonia, or buffer may also participate to some extent in ligand-exchange processes.

The selectands or the ternary complex formed between Sa and So-Me can be retained in the stationary phase. The mobile phase ligand will also be adsorbed (coated) to the solid phase and form a chiral stationary phase. Chromatographic systems similar to those with covalently bound ligands are formed when the adsorbed ligands are loaded with metal ions.

Nonchiral support materials can be used as an adsorbing stationary phase in applications of ligand-exchange chromatography using chiral eluents. Retention and stereoselectivity can easily be regulated by changes in mobile phase composition involving buffer or the organic modifier. The technique of adding a selector ligand and a metal ion to the mobile phase is straightforward, which has made this a popular chromatographic technique for resolution of chelating enantiomers. A variety of fundamental studies and applications of ligand-exchange chromatography with chiral eluents have been published. Some of these are summarized in Table IX. Lindner and colleagues introduced L-2-alkyl-4-octyl-2-diethylenetriamine (Figure 27) as a chiral selector ligand for resolution of dansylated amino acids (186,187). The separation of some DNS-amino acid enantiomers using this selector is shown in Figure 27 (187). The optically active diethylenetriamine imposed a rigidity in the ternary metal complex with the chelating selectands and gave separation factors between 1.04 (DNS-norvaline) and 1.34 (DNS-serine). The stereoselectivity changes with the structure of the alkyl group in the diethylenetriamine. The change is probably attributable to different steric and/or hydrophobic interactions caused by the alkyl groups.

TABLE IX

Selection of Chiral Mobile Phase Additives (Selectors) for Performing
Chiral Ligand-Exchange Chromatography

Chiral selector (So)	Adsorbents (stationary phase)	Metal ion	Selectand (Sa)	References
L-2-isopropyl-4-n-octyl-diethylene-triamine	RP 8	Zn^{2+}	DNS-D,L-amino acids	186
[C_3–C_8)-diene] N-octyl-L-Pro-amide	RP 8	Zn^{2+},Cd^{2+} Ni^{2+},Cu^{2+} Hg^{2+}	DNS-D,L-amino acids, dipeptides	187
N-octyl-L-Pro-amide	RP 18	Ni^{2+}	DNS-D,L-amino acids	188–190
L-Pro, D-Pro	Cation exchanger	Cu^{2+}	D,L-amino acids	191
L-Pro, D-Pro	RP 18	Cu^{2+}	D,L-amino acids	192
N-alkyl-L-OHPro	RP 18	Cu^{2+}	D,L-amino acids	193
L-Asp-L-Phe-methyl ester	RP 18	Cu^{2+},Zn^{2+}	D,L-amino acids	194,195
L-Aspartyl-alkylamide	RP 18	Cu^{2+}	D,L-amino acids	196,197
L-Pro, L-Arg, L-His	RP 18	Cu^{2+}	DNS-D,L-amino acids	198
L-Phe	RP 18	Cu^{2+}	D,L-methyldopa, D,L-Trp	199
L-Pro	Silica gel	Cu^{2+}	D,L-thyroid hormones	199
N-decyl-L-histidine	Zorbax C_8 and ODS	Cu^{2+}	D,L-amino acids	200
N,N,N',N'-tetramethyl-(R)-propanediamine-1,2	RP 18	Cu^{2+}	DOPA, Phe, Trp, mandelic acid	201
N,N-dialkyl-L-amino acids	Nucleosil C_{18}	Cu^{2+}	D,L-2-hydroxy acids	161
L-His-methyl ester	Nucleosil C_{18}	Cu^{2+}	D,L-DNS-amino acids	202
L-Phe	Supelcosil LC-18	Cu^{2+}	Dopa, methyldopa, carbidopa, Tyr	203
N,N-dialkyl-amino acids	C_{18}	Cu^{2+}	D,L-amino acids	204
N-(p-toluenesul-fonyl)-L-phenylala-nine	C_{18}	Cu^{2+}	D,L-Pro	205
(R,R)-tartaric acid mono-n-octylamide (TAMOA)	C_{18}	Cu^{2+},Ni^{2+}	D,L-amino acids	206

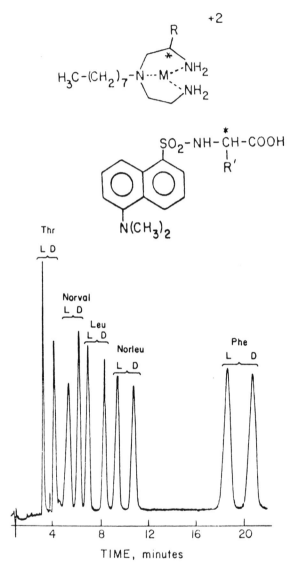

FIGURE 27.   Separation of D,L-DNS-amino acids. Column: 15 cm × 4.6 mm Hypersil 5-μm C₁₈; flow rate: 2 mL/min; conditions: 0.65 mM L-2-isopropyl-dien-Zn(II); 0.17 M NH₄Ac to pH 9.0 with aqueous NH₃; mobile phase: 35:65 acetonitrile/H₂O; temperature: 30 °C. Solutes: Thr = threonine; Norval = norvaline; Leu = leucine; Norleu = norleucine; Phe = phenylalanine. (Reprinted, with permission, from Reference 187.)

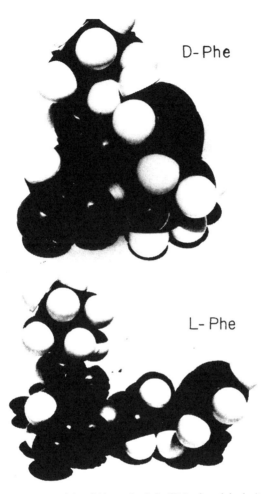

FIGURE 28. Space-filling models of (Asp-c-hex)-Cu(II)(L-phenylalanine) and (Asp-c-hex)-Cu(II)(D-phenylalanine). (Reprinted, with permission, from Reference 196.)

The sweetener aspartame (L-aspartyl-L-phenylalanine methyl ester) (194) and other amino acid amides (196,197) were introduced by Gilon and co-workers as chiral additives for separation of amino acid, with copper as the chelating metal ion. The use of the L-aspartyl-cyclohexylamide/Cu(II)complex (Asp-c-hex complex) as selector ligand has the advantage of facilitating detection. The five-membered ring that is formed produces a ternary metal complex with UV-absorbing properties near 230 nm (196). The difference in stereoselectivity obtained for amino acids with this selector ligand was interpreted to stem from differences in hydrophobic interactions between the two ligands in the metal complex (Figure 28) (196).

The model assumes a three-point contact for the D-enantiomers based on an overlap between the two alkyl groups and a two-point interaction of the L-carboxy and L-amino groups of the amino acid and the Cu(II) in the Asp-c-hex complex. The L-enantiomer is coordinated by the carboxylic and the amine function, but no additional hydrophobic interaction seems to be possible according to the molecular model. The aspartame ligand has been used in bioanalytical applications for determination of pipecolic acid enantiomers in urine from patients with hypersinemia (195). A chromatogram of the enantiomeric separation of pipecolic acid in urine is given in Figure 29 (195). A modified silica support, Nucleosil C_{18} (Macherey-Nagel & Co.), was used as the solid phase; the mobile phase was a water solution of aspartame and $CuSO_4$.

Hydrophobic interactions in reversed-phase chromatography can also be used to coat (adsorb) hydrophobic chiral ligands onto the solid phase. Modified reversed-phase materials have been coated with nonpolar N-alkyl-L-hydroxyproline (alkyl = C_7H_{15}, $C_{10}H_{21}$, and $C_{16}H_{33}$) (193). When loaded with copper or other transition metal ions, the chiral stationary phase can give stereoselective retention for ligands such as amino acids. Hydrophobic interactions in these systems are also believed to play a primary role in retention and stereoselectivity.

A similar principle was applied with L-prolyl-N-n-octylamide (nC_8 proamide) and L-prolyl-N-n-dodecylamide Ni(II) complexes as additives that gave resolution for DNS-amino acids (188,189). A coupled column technique was used to obtain high separation selectivity for amino acids and their enantiomers (Figure 30) (189). In the first column, a nonchiral system was used to separate the different

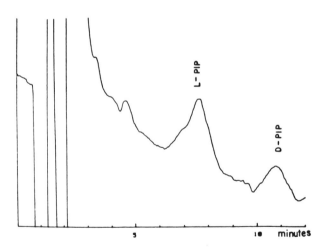

FIGURE 29.   Pipecolic acid in urine of a patient with hyperlysinemia. Column: 150 mm × 4.2 mm Nucleosil 5-μm C_{18}, was equilibrated with loading buffer containing 295 mg L-aspartame and of copper sulfate in 1 L of water; mobile phase: 100 mg of copper sulfate and 25 mL of loading buffer diluted to 1 L with water; flow rate: 1.5 mL/min. (Reprinted, with permission, from Reference 195.)

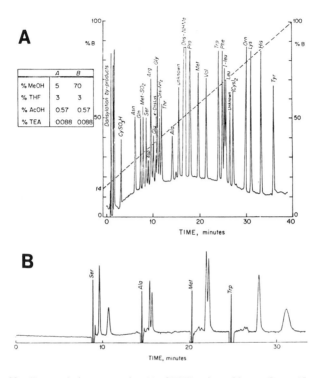

FIGURE 30.   Reversed-phase separation (a) of DNS-amino acids on a 5-μm, 15 cm × 4.6 mm C₁₈ Hypersil column; gradient conditions given in the figure; AcOH = acetic acid, TEA = triethylamine. Chiral separation (b) of four D,L-DNS-amino acids that have been first eluted from the achiral column (conditions as in Figure 30a) and the peak fractions have been transferred onto the chiral system. Other conditions same as in Figure 30a except mobile phase: 1.75×10^{-1} M ammonium acetate; flow rate: 3 mL/min. (Reprinted, with permission, from Reference 189.)

amino acids; the fractions of interest were then transferred to a second column. The second column was eluted with a mobile phase containing L-propyl-*N-n*-octylamide-Ni to promote stereoselective retention of the enantiomers of the amino acids.

DNS-amino acids have also been separated with more hydrophilic ligands, including L-proline (195,198), L-arginine, L-histidine, and L-histidine methyl ester (195,198) having copper as the chelating metal. A high carbon content and bulky groups increased the retention and stereoselectivity, indicating that hydrophobic interactions play a dominant role. The L-histindine methyl ester gives relatively high stereoselectivity and was used to determine the amino acid profile in an analysis of cerebrospinal fluid (CSF) for patients with bacterial meningitis (Figure 31) (195).

Hare and Gil-Av used a somewhat different approach for separation of amino acids using an ion-exchange resin as the stationary phase (191). Stereoselectivity was promoted by addition of L-proline and copper to the mobile phase. The amino acids were separated in underivatized form to avoid overloading effects that would affect enantioselectivity, and a postcolumn reaction of the amino acid with *o*-phthalaldehyde was needed to increase detection sensitivity.

In the form of ligand-exchange chromatography which has both the selector ligand (So) and the metal ion in the mobile phase, there is a simple way to verify whether a suspected enantioselective separation is a real separation or an artifact. The change in absolute configuration of So should give rise to reversed retention order for the enantiomers; when a racemic mixture of So is used, no resolution of the enantiomers (double peaks) should be observed (Figure 32) (191).

Besides the various applications to the resolution of amino acids and amino acid derivatives, other types of compounds (Sa) may be resolvable by chiral ligand-exchange chromatographic (CLEC) techniques, provided they are capable of forming moderately stable chelate complexes with transition metal ions. For example, this approach can be applied to the resolution of hydroxyphenyl phenyl hydantoin, a metabolite of phenytoin. The chiral mobile phase additive is nC_8 proamide, and Figure 33 shows resolution of a standard sample. This method is routinely used to screen the urine of patients in order to spot deviations in this stereospecific hydroxylation reaction (190).

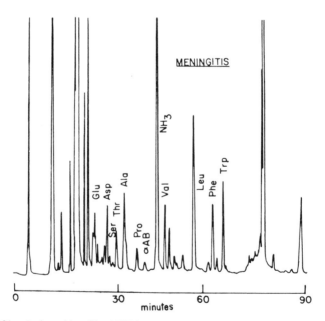

FIGURE 31.   Amino acid profile of CSF from a patient with meningitis. Column: 150 mm × 4.2 mm Nucleosil 5-μm C_{18}; mobile phase: 20% acetonitrile in a buffer containing 5 mM L-histidine methyl ester, 2.5 mM $CuSO_4$ + 5 H_2O and 2.0 g ammonium acetate/L of deionized water, pH 7.0; flow rate: 2.0 mL/min. (Reprinted, with permission, from Reference 195.)

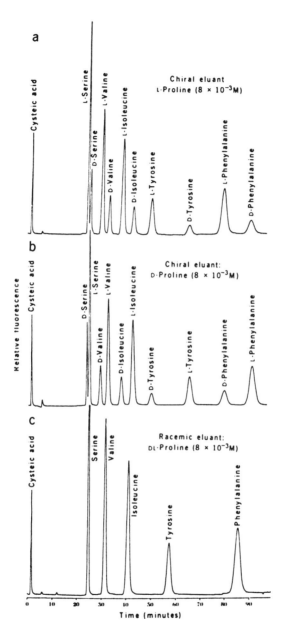

FIGURE 32.   Effect of the chiral eluent on the separation of D- and L-amino acids. Column: 120 mm × 2 mm packed with ion-exchanger DC 4a resin; mobile phase: sodium acetate buffer (0.05 M, pH 5.5) containing 4×10^{-3} M CuSO$_4$ and 8×10^{-3} M of (a) L-proline, (b) D-proline, (c) D,L-proline; temperature: 75 °C. (Reprinted, with permission, from Reference 191.)

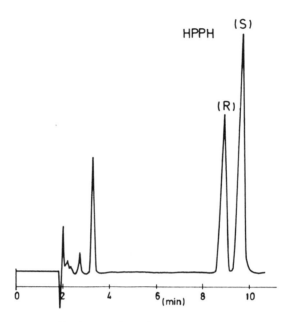

FIGURE 33.   Resolution of 5-*p*-hydroxyphenyl-5-phenyl hydantoin (HPPH). Column: 250 mm × 4.6 mm, 5-μm Spherisorb ODS; mobile phase: 0.1 M NH$_4$Ac + 1.5 mM *n*-octyl-L-prolinamide + 1.5 mM NiAc, pH 9.0, adjusted with NH$_3$(conc.) and methanol (55:45, v/v).

Chiral Resolution in Ion-Pair Chromatography

Ion-pair chromatography has been successfully applied as an efficient separation technique for compounds that can be ionized. The technique can also be used for separation of enantiomers by adding a chiral counter ion to the mobile phase, as was demonstrated by Pettersson and Schill (207). The separation of enantiomers is based on the formation of diastereomeric ion pairs with the chiral counterion; thus a mobile phase of low polarity is used to promote a high degree of ion-pair formation. Nonchiral silica supports are used as adsorbing stationary phases.

The chiral counterion and the enantiomers to be separated should have complementary structures. It seems that, besides the electrostatic attraction between the amine and the acidic function, further interactions — hydrogen-bonding, hydrophobic, and steric interactions — between the ions are necessary to yield diastereomeric ion pairs with different distribution properties (208,209). Retention and stereoselectivity can be regulated by structure and concentration of the counter ion as well as by the nature of the stationary phase. The (+)-10-camphorsulfonic acid (Figure 34a) was used in normal-phase chromatography as a chiral counterion for separation of enantiomers of β-receptor blocking agents (derivatives of 1-aryl-2-aminoalcohols) (207,208). Small changes in the structure of the solutes may have a drastic effect on stereoselectivity, as is demonstrated in Table X (208). The dis-

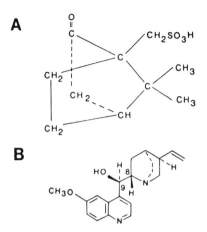

FIGURE 34. Counterion structure. (a) (+)-10-camphorsulfonic acid; (b) quinine.

TABLE X

Ion-Pair Separation of Enantiomeric Amines*

$$R_2 - \text{C}_6\text{H}_3(R_1) - R_4 - \underset{\overset{|}{OH}}{\overset{H}{C}} - (CH_2)_n - NHR_3$$

Number	n	R_1	R_2	R_3	R_4	$\alpha(-/+)$
1	1	H	$CH_2CH_2OCH_3$	$CH(CH_3)_2$	$-OCH_2-$	1.09
2	2	H	$CH_2CH_2OCH_3$	$CH(CH_3)_2$	$-OCH_2-$	1.00
3	3	H	$CH_2CH_2OCH_3$	$CH(CH_3)_2$	$-OCH_2-$	1.00
4	1	H	$CH_2CH_2OCH_3$	$CH(CH_3)_2$		1.09
5	1	H	OCH_2CHCH_2	$CH(CH_3)_2$	$-OCH_2-$	1.08
6	1	CH_2CHCH_2	H	$CH(CH_3)_2$	$-OCH_2-$	1.08
7	1	OCH_2CHCH_2	H	$CH(CH_3)_2$	$-OCH_2-$	1.00
8	1	H	OCH_3	$CH(CH_3)_2$	$-OCH_2-$	1.08
9	1	H	CH_2CH_3	$CH(CH_3)_2$	$-OCH_2-$	1.09
10	1	H	$OCH_2CH_2OCH_3$	$CH(CH_3)_2$	$-OCH_2-$	1.09
11	1	CH_2CHCH_2	H	$CH_2CH_2OC_6H_4CONH_2$	$-OCH_2-$	1.11
12	1	Cl	H	$-CH_2OC_6H_4CONH_2$	$-OCH_2-$	1.11
13	1	Br	$CH_2CH_2OCH_3$	$-CH_2OC_6H_4CONH_2$	$-OCH_2-$	1.11
14	1	CH_3	H	$CH_2CH_2CH_2C_6H_5$	$-OCH_2-$	1.00
15	1	H	H	$CH_2CH_2C_6H_4CH_3$	$-OCH_2-$	1.00

* Solid phase: LiChrosorb Diol; mobile phase: (+)-10-camphorsulfonate, 2.2×10^{-3} M, in methylene chloride/1-pentanol (199:1). Reprinted, with permission, from Reference 207.

tance between the amino and hydroxyl group is of vital importance for stereoselectivity. The loss of enantioselectivity is probably attributable to lack of a simultaneous electrostatic interaction and hydrogen bonding with the oxo group in the
counterion as the distance between the amine and hydroxyl function increases.
Substituents at the nitrogen atom may also affect stereoselectivity (Table X). Szepesi and co-workers have also used the same acid for separation of enantiomeric
alkaloids (210).

Enantiomers of sulfonic and carboxylic acids as well as *N*-blocked amino acids
can be separated with cinchona alkaloids such as quinine (Figure 34b) as the chiral
counterion (209,211). Adding quinine as the chiral counterion to the mobile phase
has, furthermore, the advantage that it may increase detection sensitivity. Indirect
detection of solutes with a UV detector is possible when a UV-absorbing probe is
present in the mobile phase (212–214). An application is given in Figure 35 for the
detection of an enantiomeric impurity of 0.7% in (+)-10-camphorsulfonic acid

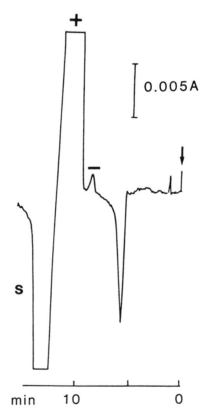

FIGURE 35.   Determination of optical impurity in (+)-10-camphorsulfonic acid. Column: 5-μm
LiChrosorb Diol; mobile phase: quinine and acetic acid 0.17 m*M* in dichloromethane/L-pentanol
(99:1); S = system peak. (Reprinted, with permission, from Reference 209.)

(209). The small positive peak is the response from 1 nmol of $(-)$-10-camphorsulfonic acid in a sample of $(+)$-10-camphorsulfonic acid that gives the main positive peak. The negative peak is the system peak, which is a peak appearing upon injection of a sample into the chromatographic system. The enantiomers of 10-camphorsulfonic acid are detected solely by changes in counterion concentration because the enantiomers have no inherent absorbing properties at 337 nm, which is the wavelength monitored by the detector.

A partial resolution of DL-tryptophan and glycyl-D,L-phenylalanine in reversed-phase chromatography with a zwitterionic ion-pair reagent, L-leucine-L-leucine-L-leucine, has been reported (215).

Enantioselective Extraction System

Diesters of tartaric acid have been used in partition studies to promote stereoselective distribution of aminoalcohols (216). The amines were distributed to the organic phase as ion pairs with a nonchiral counterion, for example, hexafluorophosphate. The ester or mixtures of the ester and halogenated hydrocarbons (dichloroethane) constituted the nonpolar phase. The chiral recognition obtained for the enantiomeric amines is suggested to stem from differences in multipoint hydrogen bonding with the chiral ester.

One ester of tartaric acid (di-5-nonyltartrate) was used at low temperature ($2\,°$–$8\,°C$) in rotation-locular countercurrent chromatography for separation of enantiomers of norephedrine (217). The principle has also been transfered to HPLC. The $(+)$-di-n-butyltartrate was used as chiral stationary phase to separate the enantiomers of aminoalcohols, including ephedrine and norephdrine (218). The ester, which is a liquid, is coated on the support (Phenyl Hypersil; Shandon Southern, Sewickley, Pennsylvania) by pumping the mobile phase (equilibrated with the ester) through the column. A chromatogram of the resolution of norephedrine with the n-butyl ester of tartaric acid as a stationary phase is given in Figure 36 (218).

Proteins as Chiral Additives in the Mobile Phase

Macromolecules bound to a rigid matrix have, as was discussed previously, been used as the chiral stationary phase in LC for resolution of enantiomers. A further way to use the stereoselective discrimination of macromolecules such as proteins is to add them to the mobile phase as complexing agents. Different interactions of the enantiomers with the protein will produce different retention times; the enantiomer with the highest affinity to the complexing agent will thus elute first from the column.

The principle has been used with albumin as the complexing agent in the mobile phase for separation of enantiomers of tryptophan (219) as well as for resolution of aromatic mono- and dicarboxylic acids (221). The technique has also been applied to determining the total affinity of, for example, furosemide, warfarin, and phenylbutazone to albumin (220). Retention of the enantiomers can in these sys-

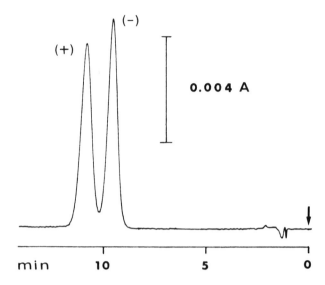

FIGURE 36.   Resolution of ($\pm$)-norephedrine. Column: Phenyl Hypersil; liquid stationary phase: (+)-di-*n*-butyltartrate; mobile phase: 90 m*M* KPF$_6$ in phosphate buffer, pH 6, μ = 0.1, total ionic strength 0.1. Peaks: ($-$)-norephedrine and (+)-norephedrine. (Reprinted, with permission, from Reference 218.)

tems be regulated by concentration of albumin, pH, or addition of an ion-pair reagent to the mobile phase. The same principle was also used for separation of enantiomeric amines with α-glycoprotein as the complexing agent (222).

A disadvantage when using albumins and similar proteins as chiral selectors — whether chemically bonded to a solid matrix or added as complexing agents to the mobile phase — is that little is known about the resolution mechanism. The nature of the binding sites responsible for the stereoselective interactions is generally unknown, so it is difficult to predict the resolution of a racemic mixture.

Miscellaneous Chiral Additives

Stereoselective retention by inclusion complexation with β-cyclodextrin can also be performed in chromatographic systems with the dextrin present in the mobile phase (223,224). The technique has been used for separation of enantiomers of a new analgesic, 1-[2-(3-hydroxyphenyl)-1-phenylethyl]-4-(3-methyl-2-butenyl)piperazine, as is illustrated in Figure 37 (225).

Recently, *N*-(2,4-dinitrophenyl)-L-alanine-*n*-dodecyl ester, a π-acceptor, was used as the chiral selector in the mobile phase (methanol/water) for separation of enantiomers of 1-aza-hexahelicene (226). Different amide derivatives have been used as chiral additives in nonpolar mobile phases. L-(+)-*N,N*-diisopropyltartramide present in the mobile phase promotes different retention of enantiomers with hydrogen-bonding functions, including hydroxycarboxylic acids, amino ac-

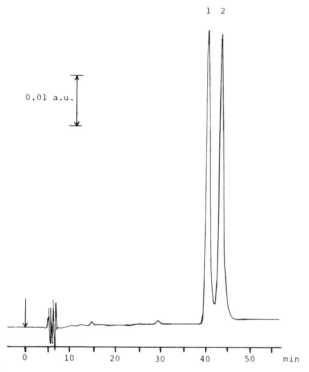

FIGURE 37.   Chromatogram of the enantiomeric resolution of (+/−)-1-(2-(3-hydroxyphenyl)-1-phenylethyl-4-(3-methyl-1-butenyl)piperazine. Column: 250 mm × 4.6 mm Develosil ODS-5; mobile phase: 0.1 *M* acetate buffer/ethanol/β-cyclodextrin (82:12:4); flow rate: 0.5 mL/min. Peaks: 1 = *S*(+)-enantiomer (15 μg); 2 = *R*(−)-enantiomer (15 μg). (Reprinted, with permission, from Reference 225.)

ids, and derivatives of 1,2-diols (227). *N*-acetyl-L-valine-*tert*-butylamide as chiral additive in the mobile phase gave chromatographic systems that could separate enantiomers of amino acid derivatives (228,229). Formation of diastereomeric adducts with the chiral amide selector is believed to be the cause of resolution. Changes in amide concentration of the mobile phase were found to affect retention as well as stereoselectivity (229).

CONCLUSION

Table XI summarizes the main applications of the various chromatographic enantioseparation techniques, focusing on drugs and biologically active compounds. Separation of amino acids is not taken into consideration here. At this time, only a few racemic drugs are resolved by LC, so it did not seem useful to

TABLE XI

LC Separations of Racemic Drugs

Type	Compound	Separation technique(s)	References
Adrenergics	Catecholamines	LEC with chiral additives	162
	Ephedrine/ norephedrine	Indirect LC; LEC with chiral additives	62, 230, 96
	Epinephrine/ norepinephrine	Indirect LC	7, 19
	1-phenyl-2-amino-propanes	Indirect LC; enantioselective charge transfer; LEC	56, 33, 93, 162
Analgesics	Piperazine derivative	Miscellaneous chiral additives	225
Antacids	Sulfoxide	Bonded proteins	150
Antibiotics	Oxfenicine	Indirect LC	35
	Penicillamine	Indirect LC	43
Anticoagulant	Warfarin	Indirect LC; bonded proteins	38, 52, 239, 147
Antidepressants	Trimipramine	Bonded proteins	222
Antiherpes	Guanine derivatives	LEC with chiral additives	244
Antihypertensives	Dopa	Chiral ion-exchangers	159
Antiinflammatories	2-Arylpropionates	Indirect LC; enantioselective charge transfer	18, 102, 91
	Ibuprofen	Enantioselective charge transfer	22, 236
	Naproxen	Indirect LC	29, 211
	Carprofen		23
Barbiturates	Hexobarbital	Polyacryl- or polymethacryl-based CSP	6, 242
	Barbiturates	Beta-cyclodextrin	129
Beta-receptors (antiadrenergics)	Beta-blockers	Indirect LC; direct resolution using chiral additives	54, 207, 208
	Propranolol	Enantioselective charge transfer; indirect LC; direct resolution using chiral additives; polymeric chiral stationary phases	95, 231, 44, 45, 40, 65, 58, 207, 208, 130
	Oxprenolol	Indirect LC	57
	Alprenolol	Indirect LC; enantioselective extraction system	232, 207, 208

TABLE XI (Continued)

	Metoprolol	Indirect LC; direct resolution using chiral additives	232, 207, 208
	Medroxolol		233
	Nadolol (diastereomers)		234, 235
Carcinogens/mutagens	Polyaromatic hydrocarbons	Enantioselective charge transfer	115
	Thalidomide	Polyacryl- or polymethacryl-based CSP	245,135
Cardiac depressants	Disopyramide	Bonded proteins; indirect LC	237, 238, 152, 222
CNS depressants	Oxazepam	Polyacryl- or polymethacryl-based CSP; bonded proteins	136, 145
Hormones	Tyroxines	Indirect LC	240, 42
	Iodinated thyronines	LEC with chiral additives	199, 241
Insecticides	Pyrethroid acids	Indirect LC; enantioselective charge transfer	21, 101
Neuroleptics	Promethazine	Bonded proteins	222
Smooth muscle relaxant	Proxyphylline		46
Miscellaneous	Omeprazole		150
	Tropic acid derivatives	Enantioselective charge transfer	103
	Mandelic acid	Indirect LC; microcrystalline triacetyl cellulose; miscellaneous chiral additives	70, 124, 242, 223, 243, 132, 133
	Various basic drugs	Bonded proteins	153
	Eburnane alkaloids	Enantioselective extraction	210
	Indole alkaloids	Beta-cyclodextrin	131

classify specially the applications by well-defined pharmaceutical terms; instead, the compounds are put in order from a chemical point of view.

This chapter has reviewed advances in the resolution of optical isomers using liquid chromatographic techniques. Direct and indirect approaches were examined, with particular emphasis (attributable to number of current applications) given to the direct separation of enantiomers. An effort was made to discuss methods development and the advantages and disadvantages involved in each technique. It is hoped that this survey will encourage further work in the field of chromatographic enantioseparations.

REFERENCES

(1) *Asymmetric Synthesis,* Volume 1, *Analytical Methods,* J. Morris, ed. (Academic Press, New York, 1983).

(2) *Stereochemistry and Biological Activity of Drugs,* A. Ariens, W. Soudijn, and P. Timmermans, eds. (Blackwell Scientific Publications, Oxford, 1983).

(3) M. Simonyi, in *Medical Research Reviews,* Volume 4 (John Wiley & Sons, New York, 1984), p. 359.

(4) S. Allenmark, *J. Biomed. Biophys. Meth.* **9,** 1 (1984).

(5) R. Audebert, *J. Liq. Chromatogr.* **2,** 1063 (1979).

(6) G. Blaschke, *Angew. Chem. Int. Ed. Engl.* **19,** 13 (1980).

(7) V. Davankov, *Adv. Chromatogr.* **18,** 139 (1980).

(8) V. Davankov, A. Kurganov, and A. Bochkov, *Adv. Chromatogr.* **22,** 71 (1983).

(9) W. Lindner, in *Chemical Derivatization in Analytical Chemistry,* Volume 2, J.F. Lawrence and R.W. Frei, eds. (Plenum Press, New York, 1982), p.145.

(10) W. Pirkle and J. Finn, in *Asymmetric Synthesis,* Volume 1, *Analytical Methods,* J. Morris, ed. (Academic Press, New York, 1983), p. 87.

(11) T. Tamegai et al., *J. Liq. Chromatogr.* **2,** 1229 (1979).

(12) B. Testa, in *Principles of Organic Stereochemistry* (Marcel Dekker, New York, 1979), p. 1.

(13) P. Salvadori, C. Rosini, and C. Bertucci, *J. Org. Chem.* **49,** 5050 (1984).

(14) W. Pirkle, H. Hyun, A. Tsipouras, B. Hamper, and B. Banks, *J. Pharmacol. Biomed. Anal.* **2,** 173 (1984).

(15) V. Davankov and A. Kurganov, *Chromatographia* **17,** 686 (1983).

(16) V. Schurig, in *Asymmetric Synthesis,* Volume 1, *Analytical Methods,* J. Morris, ed. (Academic Press, New York, 1983), p. 59.

(17) W. Möhrke, *J. Chromatogr.* **307,** 145 (1984).

(18) J.M. Maitre, G. Boss, and B. Testa, *J. Chromatogr.* **299,** 397 (1984).

(19) H. Weber, H. Spahn, E. Mutschler, and M. Möhrke, *J. Chromatogr.* **307,** 145 (1984).

(20) G. Helmchen, G. Nill, D. Flockerzi, W. Schühle, and M. Youssef, *Angew. Chem.* **91,** 64 (1979).

(21) M. Jiang and D.M. Soderlund, *J. Chromatogr.* **248,** 143 (1982).

(22) D.G. Kaiser, G.J. Vangiessen, R.J. Reischer, and W.J. Wechter, *J. Pharm. Sci.* **65** (2), 269 (1976).

(23) K.M. Kemmerer, F.A. Rubio, R.M. McClain, and B.A. Koechlin, *J. Pharm. Sci.* **68** (2), 1274 (1979).

(24) R. Oshima, Y. Yamachi, and J. Kumanotani, *Carbohydr. Res.* **107,** 169 (1982).

(25) C.G. Scott, M.J. Petrin, and T. McCorkle, *J. Chromatogr.* **125,** 157 (1976).

(26) D. Valentine, K.K. Chan, C.G. Scott, K.K. Johnson, K. Toth, and G. Saucy, *J. Org. Chem.* **41** (1), 62 (1976).

(27) J. Gaal, J. Inczedy, R.D. Gillard, P. O'Brien, and S.E. Turgoose, *J. Chromatogr.* **174,** 212 (1979).

(28) G. Helmchen, H. Völter, and W. Schühle, *Tetrahedron Lett.* **16,** 1417 (1977).

(29) J. Goto, N. Goto, and T. Nambara, *J. Chromatogr.* **239,** 559 (1982).

(30) J.K. Whiretell and D. Reynolds, *J. Org. Chem.* **48,** 3548 (1983).

(31) G. Helmchen and W. Strubert, *Chromatographia* **7,** 713 (1974).

(32) J. Goto, M. Hasegawa, S. Nakamura, K. Shimada, and T. Nambara, *J. Chromatogr.* **152,** 413 (1978).

(33) J. Goto, N. Goto, A. Hikichi, and T. Nambara, *J. Liq. Chromatogr.* **2,** 1179 (1979).

(34) S.K. Balani, D.R. Boyd, E.S. Cassidy, G.I. Devine, J.F. Malone, K.M. McCombe, N.D. Scharma, and W.B. Jennings, *J. Chem. Soc. Perkin Trans.* **11,** 2751 (1983).

(35) M.W. Coleman, *Chromatographia* **17,** 23 (1983).

(36) J.A. Dale, D.L. Dull, and H.S. Mosher, *J. Org. Chem.* **34,** 2543 (1969).

(37) R.E. Doolittle and R.R. Heath, *J. Org. Chem.* **49,** 5041 (1984).

(38) C. Banfield and M. Rowland, *J. Pharm. Sci.* **72,** 921 (1983).

(39) M. Goodman, P. Keogh, and H. Anderson, *Bioorg. Chem.* **6,** 239 (1977).

(40) J. Hermansson, T. Iversen, and U. Lindquist, *Acta Pharm. Suec.* **19,** 199 (1982).

(41) A.R. Mitchell, S.B.H. Kenz, I.S. Chu, and R.B. Merrifield, *Anal. Chem.* **50,** 637 (1978).

(42) E.P. Lankmayr, W. Budna, and F. Nachtmann, *J. Chromatogr.* **198,** 471 (1980).

(43) F. Nachtmann, *Int. J. Pharm.* **4,** 337 (1980).

(44) J.A. Thompson, M. Tsuru, C.L. Lerman, and J.L. Holtzman, *J. Chromatogr.* **238,** 470 (1982).

(45) B. Silber and S. Riegelman, *J. Pharmacol. Exp. Ther.* **215,** 643 (1980).

(46) K. Selvig, M. Ruud-Christensen, and A.J. Aasen, *J. Med. Chem.* **26,** 1514 (1983).

(47) R.W. Souter, *Chromatographia* **9** *(12),* 635 (1976).

(48) H. Furukawa, Y. Mori, Y. Takeuchi, and K. Ito, *J. Chromatogr.* **136,** 428 (1977).

(49) R.C. Clark and J.M. Barksdale, *Anal. Chem.* **56,** 958 (1984).

(50) J. Jurczak and Z. Krawczyk, *Pol. J. Chem.* **55,** 2625 (1981).

(51) S.K. Yang, H.V. Gelboin, J.D. Weber, V. Sankaran, D.L. Fischer, and J.F. Engel, *Anal. Biochem.* **78,** 520 (1977).

(52) G.L. Jeyaraj and W.R. Porter, *J. Chromatogr.* **315,** 378 (1984).

(53) N. Bodor, K. Knutson, and T. Sato, *J. Am. Chem. Soc.* **102,** 3969 (1980).

(54) W. Lindner, Ch. Leitner, and G. Uray, *J. Chromatogr.* **316,** 605 (1984).

(55) W.D. Landen and D.S. Caine, *J. Pharm. Sci.* **8,** 1039 (1979).

(56) K.J. Miller, J. Gal, and M.M. Ames, *J. Chromatogr.* **307,** 335 (1984).

(57) W. Dieterle and J.W. Faigle, *J. Chromatogr.* **259,** 301 (1983).

(58) M.J. Wilson and T. Walle, *J. Chromatogr.* **310,** 424 (1984).

(59) G. Gübitz and S. Mihellyes, *J. Chromatogr.* **314,** 462 (1984).

(60) W.H. Pirkle and M.S. Hoekstra, *J. Org. Chem.* **39,** 3904 (1974).

(61) N. Nimura, Y. Kasahara, and T. Kinoshita, *J. Chromatogr.* **213,** 327 (1981).

(62) J. Gal, *J. Chromatogr.* **307,** 220 (1984).

(63) N. Nimura, H. Ogura, and T. Kinoshita, *J. Chromatogr.* **202,** 375 (1980).

(64) T. Kinoshita, Y. Kasahara, and N. Nimura, *J. Chromatogr.* **210,** 77 (1981).

(65) A.J. Sedman and J. Gal, *J. Chromatogr.* **278,** 199 (1983).

(66) J.S. Davies and A.K.A. Mohammed, *J. Chem. Soc. Perkin Trans.* **2,** 1723 (1984).

(67) J.S. Davies and E. Hakeen, *18th EPS Stockholm* (1984), in press.

(68) C.R. Noe, *Chem. Ber.* **115,** 1591 (1982).

(69) J. Goto, N. Goto, and T. Nambara, *Chem. Pharm. Bull.* **30,** 4597 (1982).

(70) I.W. Wainer, *J. Chromatogr.* **202,** 478 (1980).

(71) D.M. Johnson, A. Reuter, J.M. Collins, and G.F. Thompson, *J. Pharm. Sci.* **68,** 112 (1979).

(72) J. Michael, *J. Chromatogr.* **176,** 440 (1979).

(73) W.H. Pirkle and K.A. Simmons, *J. Org. Chem.* **48,** 2520 (1983).

(74) D.A. Aswad, *Anal. Biochem.* **137,** 405 (1984).

(75) R.H. Buck and K. Krummen, *J. Chromatogr.* **315,** 279 (1984).

(76) K.D. Ballard, T.D. Eller, and D.R. Knapp, *J. Chromatogr.* **275,** 161 (1983).

(77) J. Köhler and G. Schomburg, *Chromatographia* **14,** 559 (1981).

(78) J. Köhler, A. Deege, and G. Schomburg, *Chromatographia* **18,** 119 (1984).

(79) E. Gil-Av, B. Feibush, and R. Charles-Sigler, *Tetrahedron Lett.* **10,** 1009 (1966).

(80) V. Schurig, *Angew. Chem. Int. Ed. Engl.* **23,** 747 (1984).

(81) S. Hara and A. Dobashi, *HRC&CC, J. High Resolut. Chromatogr. Chromatogr. Commun.* **2,** 531 (1979).

(82) S. Hara and A. Dobashi, *J. Chromatogr.* **186,** 543 (1979).

(83) J.A. Akanya, S.M. Hitchen, and D.R. Taylor, *Chromatographia* **16,** 224 (1982).

(84) C. Facklam, H. Pracejus, G. Oehme, and H. Much, *J. Chromatogr.* **257,** 118 (1983).

(85) A. Dobashi, K. Oka, and S. Hara, *J. Am. Chem. Soc.* **102**, 7123 (1980).

(86) N. Oi and H. Kitahara, *J. Chromatogr.* **285**, 198 (1984).

(87) W.H. Pirkle and D.L. Sikkenga, *J. Org. Chem.* **40**, 3430 (1975).

(88) W.H. Pirkle and D.L. Sikkenga, *J. Chromatogr.* **123**, 400 (1976).

(89) W.H. Pirkle and D.W. House, *J. Org. Chem.* **44**, 1957 (1979).

(90) W.H. Pirkle, D.W. House, and J.M. Finn, *J. Chromatogr.* **192**, 143 (1980).

(91) I.W. Wainer and T.D. Doyle, *LC, Liq. Chromatogr. HPLC Mag.* **2**, 88 (1984).

(92) T.D. Doyle and I.W. Wainer, *HRC&CC, J. High Resolut. Chromatogr. Chromatogr. Commun.* **7**, 38 (1984).

(93) I.W. Wainer and T.D. Doyle, *J. Chromatogr.* **259**, 465 (1983).

(94) W.H. Pirkle, C.J. Welch, and G.S. Mahler, *J. Org. Chem.* **49**, 2504 (1984).

(95) I.W. Wainer, T.D. Doyle, K.H. Donn, and J.R. Powell, *J. Chromatogr.* **306**, 405 (1984).

(96) I.W. Wainer, T.D. Doyle, Z. Hamidzadeh, and M. Aldridge, *J. Chromatogr.* **268**, 107 (1983).

(97) I.W. Wainer, T.D. Doyle, Z. Hamidzadeh, and M. Aldridge, *J. Chromatogr.* **261**, 123 (1983).

(98) M. Kasai, C. Froussios, and H. Ziffer, *J. Org. Chem.* **48**, 459 (1983).

(99) M. Kasai and H. Ziffer, *J. Org. Chem.* **48**, 712 (1983).

(100) W.H. Pirkle and J.M. Finn, *J. Org. Chem.* **47**, 4037 (1982).

(101) R.A. Chapman, *J. Chromatogr.* **258**, 175 (1983).

(102) I.W. Wainer and T.D. Doyle, *J. Chromatogr.* **284**, 117 (1984).

(103) I.W. Wainer, T.D. Doyle, and C.D. Breder, *J. Liq. Chromatogr.* **7**, 731 (1984).

(104) H.B. Weems and S.K. Yang, *Anal. Biochem.* **125**, 156 (1982).

(105) T.M. Brown and J.R. Grothusen, *J. Chromatogr.* **294**, 390 (1984).

(106) N. Oi, M. Nagase, Y. Inda, and T. Doi, *J. Chromatogr.* **259**, 487 (1983).

(107) N. Oi, M. Nagase, and T. Doi, *J. Chromatogr.* **257**, 111 (1983).

(108) N. Oi and H. Kitahara, *J. Chromatogr.* **265**, 117 (1983).

(109) N. Oi, N. Nagase, Y. Inda, and T. Doi, *J. Chromatogr.* **265**, 111 (1983).

(110) I.W. Wainer, C.A. Brunner, and T.D. Doyle, *J. Chromatogr.* **264**, 154 (1983).

(111) R. Däppen, V. Meyer, and H. Arm, *J. Chromatogr.* **295**, 367 (1984).

(112) W. Lindner and I. Hirschböck, *J. Pharm. Biomed. Anal.* **2**, 183 (1984).

(113) F. Mikes, G. Boshart, and E. Gil-Av, *J. Chromatogr.* **122**, 205 (1976).

(114) F. Mikes and G. Boshart, *J. Chromatogr.* **149**, 455 (1978).

(115) Y.H. Kim, A. Tishbee, and E. Gil-Av, *J. Chem. Soc. Chem. Commun.* 75 (1981).

(116) Y.H. Kim, A. Tishbee, and E. Gil-Av, *J. Am. Chem. Soc.* **102**, 5915 (1980).

(117) C.H. Lochmüller and R. Ryall, *J. Chromatogr.* **150**, 511 (1978).

(118) Y.H. Kim, A. Balan, A. Tishbee, and E. Gil-Av, *J. Chem. Soc. Chem. Commun.* 1336 (1982)

(119) G.W. Gokel and H.D. Durst, *Aldrichimica Acta* **9**, 3 (1976).

(120) L.R. Sousa, G.D.Y. Sogah, D.H. Hoffman, and D.J. Cram, *J. Am. Chem. Soc* **100**, 4569 (1978).

(121) W. Bussman and W. Simon, *Helv. Chim. Acta* **64**, 2101 (1981).

(122) Y. Okamoto, I. Motoshi, H. Hatada, and Y. Heimei, *Polym. I (Tokyo)* **15** (11), 851 (1983).

(123) H. Yuki, Y. Okamoto, and I. Okamoto, *J. Am. Chem. Soc.* **102**, 6356 (1980).

(124) G. Hesse and R. Hagel, *Chromatographia* **6**, 277 (1973).

(125) G. Blaschke and H. Markgraf, *Arch. Pharm.* (Weinheim, Ger.) **317**, 465 (1984).

(126) M. Mintas and A. Mannschreck, *J. Chem. Soc. Chem. Commun.* **14**, 602 (1979).

(127) K. Schlögl and M. Widhalm, *Chem. Ber.* **115**, 3042 (1982).

(128) H. Ahlbrecht, G. Becher, J. Blecher, H.-O. Kalinowski, W. Raab, and A. Mannschreck, *Tetrahedron Lett.* **24**, 2265 (1979).

(129) Y. Okamoto, M. Kawashima, K. Yamamoto, and K. Hatada, *Chem. Lett.* **5**, 739 (1984).

(130) D.W. Armstrong, *J. Liq. Chromatogr.* **7**, 353 (1984).

(131) B. Zsadon, L. Decsei, M. Szilasi, F. Tüdos, and J. Szejtli, *J. Chromatogr.* **270**, 127 (1983).

(132) K.G. Feitsma, B.F.H. Drenth, and R.A. De Zeeuw, *HRC&CC, J. High Resolut. Chromatogr. Chromatogr. Commun.* **7**, 147 (1984).

(133) G. Blaschke and F. Donow, *Chem. Ber.* **116,** 3611 (1983).

(134) A.-D. Schwanghart, W. Backmann, and G. Blaschke, *Chem. Ber.* **110,** 778 (1977).

(135) G. Blaschke, H.P. Kraft, and H. Markgraf, *Chem. Ber.* **113,** 2318 (1980).

(136) G. Blaschke and H. Markgraf, *Chem. Ber.* **113,** 2031 (1980).

(137) G. Blaschke, H.P. Kraft, and A.-D. Schwanghart, *Chem. Ber.* **111,** 2732 (1978).

(138) Y. Okamoto, I. Okamoto, and H. Yuki, *Chem. Lett.* 835 (1981).

(139) Y. Okamoto, S. Honda, I. Okamoto, H. Yuki, S. Murata, R. Noyori, and S. Takaya, *J. Am. Chem. Soc.* **103,** 6971 (1981).

(140) L. Pauling, *Chem. Eng. News* **27,** 913 (1949).

(141) H. Bartels and B. Prijs, in *Advances in Chromatography,* Volume 11 (M. Dekker, New York, 1974), p. 115.

(142) G. Wulff and W. Vesper, *J. Chromatogr.* **167,** 171 (1978).

(143) T. Alebic-Kolbah, S. Rendic, Z. Fuks, V. Sunjic, and F. Kajfez, *Acta Pharm. Jugoslav.* **29,** 53 (1979).

(144) K.K. Stewart and R.F. Doherty, *Proc. Nat. Acad. Sci. USA* **70,** 2850 (1973).

(145) I. Fitos, M. Simonyi, Z. Tegyey, L. Ötvös, J. Kajtar, and M. Kajtar, *J. Chromatogr.* **259,** 494 (1983).

(146) C. Lagercrantz, T. Larsson, and H. Karlsson, *Anal. Biochem.* **99,** 352 (1979).

(147) C. Lagercrantz, T. Larsson, and I. Denfors, *Comp. Biochem. Physiol.* **69C,** 375 (1981).

(148) S. Allenmark, *Chimica Scripta,* **20,** 5 (1982).

(149) S. Allenmark, B. Bomgren, and H. Boren, *J. Chromatogr.* **264,** 63 (1983).

(150) S. Allenmark, B. Bomgren, H. Boren, and P.-O. Lagerström, *Anal. Biochem.* **136,** 293 (1984).

(151) J. Steinhardt and J.A. Reynolds, *Multiple Equilibria in Proteins* (Academic Press, New York, 1969).

(152) J. Hermansson, *J. Chromatogr.* **269,** 71 (1983).

(153) J. Hermansson, *J. Chromatogr.* **298,** 67 (1984).

(154) R. Mertens, M. Stuening, E.W. Weiler, *Naturwissenschaften* **69,** 595 (1982).

(155) W. Rieman III and H.F. Walton, *Ion Exchange in Analytical Chemistry* (Pergamon Press, New York, 1970), p. 254.

(156) N. Grubhofer and L. Schlieth, *Z. Physiol. Chem.* **296,** 262 (1954).

(157) H. Suda and R. Oda, *Mem. Fac. Tech. Kanazawa Univ.* **2,** 215 (1960).

(158) J.A. Lott and W. Rieman III, *J. Org. Chem.* **31,** 561 (1966).

(159) R.J. Baczuk, G.K. Landram, R.J. Dubois, and H.C. Dehm, *J. Chromatogr.* **60,** 351 (1971).

(160) S. Anpeiji, Y. Toritani, K. Kawada, S. Kondo, S. Murai, H. Okai, H. Yoshida, and H. Imai, *Bull. Chem. Soc. Jpn.* **56,** 2994 (1983).

(161) I. Benecke, *J. Chromatogr.* **291,** 155 (1984).

(162) L.R. Gelber, B. Karger, J. Neumeyer, and B. Feibush, *J. Am. Chem. Soc.* **106,** 7729 (1984).

(163) R. Snyder, R.J. Angelici, and R.B. Meck, *J. Am. Chem. Soc.* **94,** 2660 (1972).

(164) V. Davankov and A. Kurganov, *Chromatographia* **13** (6), 339 (1980).

(165) B. Lefebre, R. Audebert, and C. Quivoron, *J. Liq. Chromatogr.* **1**(6), 761 (1978).

(166) G. Gübitz, W. Jellenz, G. Löffler, and W. Santi, *HRC&CC, J. High Resolut. Chromatogr. Chromatogr. Commun.* **2,** 145 (1979).

(167) K. Sudgen, C. Hunter, and G. Lloyd-Jones, *J. Chromatogr.* **192,** 228 (1980).

(168) G. Gübitz, W. Jellenz, and W. Santi, *J. Chromatogr.* **203,** 377 (1981).

(169) G. Gübitz, W. Jellenz, and W. Santi, *J. Liq. Chromatogr.* **4,** 701 (1981).

(170) G. Gübitz, F. Juffman, and W. Jellenz, *Chromatographia* **16,** 103 (1982).

(171) A. Foucault, M. Caude, and L. Oliveros, *J. Chromatogr.* **185,** 345 (1979).

(172) W. Lindner, *J. Liq. Chromatogr.,* manuscript in preparation.

(173) W. Lindner, *Naturwissenschaften* **67,** 354 (1980).

(174) W. Lindner, *Chimia,* in press.

(175) H. Engelhardt and S. Kromidas, *Naturwissenschaften* **67,** 353 (1980).

(176) J. Boué, R. Audebert, and C. Quivoron, *J. Chromatogr.* **204**, 185 (1981).

(177) A. Zolotarev, N. Myasoedov, V. Penkina, I. Dostovalov, O. Petrenik, and V. Davankov, *J. Chromatogr.* **207**, 231 (1981).

(178) B.B. Berezin, I.A. Ymaskov, and V.A. Davankov, *J. Chromatogr.* **261**, 301 (1983).

(179) N. Watanabe, *J. Chromatogr.* **260**, 75 (1983).

(180) H. Engelhardt, T. Koenig, and S. Kromidas, *Fresenius' Z. Anal. Chem.* **317**, 670 (1984).

(181) P. Roumeliotis, K.K. Unger, A.A. Kurganov, and V.A. Davankov, *J. Chromatogr.* **255**, 51 (1983).

(182) P. Roumeliotis, A.A. Kurganov, and V.A. Davankov, *J. Chromatogr.* **266**, 439 (1983).

(183) W. Lindner and J. Hirschböck, manuscript in preparation.

(184) B. Feibush, M.J. Cohen, and B.L. Karger, *J. Chromatogr.* **282**, 3 (1983).

(185) V.A. Davankov, Yu. A. Zolotarev, and A.A. Kurganov, *J. Liq. Chromatogr.* **2**, 1191 (1979).

(186) W. Lindner, J. LePage, G. Davies, D. Seitz, and B.L. Karger, *Anal. Chem.* **51**, 433 (1979).

(187) W. Lindner, J. LePage, G. Davies, D. Seitz, and B.L. Karger, *J. Chromatogr.* **185**, 323 (1979).

(188) B.L. Karger, Y. Tapuhi, W. Lindner, and J. LePage, paper presented at the 13th International Symposium on Chromatography (Cannes, France, 1980).

(189) Y. Tapuhi, N. Miller, and B.L. Karger, *J. Chromatogr.* **205**, 325 (1981).

(190) W. Lindner and A. Küpfer, unpublished results, paper in preparation.

(191) P.E. Hare and E. Gil-Av, *Science* **204**, 1226 (1979).

(192) E. Gil-Av, A. Tishbee, and P.E. Hare, *J. Am. Chem. Soc.* **102**, 5115 (1980).

(193) V.A. Davankov, A.S. Bochkov, A.A. Kurganov, P. Roumeliotis, and K.K Unger, *Chromatographia* **13**, 677 (1980).

(194) C. Gilon, R. Leshem, Y. Tapuhi, and E. Grushka, *J. Am. Chem. Soc.* **101**, 7612 (1979).

(195) S. Lam, *J. Chromatogr. Sci.* **22**, 416 (1984).

(196) C. Gilon, R. Leshem, and E. Grushka, *Anal. Chem.* **52**, 1206 (1980).

(197) C. Gilon, R. Leshem, and E. Grushka, *J. Chromatogr.* **203**, 365 (1981).

(198) S. Lam, F. Chow, and A. Karmen, *J. Chromatogr.* **199**, 295 (1980).

(199) E. Oelrich, H. Preusch, and E. Wilhelm, *HRC&CC, J. High Resolut. Chromatogr. Chromatogr. Commun.* **3**, 269 (1980).

(200) V.A. Davankov, A.S. Bochkov, and Yu. P. Belov, *J. Chromatogr.* **218**, 547 (1981).

(201) A.A. Kurganov and V.A. Davankov, *J. Chromatogr.* **218**, 559 (1981).

(202) S. Lam and A. Karmen, *J. Chromatogr.* **289**, 339 (1984).

(203) L.R. Gelber and J.L. Neumeyer, *J. Chromatogr.* **257**, 317 (1983).

(204) S. Weinstein, *Angew. Chem. Int. Ed. Engl.* **21**, 218 (1982).

(205) N. Nimura, A. Toyama, and T. Kinoshita, *J. Chromatogr.* **234**, 482 (1982).

(206) I. Hirschböck and W. Lindner, paper presented at the 14th International Symposium on Chromatography (London, 1982).

(207) C. Pettersson and G. Schill, *J. Chromatogr.* **204**, 179 (1981).

(208) C. Pettersson and G. Schill, *Chromatographia* **16**, 192 (1982).

(209) C. Pettersson and K. No, *J. Chromatogr.* **282**, 671 (1983).

(210) G. Szepesi, M. Gazdag, and R. Ivancsics, *J. Chromatogr.* **244**, 33 (1982).

(211) C. Pettersson, *J. Chromatogr.* **316**, 553 (1984).

(212) N. Parris, *Anal. Biochem.* **100**, 260 (1979).

(213) B.A. Bidlingmeyer, *J. Chromatogr. Sci.* **18**, 525 (1980).

(214) L. Hackzell, T. Rydberg, and G. Schill, *J. Chromatogr.* **282**, 179 (1983).

(215) J.H. Knox and J. Jurand, *J. Chromatogr.* **234**, 222 (1982).

(216) V. Prelog, Z. Stojanac, and K. Kovacévic, *Helv. Chim. Acta* **65**, 377 (1982).

(217) B. Domon, K. Hostettmann, K. Kovacévic, and V. Prelog, *J. Chromatogr.* **250**, 149 (1982).

(218) C. Pettersson and H.W. Stuurman, *J. Chromatogr. Sci.* **22**, 441 (1984).

(219) B. Sebille and N. Thuaud, *J. Liq. Chromatogr.* **3**, 299 (1980).

(220) B. Sebille, N. Thuaud, and J.P. Tillement, *J. Chromatogr.* **204**, 285 (1981).

(221) C. Pettersson, T. Arvidsson, A.-L. Karlsson, and I. Marle, *J. Pharm. Biomed. Anal.* (1984), in press.

(222) J. Hermansson, *J. Chromatogr.* **316,** 537 (1984).

(223) J. Debowski, D. Sybilska, and J. Jurczak, *J. Chromatogr.* **237,** 303 (1982).

(224) J. Debowski, D. Sybilska, and J. Jurczak, *Chromatographia* **16,** 198 (1982).

(225) Y. Nobuhara, S. Hirano, and Y. Nakanishi, *J. Chromatogr.* **258,** 276 (1983).

(226) C.H. Lochmüller and E.C. Jensen, *J. Chromatogr.* **216,** 333 (1981).

(227) Y. Dobashi, A. Dobashi, and S. Hara, *Tetrahedron Lett.* **25,** 329 (1984).

(228) A. Dobashi and S. Hara, *Tetrahedron Lett.* **24,** 1509 (1983).

(229) A. Dobashi and S. Hara, *J. Chromatogr.* **267,** 11 (1983).

(230) G.K.C. Low, P.R. Haddad, and A.M. Duffield, *J. Liq. Chromatogr.* **6,** 311 (1983).

(231) M. Zief, L.J. Crane, and J. Horvath, *J. Liq. Chromatogr.* **7,** 709 (1984).

(232) J. Hermansson and C. Bahr, *J. Chromatogr.* **227,** 113 (1982).

(233) A.A. Hancock, *Clin. Exp. Hypertens. AG,* 659 (1984).

(234) E. Matsutera, Y. Nobuhara, and Y. Nakanishi, *J. Chromatogr.* **216,** 374 (1981).

(235) V.K. Piotrovskii, Yu.A. Zhirkov, and V.I. Metelitsa, *J. Chromatogr.* **309,** 421 (1984).

(236) J.B. Crowther, T.R. Covey, E.A. Dewey, and J.D. Henion, *Anal. Chem.* **56,** 2921 (1984).

(237) J. Hermansson, M. Eriksson, and O. Nyquist, *J. Chromatogr.* **336,** 321 (1984).

(238) T.R. Burke, W.L. Nelson, M. Mangion, G.J. Hite, C.M. Mokler, and P.C. Ruenitz, *J. Med. Chem.* **23,** 1044 (1980).

(239) A. Yacobi, C.M. Lai, and G. Levy, *J. Pharmacol. Exp. Ther.* **231,** 72 (1984).

(240) E. Lankmayr, B. Maichin, and G. Knapp, *Fresenius' Z. Anal. Chem.* **301,** 187 (1980).

(241) I.D. Hay, T.M. Annesley, N.S. Jiang, and C.A. Gorman, *J. Chromatogr.* **226,** 383 (1981).

(242) G. Hesse and R. Hagel, *Liebigs Ann. Chem.* 996 (1976).

(243) J. Debowski, J. Jurczak, and D. Sybilska, *J. Chromatogr.* **282,** 83 (1983).

(244) U. Forsman, *J. Chromatogr.* **303,** 217 (1984).

(245) G. Blaschke, H.P. Kraft, K. Fickentscher, and F. Köhler, *Drug. Res.* **29** (II), 1640 (1979).

SAMPLE PREPARATION FOR PHARMACEUTICALS VIA SOLID-PHASE EXTRACTION

Morris Zief

Research Laboratory
J.T. Baker Chemical Company
Phillipsburg, NJ 08865

Detection and quantitation of a specific analyte in a pharmaceutical preparation present few problems today. Removal of matrix interferences, however, has received considerably less attention in the literature even though sample cleanup is frequently the most critical aspect of an analysis.

The need for sample cleanup as a prelude to modern instrumental analysis was reviewed several years ago (1). Two important reasons for sample cleanup are the removal of impurities that interfere with measurement of the analyte by the analytical method of choice and sample concentration when the solution of the sample is too dilute for direct measurement of the analyte. In addition, impurities that can damage expensive analytical equipment must be removed before sample introduction into the instrument. Components, for example, that can damage LC or GC columns must be eliminated before sample injection.

Sample preparation procedures for antibiotics, vitamins, and steroids traditionally have included sample dissolution followed by liquid-liquid extraction (2). Within the past five years, solid-phase extraction has replaced many solvent extraction procedures, and it is expected that this new technique will supplant the older assay methods in the forthcoming volumes of the *United States Pharmacopeia* (3).

Traditional liquid-liquid extractions are performed in separatory funnels, and frequently they are tedious, time-consuming, and expensive. These methods not only require several sample-handling steps but may also present the following problems to the analyst: phase emulsions, evaporation of large solvent volumes, disposal of toxic and flammable solvents, impure and wet extracts, and nonquantitative and irreproducible extractions. Not only are all of these problems eliminated by solid-phase extraction, but the time for sample preparation is reduced by factors of six to nine.

133

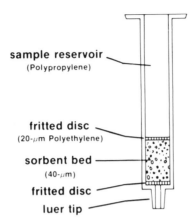

sample reservoir
(Polypropylene)

fritted disc
(20-μm Polyethylene)

sorbent bed
(40-μm)

fritted disc

luer tip

FIGURE 1.   Solid-phase extraction column.

SOLID-PHASE EXTRACTION COLUMNS

Figure 1 illustrates how the concept of solid-phase extraction has been reduced to practice in one of several types of disposable columns that are now available (3). The polypropylene columns are packed with 100, 200, or 500 mg of an absorbent sandwiched between two 20-μm polyethylene frits and have capacities of 1, 3, or 6 mL, respectively. Adsorbents with average particle diameters of 40 μm and possessing a narrow size distribution of 32–60 μm produce the optimum efficiency. For samples with molecular weights of less than 2000, the standard pore size of the adsorbent is 60–100 Å. Large volumes of samples (greater than 100 mL) can be added to the larger columns through disposable reservoirs.

EXTRACTION AND ELUTION OF SAMPLES

A sample solution can be forced through a single column by a syringe; alternatively, the solution can be aspirated through an appropriate adsorbent by vacuum. A valve with six functional ports simplifies the repetitive use of single columns (Figure 2). Ports A, B, C, and D introduce conditioning, sample, wash, and eluting solutions into the syringe or expel wastes from the syringe by positioning the indicating valve to the selected port. When the proper solution has been introduced into the syringe, the indicating valve is positioned to the bottom port. The solution in the syringe is then slowly forced through the extraction column. Air is drawn into the syringe to force the remaining solution in the column through the sorbent.

Up to ten extraction columns can be handled simultaneously with a specially designed vacuum manifold (Figure 3). The columns are first inserted into the Luer fittings of the manifold and washed with an appropriate solvent to remove impuri-

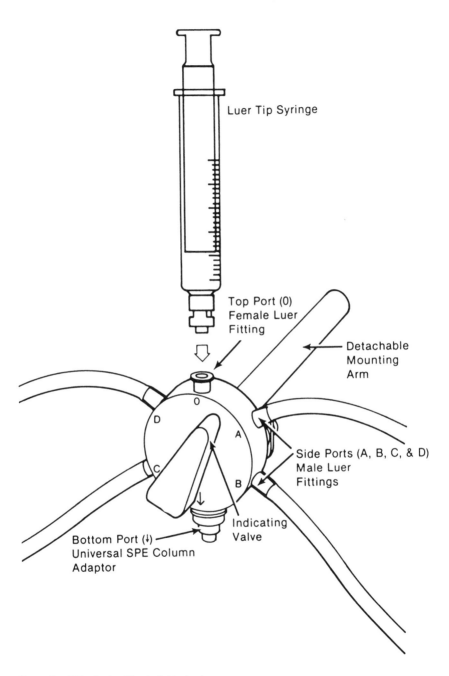

FIGURE 2.   Valve for handling individual columns.

ties and wet the surface of the adsorbent. Sample solutions are then aspirated through the columns by a water-aspirator vacuum at 10–20 in. Hg. Column eluates and washes are collected in a trap inserted between the manifold and the water aspirator. After the washes are completed, collection tubes are placed under each column in a specially designed rack.

Appropriate elution solvents are added to the columns; sample eluates are collected and used directly. Typically two 10-μL rinses, two 200-μL rinses, and two 500-μL rinses of eluting solvent remove the analyte quantitatively from the 1, 3, and 6 mL columns, respectively. When standard solutions containing from 0.001 mg to 2 mg of hydrocortisone in 20:80 methanol/water were added to 3 mL

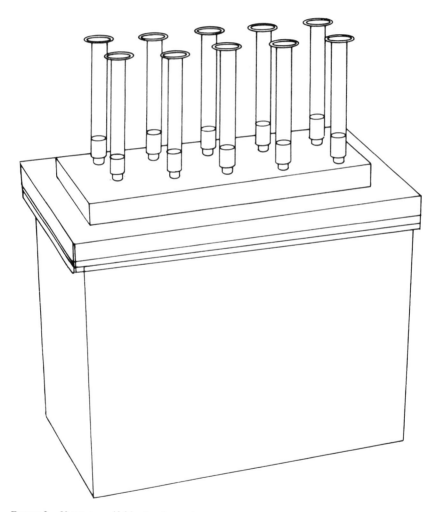

FIGURE 3.   Vacuum manifold extraction system.

octadecyl-silanized silica columns, quantitative recovery of hydrocortisone was essentially obtained in all cases by elution with two 500-μL rinses of methanol (4).

Sample processors for eight (5) and twelve (6) samples with other column configurations are also available.

SELECTION OF THE COLUMN ADSORBENT

Until 1960, silica and alumina were used almost exclusively as adsorbents for the separation of polar compounds from relatively nonpolar mobile phases such as hexane, ether, ethyl acetate, and methylene chloride. Dissolution of the sample in a solvent less polar than a support such as silica was indeed considered the normal mode of solid-phase extraction. In the late 1960s the idea of treating the free hydroxyl groups of silica with mono-, di-, and trihalo or alkoxyl silyl derivatives to form siloxanes was conceived. A typical reaction of silica with a trihalo-substituted silane proceeds as follows:

| | Cl | | OH |
| Silica | Trichlorosilane | Siloxane | |

The R group can have nonpolar, polar, or ion-exchange functional groups. When R is octadecyl, the polar silica is converted into a nonpolar silica bonded phase that is more useful for separating nonpolar analytes. The use of a nonpolar adsorbent such as the octadecyl siloxane gave rise to the term reversed-phase extraction, because in this case a nonpolar adsorbent extracts a nonpolar sample from a more polar solvent such as acetonitrile, methanol, or water. By varying the R group of the siloxane, adsorbents with substantial differences in polarity can be prepared.

Table I lists adsorbents in order of increasing polarity from the nonpolar octadecyl group to the strongly polar sulfonic acid and quaternary functional groups. The octadecyl, octyl, butyl, cyclohexyl, and phenyl groups confer nonpolar character to the support. If an analyte is retained too strongly on the octadecyl siloxane, columns with the octyl or butyl bonded phases can frequently be substituted with promising results. The remaining 12 adsorbents are polar, with the polarity increasing from the weakly polar CN-derivative to the highly polar cationic and anionic derivatives.

Although Table I emphasizes silica-based adsorbents, in special cases other sorbents can be substituted. Amberlite XAD polymeric adsorbents (Rohm and Haas,

TABLE I

Polarity Scale of Siloxane Adsorbents

		R*
Nonpolar	octadecyl	$-C_{18}H_{37}$
	octyl	$-C_8H_{17}$
	butyl	$-C_4H_9$
	cyclohexyl	$-CH\langle{}^{CH_2CH_2}_{CH_2CH_2}\rangle CH_2$
	phenyl	$-C_6H_5$
Polar	cyanopropyl Kieselguhr silica gel	$-CH_2CH_2CH_2CN$
	diol	$-(CH_2)_3-OCH_2CH-CH_2$
		OH OH
	aminopropyl Florisil alumina	$-CH_2CH_2CH_2NH_2$
Anionic	diamino	$-(CH_2)_3NHCH_2CH_2NH_2$
	polyethylenimine	$-(CH_2)_3NH(CH_2CH_2NH)_xH$
	quaternary amine	$-(CH_2)_3N^+(CH_3)_3Cl^-$
Cationic	carboxylic acid	$-(CH_2)_3COOH$
	sulfonic acid	$-C_6H_4-SO_2OH$

(right margin, vertical: Increasing Polarity ↓)

* R represents the organic substitutents of siloxane bonded phases (see structure in text, above). Florisil = activated magnesium silicate; Kieselguhr = biogenic amorphous silica; silica gel = synthetic surface-hydroxylated, amorphous silica.

Philadelphia, Pennsylvania), for example, extract hydrophobic solutes from polar solvents. These sorbents are cross-linked poly(styrene-divinylbenzene) polymers without any ionic functional groups incorporated onto the resin structure. The highly aromatic structure thus extracts aromatic hydrocarbons from aqueous solutions. These nonpolar resins are positioned between the phenyl and cyanopropyl bonded siloxanes in Table I.

The most polar bonded silicas in Table I adsorb acids and bases from aqueous solutions according to classic ion-exchange theories. The retention of ionic compounds is achieved by promoting ionization in the aqueous sample solution; for example, the pH is increased for acidic analytes and decreased for basic analytes. Acidic compounds are extracted on an anionic $[-N^+(CH_3)_3]$ solid-phase extraction column from a sample solution having a pH two units higher than the pK_a of the analyte (pH = pK_a + 2). The analytes are eluted with an acidic solution (pH = pK_a − 2).

Basic compounds are extracted on a cationic ($-C_6H_5SO_3^-$) column from a sample solution having a pH two units lower than the pK_a of the analyte (pH = pK_a − 2). The analytes are eluted with an eluting solvent where pH = pK_a + 2. Eluting solvents of required pH are easily prepared from 0.01–0.1 M buffer solutions.

To optimize solid-phase extraction, the interaction of analyte with the sorbent and solvent matrix must be carefully evaluated in all cases. The situation illustrated in Figure 4 must prevail; that is, the analyte must be more strongly attracted to the sorbent than to the solvent. If the compound of interest interacts more readily with the solvent, the sample solution should be rendered weaker or a stronger sorbent should be selected.

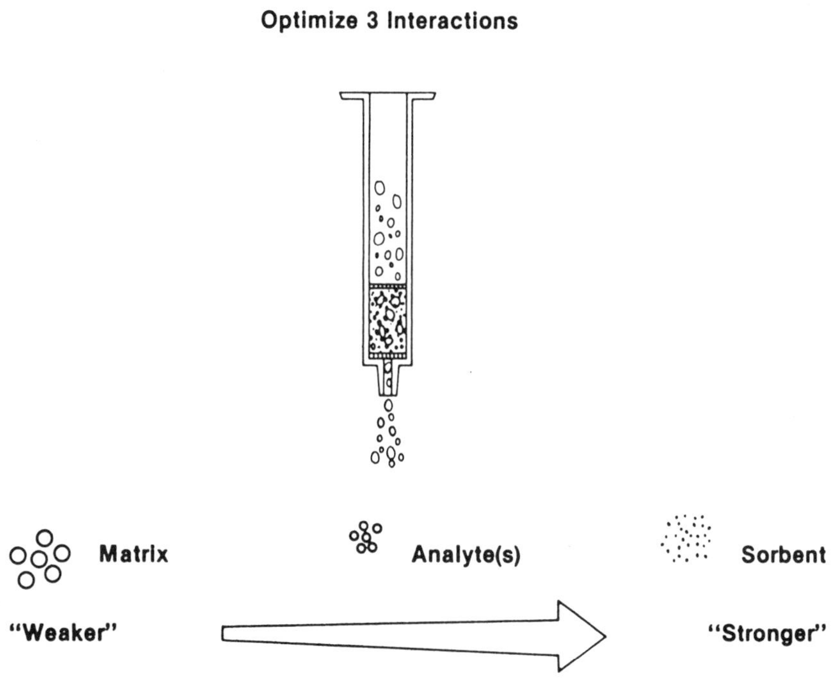

Optimize 3 Interactions

FIGURE 4.   Optimization of solid-phase extraction.

SELECTION OF SOLVENTS

To elute the analyte from the column, the conditions of Figure 4 must be reversed; the eluting solvent must now overcome the attraction between analyte and sorbent. Selection of a solvent with the optimum eluting strength is simplified by an examination of Table II, where the polarities of fifteen common solvents are listed according to $E°$, the eluotropic strength of adsorption on silica (7). For solid-phase extraction the arrangement of solvents in order of increasing strength for removal of solutes from a particular adsorbent is more useful than P', the polarity index, based on classification of liquids according to their solubility for a number of test solutes (8).

Fine adjustments of $E°$ can be obtained by a mixture of solvents, because solvent strength changes continuously with composition. The change in $E°$ from 0 to 0.73 can be easily controlled with binary mixtures of hexane, methylene chloride, ethyl ether, acetonitrile, and methanol. Selection of an appropriate $E°$ can be obtained from the plot for solvent strength for various binaries as a function of binary composition (8). Solvent strength does not vary linearly in this plot but makes for a good approximation in developing new solid-phase extraction methods. Examples of $E°$ from 0.63 to 0.67 with methanol mixtures are presented in Table II.

Experience has shown that solvents with $E° < 0.38$ are preferred as diluents for the adsorption of moderately polar compounds on polar sorbents such as silica and its polar derivatives. Mixtures of hexanol and phenethyl alcohol, for example, are dissolved in hexane/ethyl acetate (3:1) for adsorption on silica. Binary mixtures with $E° > 0.5$ are generally considered first for the dissolution of nonpolar solutes before adsorption on nonpolar adsorbents such as octyl and octadecyl bonded sili-

TABLE II

Solvent Polarity and Eluotropic Strength

Solvent	$E°*$
Hexane	0.00
Isooctane	0.01
Carbon tetrachloride	0.11
Chloroform	0.26
Methylene chloride	0.32
Tetrahydrofuran	0.35
Ethyl ether	0.38
Ethyl acetate	0.38
Acetone	0.47
Dioxane	0.49
Acetonitrile	0.50
Isopropanol	0.63
Methanol/methylene chloride (20:80)	0.63
Methanol/ethyl ether (20:80)	0.65
Methanol/acrylonitrile (40:60)	0.67
Methanol	0.73
Water	>0.73
Acetic acid	>0.73

* $E°$ = eluotropic strength (eluting solvent strength on silica).

cas. Cholesterol is dissolved in methanol/chloroform (10:1) when adsorption on the octadecyl derivative is the objective. The cholesterol is easily removed from the column by using the less polar chloroform alone. The latter example shows that polar solvents tend to have weak solvent strength while nonpolar solvents are strong eluting solvents for reversed-phase extraction. Water acts as a relatively weak solvent, while hydrocarbons, such as hexane, are strong solvents for reversed-phase conditions.

When a sample is extracted from a solvent by an adsorbent, selective removal of the impurities can frequently be accomplished by changing the polarity of the eluate. Hydrocortisone in urine or plasma, for example, was adsorbed onto an octadecyl bonded silica. The adsorbed sample was then washed with 20:80 acetone/water to remove impurities (4). The hydrocortisone, free from impurities, was subsequently eluted with methanol. In the previous example, the impurities were more polar than hydrocortisone. The strongly polar constituents in urine or plasma pass through the column with the matrix. Compounds that are less polar, but that are more polar than hydrocortisone itself, are removed by the acetone/water wash.

If the hydrocortisone is still contaminated with components in the urine that are less polar, then the methanol eluate containing the hydrocortisone is concentrated to dryness and the residue is dissolved in methylene chloride. The resulting solution is then passed through a silica column that binds the hydrocortisone more strongly than the less polar impurities. Experiments with methanol/methylene chloride washes will quickly yield information concerning the proper ratio of solvents for removal of the less polar materials. Finally, the pure hydrocortisone is removed with 100% methanol. Earlier work with hydrocortisone showed that this steroid is adsorbed by the octadecyl, cyano, silica, and diol as well as the amine columns (4). Although prednisone, cortisone, and deoxycorticosterone are adsorbed only onto the octadecyl, cyano, and silica columns, impurities more or less polar than any of these steroids can be separated by the same strategy used with the hydrocortisone sample.

The choice of an eluting solvent is determined by the relationship between $E°$ and the polarity of the analyte. The high $E°$ of methanol, 0.73, is the basis for its selection as an eluate for the removal of moderately polar and strongly polar analytes from polar adsorbents. Methylene chloride, $E° = 0.32$, is used frequently to elute nonpolar analytes from nonpolar bonded phases. Although the principle that like dissolves like — selection of a polar solvent to elute a polar analyte from the sorbent — underlies all experimental work, evaluation of the proper solvent is still more of an art than a science. Methanol, for example, possesses a high $E°$, yet it is frequently an excellent eluting solvent for nonpolar bonded phases.

COLUMN CAPACITY

Bidlingmeyer pointed out that mass overload is unlikely in solid-phase extraction because the columns are usually used in trace analysis (9). Because analysts have expressed legitimate concern about the maximum allowable concentrations for solid-phase extraction, however, the capacity of 200-mg columns for several analytes of practical interest was investigated. Table III lists the capacities of the pure compound before breakthrough (the point at which the adsorbent becomes saturated and can no longer retain the analyte). If other compounds in the sample are retained by the column, the capacity for the analyte is reduced proportionally to the amount of competing compound present.

Capacities from 3 mg to 12 mg in the documented examples are far above the analyte concentrations in practical pharmaceutical analyses. Drug levels in urine and serum are usually at the $\mu g/mL$ and ng/mL levels. The 1-mL columns containing 100 mg of sorbent have been found adequate for nanogram levels of analyte.

In Table III, the adsorption of cholesterol palmitate on the nonpolar C_{18} group is easily predictable. This analyte contains no functional group that is capable of interacting with polar adsorbents. When the ester is cleaved to liberate cholesterol containing a free hydroxyl group, the nonpolar character of the molecule predominates, and the octadecyl bonded phase is the column of choice. When pendant groups on the steroid skeleton predominate — for example, in hydrocortisone — silica or the cyanopropyl bonded phase becomes a good candidate for adsorption.

TABLE III

Capacity of Solid-Phase Extraction Columns

Analyte	Column (200 mg)	Adsorption without breakthrough (mg/200 mg)	Solvent
Hydrocortisone	octadecyl	8.0	water (pH 7.0)
	silica	6.0	methylene chloride
	cyanopropyl	5.0	methylene chloride
Cholesterol	octadecyl	5.0	methanol/chloroform (9:1)
Cholesterol palmitate	octadecyl	5.0	methanol/chloroform (9:1)
Adenosine-5'- monophosphate	quaternary amine	3.0	water (pH 7.0)
Adenosine-5'- diphosphate trisodium salt	quaternary amine	>10.0	water (pH 7.0)
Lead, mercury	diamino	>5.0	water (pH 4.0)
Gold	diamino	12.0	water (pH 1.0)

APPLICATIONS

Fat-Soluble Vitamins

Comparison of official methods of analysis for vitamins A, D, and E in multivitamin tablet preparations (10) with a technique developed on octadecyl extraction columns (11) identifies the advantages of the latter method. The official method requires a time-consuming liquid-liquid extraction of saponified analytes that may take 4 h. Ten samples can be processed by solid-phase extraction in 20 min with average recoveries for vitamins A_1, D_2, and E of 99.8%, 96%, and 99.7%, respectively.

Ground multivitamin tablet powder containing 5000 IU vitamin A acetate, 400 IU vitamin D_2, and 15 IU vitamin E acetate was dissolved in 1% acetic acid/2-propanol (1:2). Two milliliters of the sample solution was aspirated through a 3-mL octadecyl-silanized silica column after conditioning with one column volume of methanol. The column was then washed with 1 mL of water/2-propanol (45:55) to remove interfering compounds. The three vitamins were then eluted with two 0.5-mL aliquots of methylene chloride. The eluate was then analyzed using HPLC with a 5-μm C_8 analytical column and a mobile phase of acetonitrile/methanol/water (47:47:6).

Water-Soluble Vitamins

Reversed-phase, ion-pair extraction has been applied to analysis of water-soluble vitamins. In this technique ionic compounds in solution are extracted with an ion-pair reagent of opposite charge. When a solution containing pyridoxine, for example, and an ion-pair reagent such as heptanesulfonic acid is aspirated through an octadecyl column, the extraction proceeds as follows:

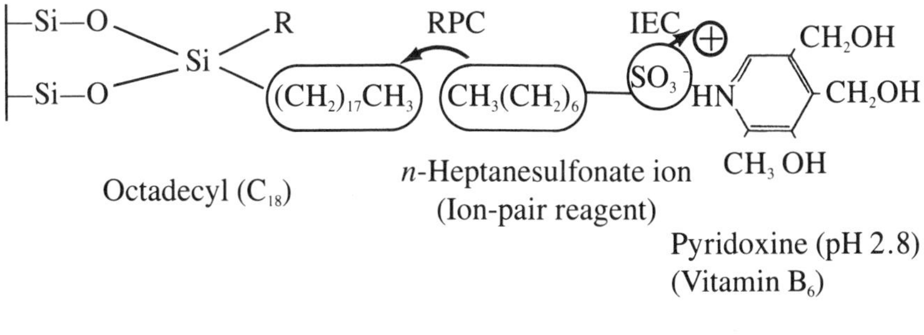

Octadecyl (C_{18}) n-Heptanesulfonate ion CH_3 OH
                              (Ion-pair reagent)

                                        Pyridoxine (pH 2.8)
                                        (Vitamin B_6)

The nonpolar heptyl group of the ion-pair reagent partitions into the adsorbent by interacting with the nonpolar octadecyl pendant group of the sorbent. Simultaneously the negative sulfonic acid portion of the ion-pair reagent interacts with the protonated nitrogen group of pyridoxine. The concurrent interactions optimize this reversed-phase extraction. Usually 0.005–0.01 M solutions of the ion-pairing reagent are used. When the analyte is an acid, a tetraalkyl ammonium compound — a tetrabutyl ammonium derivative, for example — becomes an appropriate ion-pairing reagent. Ion-pairing is a good alternative to ion-exchange extraction in cases in which the analyte must be eluted with organic solvents.

Prenatal vitamin and mineral tablet powder containing <6 mg of riboflavin was dissolved in 50 mL of 0.01 M sodium heptanesulfonate in 99:1 water/acetic acid. Two milliliters of the sample solution was added to a methanol-conditioned 3-mL octadecyl extraction column inserted into a vacuum manifold such as the one illustrated in Figure 2; the sample was then aspirated onto the packing. After the column was washed with 0.5 mL of the extracting solution, the column was eluted three times with 0.5-mL aliquots of methanol. The effluent was collected in a 10-mL volumetric flask, diluted to volume, then injected onto a C_8 analytical HPLC column. The average recoveries of niacinamide, pyridoxine, riboflavin, and thiamine ranged from 97.1% to 100% of the expected values in standard addition experiments (12). The method successfully separated these vitamins from tablet excipients that could interfere in the final HPLC analysis. In addition, the vitamins were found to be stable for a minimum of 3 h on the solid-phase support.

Two columns, the quaternary amine and the phenyl bonded phases, have been used in series for the rapid extraction of vitamin B_{12} (cyanocobalamin) (13). In this example, a quaternary amine column was inserted, using an adapter, on top of a phenyl column. When an aliquot of sample solution [1 tablet weight of multivitamin tablet powder in 10 mL of extracting solution (1.3 g of Na_2HPO_4, 1.2 g of citric acid monohydrate, and 1 g of sodium metabisulfite dissolved in 100 mL of distilled water)] was aspirated through this system, some impurities were retained by the quaternary amine column. Other impurities were removed from the phenyl column with hexane, methylene chloride, and acetonitrile washes. Vitamin B_{12} was finally eluted from the phenyl column with two 0.5-mL aliquots of 90:10 methanol/water. Experience has taught that two or more small aliquots invariably afford higher recoveries than one large aliquot in all extraction procedures.

Antibiotics

For the solid-phase extraction of polar antibiotics such as bacitracin (a polypeptide) and chlortetracycline hydrochloride, the strongly polar diol bonded phase is the preferred adsorbent. Two milliliters of methylene chloride was added to 200 mg of bacitracin ointment in a small stoppered vial. The ointment base was dissolved with gentle heating and shaking. The resulting suspension containing the insoluble bacitracin was quantitatively transferred to a diol column and aspirated through the column. The vial and stopper were then washed with 2-mL aliquots of methylene chloride that were added to the column. The bacitracin was then eluted from the column with two 1-mL aliquots of 0.1 N HCl into a 2-mL volumetric flask and diluted to volume. The eluate was then analyzed by HPLC (14).

Chlortetracycline hydrochloride ointment was extracted in a similar manner; 50 mg was shaken in a vial with 2 mL of hexane, then added to a diol column. After a few hexane washes, the sample was eluted with two 1-mL aliquots of methanol/ 0.1 N HCl (50:50), diluted to volume as above, then analyzed by HPLC (15).

Benzylalkonium Chloride

The analysis of benzylalkonium chloride, a preservative in commercial eye washes, illustrates the relative ease with which fairly large volumes of a polar sample can be handled with a 6-mL cyano column. A 75-mL reservoir was attached to the top of the cyano column with an adapter, and a 50-mL sample of eye wash containing 0.005% benzalkonium chloride was aspirated through the column at flow rates of approximately 5 mL/min. After the column was washed with 2 mL of 1.5 N HCl, the analyte was eluted with 2-mL aliquots of methanol/1.5 N HCl (80:20) (16). Quaternary amine compounds are readily extracted by cyano and diol sorbents. This class of compound is more easily eluted from these sorbents than from classic sulfonic acid and carboxylic acid ion-exchange resins.

Parabens

Parabens, which are a class of benzoic acid esters, are also used as preservatives in pharmaceuticals. In these compounds, because of the aromatic ester structure, use of the octadecyl bonded phase is suggested for extraction. Commercial samples containing parabens (0.1–1.0 g) were extracted with 5–10 mL of methanol. A 100-μL aliquot of the supernatant was added to a 2-mL volumetric flask and diluted to the mark with distilled water. The 2-mL sample was added to a column containing an octadecyl bonded phase. After the column was washed with water to remove impurities, the analyte was eluted with two 500-μL aliquots of methanol (17).

Hydrocortisone

Hydrocortisone, unlike most analytes, can be adsorbed readily on polar as well as nonpolar adsorbents, as is indicated in Table III. The analysis of this steroid in commercially available 0.5% hydrocortisone cream was carried out conveniently with a silica column. One gram of cream was weighed into a 20-mL vial. Then 10 mL of hexane/ethyl acetate (50:50) was added, followed by agitation on a vortex mixer. The supernatant was decanted into a 50-mL volumetric flask. The extraction was repeated once more; the supernatants were combined and diluted to the mark with the extracting solution. One milliliter of the diluted solution was added to a silica column previously conditioned with hexane/acetone (80:20). After the analyte on the column was washed with the same conditioning mixture, the analyte was eluted with two 500-μL aliquots of methanol (18).

CONCLUSIONS

The bulk of literature references report solid-phase extraction with silica gel or the octadecyl-siloxane derivatized silica — an understandable tendency because silica gel was used almost exclusively as an adsorbent until 1960; the later reversed-phase extraction for nonpolar materials was dominated by the octadecyl derivative because it was the first bonded silica available. With the wide selection of adsorbents available today, however, a basic understanding of the mechanisms of liquid chromatography is essential in order to select the proper column. A complete characterization of the sample can reduce trial-and-error experiments in the laboratory. Methods development is simplified by determining the following: sample classification (including impurities or interferences), sample solubility, sample polarity, selection of sorbent and elution solvents, assessment of matrix solution strength, selection of stronger sorbent, and selection of stronger elution solvent. The choice of a solid-phase extraction column for the sample is the final step in methods development.

REFERENCES

(1) J.W. Dolan, in *Introduction to Modern Liquid Chromatography,* L.R. Snyder and J.J. Kirkland (John Wiley and Sons, New York, 1979), Chapter 17.

(2) G. Zweig and J. Sherma, *Handbook of Chromatography* (CRC Press, Cleveland, Ohio, 1972), Vol. 2, pp. 191–254.

(3) Baker-10 SPE Applications Guide, J.T. Baker Chemical Co. (Phillipsburg, New Jersey), Vol. 1 (1982) and Vol. 2 (1984).

(4) M. Zief, L.J. Crane, and J. Horvath, *Am. Lab.* **15**(5), 120–126 (1982).

(5) Sep-Pak Cartridge Rack, Waters Associates, Inc. (Milford, Massachusetts).

(6) Prep I, DuPont Company, Clinical and Instrument Systems Division (Wilmington, Delaware).

(7) L.R. Snyder, *Principles of Adsorption Chromatography* (Marcel Dekker, New York, 1968), Chapter 8.

(8) L. Rohrschneider, *Anal. Chem.* **45**, 1241–1247 (1973).

(9) B.A. Bidlingmeyer, *LC, Liq. Chromatogr. HPLC Mag.* **2**, 578–580 (1984).

(10) *The United States Pharmacopeia,* 21st rev. (Easton, Pennsylvania, Mack Publishing Co., 1984), pp. 1118, 1215.

(11) G. Wachob, *LC, Liq. Chromatogr. HPLC Mag.* **1**, 428–430 (1983).

(12) G. Wachob, *LC, Liq. Chromatogr. HPLC Mag.* **1**, 110–112 (1983).

(13) Baker-10 SPE Applications Guide, J.T. Baker Chemical Co. (Phillipsburg, New Jersey), Vol. 1 (1982), p. 40.

(14) *Ibid.,* Vol. 2 (1984), p. 146.

(15) *Ibid.,* p. 148.

(16) *Ibid.,* p. 150.

(17) *Ibid.,* Vol. 1 (1982), p. 44.

(18) *Ibid.,* p. 46.

B: OPTIMIZATION OF MOBILE PHASES

THE THEORY OF RETENTION IN REVERSED-PHASE HIGH PERFORMANCE LIQUID CHROMATOGRAPHY

Henri Colin

Varex Corporation
Rockville, MD 20852

Ante M. Krstulović

Laboratoires d'Etudes et
de Recherches Synthelabo
Paris, France

The use of a nonpolar stationary phase and a polar mobile phase was first mentioned in the literature in the work of Boscott (1) and, subsequently, Howard and Martin (2), who treated kieselguhr with dimethyldichlorosilane vapor. The technique was termed reversed phase (RP). Although reversed-phase liquid chromatography (RPLC) has been successfully used in many fields since its inception, its true ascent had to await the introduction of pellicular hydrocarbonaceous bonded silica phases developed by Kirkland and De Stefano in 1970 (3). Since then, many theoretical and technological advances have been made in this field, and it is now estimated that more than 80% of the LC separations (except gel permeation) are carried out using this chromatographic mode.

Although in normal-phase liquid chromatography (NPLC) the stationary phase interactions play a key role in overall retention, the opposite is true in RPLC, in which interactions in the stationary phase represent a weak, although not negligible, contribution. The high polarity of RPLC solvents (most often hydroorganic mixtures) makes possible the existence of polar interactions such as those encountered in ion-pair formation, ionization control, and, more generally, specific complexation between solute molecules and complexing agents. In addition, reversed-phase systems afford higher column stabilities, low equilibration times, and higher retention reproducibility than normal-phase or ion-exchange systems.

149

This chapter will discuss some important aspects of RPLC: the theory of retention, the role of sample, and mobile phase composition [water content, nature of the organic solvent(s), effect of exotic additives, ionic strength and pH, adsorption from the mobile phase, and dead volume determination]. Some extrathermodynamic relationships pertaining to RPLC systems will also be examined.

RETENTION MECHANISM

Since the development of chemically bonded RP packings, the separation mechanism has been a subject of continuing controversy. Basically, three modes of interaction have been proposed: a partitioning process between the mobile phase and a liquidlike stationary phase (3–8), adsorption on a solidlike stationary phase (9–23), and a mixed adsorption/partition mechanism (24–26). Before discussing these concepts, it is necessary to recall the difference between adsorption and partition (14).

In *adsorption,* the stationary phase interacts with the solute or solvent molecules at the external surface of the adsorbent. The separation surface between mobile and stationary phases is arbitrarily chosen in a position where the Gibbs' surface excess for the solvent is zero. In a *partition* process, the support is a substrate with enough porous volume to hold the stationary liquid phase. Direct interactions between the solute and the support do not occur or are kept as small as possible. The basic difference between adsorption and partition is that the stationary region contains only the stationary phase and solute molecules in partition (bulk stationary phase), whereas it contains also solvent molecules in adsorption (interface). The pertinence of each approach has been a subject of many publications (for example, References 7 and 21–23) and will be briefly discussed here.

Two different processes of liquid partitioning can be considered. In the first one, the stationary phase is composed of nonpolar material alone. Polymeric phases, for instance, can be considered (at least quantitatively) to be a silica matrix covered with a film of nonpolar liquid. In this case, it is likely that solute retention originates from a partitioning process between a polar (eluent) and a nonpolar (sorbent) phase (3,8). This case will not be further discussed, however, because polymeric materials are not as popular as monomeric phases. In the case of monomeric phases, it is more difficult to conceive such a description of the stationary phase (7,21) because it is clear that a monomolecular layer of bonded alkyl chains undoubtedly has different properties than a nonpolar liquid phase. This point will be further discussed below.

In the second type of partitioning system, the stationary phase is actually composed of molecules in the organic component(s) of the mobile phase that have been selectively extracted by the modified silica surface (5,8). Depending on the eluent composition, a variable amount of organic solvent can be extracted, yielding a stationary phase of variable thickness. In the case of a monolayer, distinction must be made between adsorption on the nonpolar moieties with associated displacement of solvent molecules in the adsorbed monolayer, or adsorption on the solvent monolayer with no direct contact between sample molecules and bonded moieties. According to Spanjer and De Ligny (27), such a mechanism is highly improbable, at least with methanol/water mixtures.

According to the lattice model developed by Martire and Boehm (7), the bonded chains at the surface of a monomeric phase collapse in most commonly used solvent conditions (aqueous mixtures of methanol or acetonitrile), except in tetrahydrofuran (THF). These collapsed chains act as a stationary liquid in which sample molecules dissolve. The distribution process resembles that of classical liquid-liquid chromatography, the differences originating from different solute configuration and entropy in the respective stationary phases (7). The contribution of adsorption at the interface is negligible in most cases. Martire and Boehm have used their model to justify the characteristic features of RP systems, such as behavior of homologous series, role of solvent composition, surface coverage, and chain length (7). It must be noted, however, that such characteristic properties can be explained as well in terms of pure adsorption. Moreover, the assumptions made in the derivation of the model contradict the data of Kováts and associates (19), who concluded that the adsorbed solvent layer is more than a molecule thick, and those of Bogar and co-workers (28), who found that, in the presence of wetting mobile phases (such as methanol/water mixtures containing a *minimum* amount of methanol), the bonded chains are straightened because of the sorption of organic modifier. This assumption of more or less straight bonded bristles under wetting conditions is also suggested by the concept of critical chain length introduced by Berendsen and de Galan (29), as well as the results of Colin et al. on homologous series (30). The knowledge of the configuration of the stationary phase is of great importance (22,31). According to Lochmüller (16), the bonded alkyl chains are gathered in bundles at the surface of the modified silica for solvophobic reasons, creating greasy patches. The solute molecules penetrating these bushes undergo three-dimensional interactions with the alkyl chains similar to those in a liquid medium (31). This concept of three-dimensional interactions is also suggested by Bogar and co-workers (28), who used spectroscopy of pyrene excimers to conclude that the bonded chains do not behave as solid material but rather have a microviscosity close to that of ethylene glycol. The authors concluded that solvated reversed phases are a unique state of matter with a highly anisotropic character to which bulk-phase thermodynamic properties may not apply.

Various treatments of adsorption on RP materials have been proposed (11,12,14,20,32). Critical parameters in adsorption models are the molecular surface areas and interfacial tensions. The capacity ratio of a solute, 1, which is distributed between an interfacial layer, 2/3, separating a mobile eluent, 2, and a stationary phase, 3, can be described by the following equation (32):

$$\ln k' = \ln (\gamma_{1,2}^{\infty}/n_2) - \ln(\gamma_{1,2/3}^{\infty}/\bar{n}_2) \qquad [1]$$

where n_2 and $\bar{n}_2$ are the numbers of solvent molecules in the stagnant interfacial phase and the mobile phase, respectively, and $\gamma_{1,2}^{\infty}$ and $\gamma_{1,2/3}^{\infty}$ are the chemical potentials of the sample at infinite dilution in bulk eluent and at the interface, respectively.

According to the displacement model proposed by Locke (11), Equation 1 can be rewritten in the following form:

$$\ln k' = A + \ln \gamma_{1,2}^{\infty} - \ln \bar{\gamma}_{1,2/3}^{\infty} + A_1(\sigma_{2,3}^{\circ} - \sigma_{1,3}^{\circ})/RT \qquad [2]$$

where A is a constant, $\bar{\gamma}_{1,2/3}^{\infty}$ is $\gamma_{1,2/3}^{\infty}$ assuming the standard state corresponds to a monolayer of pure sample, A_1 is the molecular surface area of the solute, and $\sigma_{i,j}^{\circ}$ is the interfacial tension between the species i and j. The change in k' when going from a solvent, 2, to a solvent, 2', is thus:

$$\ln(k'_{1,2/3} k'_{1,2'/3}) = A' + \ln(\gamma_{1,2}^{\infty} \bar{\gamma}_{1,2'/3}^{\infty}/\gamma_{1,2'}^{\infty} \bar{\gamma}_{1,2/3}^{\infty}) \qquad [3]$$
$$+ A_1 N(\sigma_{2,3}^{\circ} - \sigma_{2',3}^{\circ}) RT$$

The second term on the right-hand side of Equation 3 is the sum of two contributions, $\ln(\gamma_{1,2}^{\infty}/\gamma_{1,2'}^{\infty})$ and $\ln(\bar{\gamma}_{1,2'/3}^{\infty}/\bar{\gamma}_{1,2/3}^{\infty})$, which characterize a mobile phase effect and an interface contribution, respectively.

Colin and Guiochon (14) have simplified Equation 3 according to:

$$\ln(k'_{1,2/3}/k'_{1,2'/3}) = A' + A_1 N(\sigma_{2,3}^{\circ} - \sigma_{2',3}^{\circ})/RT \qquad [4]$$

This simplification makes it easier to predict retention when changing solvent composition, but the assumptions made are only valid for a given class of samples (see reference 27 for a discussion).

Another adsorption model was proposed by Horváth and associates (12,18,33). It is based on Sinanoğlu's solvophobic interaction theory, which describes the various interactions in terms of macroscopic properties (34,35). According to this approach, nonpolar moieties associate in a polar medium not because of any particular attraction between the species but rather because of the polar character of the solvent. As discussed by Karger and co-workers (37), this forced association often results in a much larger structural selectivity than that originating from dispersion forces. The effect is well known for water as a solvent, and it has been suggested that the large excess entropy accompanying the dissolution of alkanes (or alkyl chains) in water is attributable to reordering of water molecules with creation of iceberg-like structures (36). It has been shown that the hydrophobic effect is not unique to water (37,38). The hydrophobic effect between the solute molecules in an aqueous medium results from van der Waals interactions that are operative between low surface energy materials (or sites) immersed in high energy liquids such as water (39). In the process, some more-organized interstitial water molecules are expelled into the bulk liquid and become more randomized, thus often giving rise to an increase in entropy. This is an accompanying — and occasionally even a contributing — effect, but not a cause of the bonding. Although it is generally believed that this process is accompanied by an increase in hydrogen bonding, it has been shown recently that the solvophobic effect is also operative with aprotic solvents that cannot form hydrogen bonds. Thus, a mechanism must exist by which entropy can decrease without involving hydrogen bonds.

Although the hydrophobic effect is widely accepted (39) it has been questioned by Cramer (40), who has shown that the free energy of partitioning of some compounds — such as rare gases, light hydrocarbons, and several other solutes — between water and *n*-octanol cannot be explained simply in terms of unique properties of water. Cramer raised several points against hydrophobic effects, including: increased aqueous solubility of rare gases with solute bulk; large dependence of the energy of transfer of a $-CH_2-$ moiety from water to a lipid phase; and a remarkably small entropy of solvation in water of an incremental $-CH_2-$ group (around 3–4 cal/mol degree calculated from the series methane-butane). Cramer concluded his publication by noting that "the preceding analysis of the partitioning process leaves little if any experimental evidence remaining to support the notion of a specifically hydrophobic effect" (40).

The model developed by Horváth and others (12,18,33) can be summarized as follows: a nonionic molecule (S) binds to the hydrocarbon ligand (L) in a reversible reaction, forming the complex SL according to Equation 5:

$$S + L \rightleftharpoons SL \qquad\qquad [5]$$

The free energy change, $\Delta G°$, associated with this equilibrium can be decomposed into various discrete hypothetical processes:
- formation of the complex SL in the gaseous phase
- creation of a cavity in the solvent to accommodate the complex
- transfer of the complex from the gaseous to the liquid phase
- interaction of the complex SL with the solvent.

This is schematized in Figure 1.

Basically, the energies of the retention process can be divided into three contributions:
- the solvation in the mobile phase (a). Increasing the solvation favors solute/mobile-phase interactions and thus decreases retention. Stated differently, increasing sample polarity decreases retention. The energy of solvation depends obviously on solvent composition.
- the interactions in (or on) the stationary phase (b). Nonspecific in nature, the intensity of these interactions increases with increasing size of the hydrocarbonaceous skeleton of the sample molecule, independent of solvent composition.
- the stabilization energy stemming from a decrease in area of the total solvent cavity upon binding of the solute with the bonded ligands (c). This energy depends on the sample, stationary phase, and solvent composition. Both contributions b and c tend to increase retention.

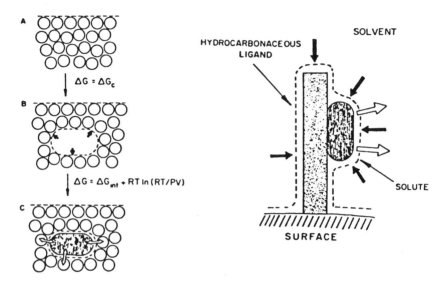

FIGURE 1.   Schematic illustration of placing a solute into a solvent shown in A (left). The energy required to form a suitable cavity depicted in B is given by $\Delta G°_{cav}$, the magnitude of which is essentially determined by the cavity surface area and the appropriate surface tension of the solvent, and it is greater than zero. The solute in the cavity interacts with the solvent, as shown in C. The energy of interaction is $\Delta G_{int} + RT\ln(RT/PV)$, which is smaller than zero.

Association between the solute and nonpolar ligand at the surface of the stationary phase in reversed-phase chromatography (right). The binding is essentially a result of solvent effects. As suggested by the solid arrows, the association of the two species is facilitated by the decrease in the molecular surface area exposed to the solvent upon complex formation. Attractive interactions with the solvent, which are symbolized by open arrows, have a countervailing effect. The magnitude of the nonpolar interaction between the solute and ligand, which determines solute retention, is given by the difference between the two effects. Polar functions in the solute enhance interaction with polar solvents and therefore reduce binding. The binding is stronger when the contact area in the complex is greater and/or the surface tension of the solvent is higher. (Reproduced from Reference 18 with permission.)

$\Delta G°$ of the overall process is given by:

$$\Delta G° = \Delta G°_{gas} + \Delta G°_{cav} + \Delta G°_{vdw} + \Delta G°_{es} - RT\ln(RT/P_oV) \quad \quad [6]$$

where

$\Delta G°_{gas}$ = free energy change associated with interaction in the gaseous phase

$\Delta G°_{vdw}$ = free energy change associated with van der Waals interactions in the solvent

$\Delta G°_{es}$ = free energy change associated with the electrostatic interactions in the solvent

$\Delta G°_{cav}$ = free energy change associated with the formation of the cavity in the solvent

$RT \ln(RT/P_oV)$ = change in free volume during the transfer from the gaseous to the liquid phase.

By rewriting Equation 6 in terms of k' and by expressing the various $\Delta G°$ contributions in terms of measurable physicochemical parameters, one obtains Equation 7:

$$\ln k' = \ln \phi + \frac{1}{RT}[\Delta A(N\gamma + a) + NA_s\gamma(\chi^c - 1) + W - \Delta Z/\varepsilon] \qquad [7]$$
$$+ \ln(RT/P_oV)$$

where

ϕ = phase ratio

ΔA = reduction in surface of the solvent cavity upon formation of the solute/ stationary phase complex

N = Avogadro's number

γ = solvent surface tension

a = parameter involving physicochemical constants of the solvent

A_s = solvent molecular area

χ^c = factor for converting bulk to molecular dimensions properties (may be defined as the ratio of the energy required to extend the polar surface of the solvent to the surface area of added solute molecule)

W = parameter involving physicochemical constants of the solvent

ΔZ = reduction in charge after formation of the solute/stationary-phase complex

ε = dielectric constant of solvent.

ΔA and ΔZ are given by:

$$\Delta A = A_{SL} - A_S - A_L \qquad [8]$$

$$\Delta Z = Z_{SL} - Z_S - Z_L \qquad [9]$$

where subscripts SL, S, and L represent the solute/ligand complex, solute, and ligand, respectively.

If it can be assumed that the length of the alkyl chain bonded to the surface is not a limiting factor in the solute/ligand association, then it results that:

$$A_{SL} = A_L + \theta A_S \qquad [10]$$

where $(1 - \theta)$ is the fraction of the surface area of the solute in contact with the ligand. It follows that ΔA is proportional to A_S. This relationship has been verified by Horváth and colleagues (12) and Wells and Clark (41–45).

The solvophobic approach has been used by many authors to interpret and justify their experimental results (41–45). It has also been shown that it can be used

for various techniques, such as ion pairing (46–48), as well as to calculate physicochemical parameters from chromatographic data (49). According to Hammers
and co-workers (32), however, a fundamental shortcoming of Horváth's model is
that it fails to account for the contribution of solute/solvent interactions in the interfacial layer. As it will be seen in the following section, the contribution of these
interactions sometimes may indeed be very significant.

Another interesting approach to describing retention in chromatography is the
used of solubility parameters (50–52). According to this model, the capacity ratio
of solute, 1, distributed between a mobile, 2, and a stationary, 3, phase is given by:

$$\ln k'_1 = \overline{V}_{1/RT}[(\delta_1 - \delta_2)^2 - (\delta_1 - \delta_3)^2] + \ln(n_2/n_3) \qquad [11]$$

where

δ_i = the Hildebrand solubility parameter (square root of the cohe-
sive energy density)

$\overline{V}_1$ = the sample molar volume

n_2 and n_3 = mole numbers of mobile and stationary phases, respectively.

The direct use of Equation 11 to calculate retention is difficult because δ_3 and n_3
cannot easily be assessed for bonded phases. It has also been shown that this method gives a four- to fivefold overestimation of capacity ratios (52). Nevertheless,
the solubility parameter concept is a useful tool for understanding of the chromatographic process and, particularly, the role of the mobile phase.

Although the stationary phase may in certain conditions control a separation, it
is generally accepted that the properties of a reversed-phase system, especially
the selectivity, are mainly controlled by the mobile phase. Using this assumption,
Jandera and associates (53–55) developed a semiempirical model based on interaction indices. According to this model, the capacity ratio is given as:

$$\ln(k'_i/V_i) - \ln(\phi/V_i) = (\alpha_o - \alpha_2 x) - (\beta_o - \beta x)I_i + \delta x^2 \qquad [12]$$

where V_i, I_i, and ϕ are the solute molar volume, interaction index, and the phase ratio, respectively; α_o and β_o are constants; α and β are also constants that depend on
the nature of the organic solvent in the mobile phase; and x is the volume fraction
of organic solvent in the eluent for a binary mixture. The definition of interaction
indices is described in detail elsewhere (55).

Retention in RPLC has also been correlated with solute solubility in the mobile
phase. According to Hennion and co-workers (8), the hydrophobic hydrocarbonaceous chains serve only as a support for a layer of organic solvent; there are no
specific interactions between the bonded ligands and the solute. These authors
have reported a high degree of correlation between k' and sample solubility in the
mobile phase (s) and in the pure organic solvent (s_o). The relationship between k',
s, and s_o is given by:

$$k' = C's_o/s \qquad [13]$$

where C' is a constant that apparently does not depend on the solute but is equivalent to a phase ratio. The data given in Reference 8 are, however, quite intriguing because the authors report a linear relationship between log k' and NS (N being the number of bonded bristles having a surface area S), whereas, on the basis of the retention mechanism proposed, one would expect C' (and thus k') to be proportional to NS. Moreover, if log k' is linearly related to NS, the selectivity should be an exponential function of NS whereas a linear dependence has been reported. Killer and co-workers (56) have also investigated the relationship between k' and the solubility in the mobile phase, and their results indicate that these two parameters are related according to:

$$k' = b(1/C_{sat})^{2/3} \qquad [14]$$

where C_{sat} is the sample concentration in the mobile phase. It seems difficult to correlate Equations 13 and 14.

In conclusion, retention behavior in RPLC can be summarized as follows:

- it is governed by mobile phase interactions
- the amount of organic material extracted by the stationary phase depends on mobile phase composition and, in most cases, has a marked influence on retention
- the bonded bristles are likely to interact with the solute through dispersive forces
- the contribution of the nonpolar phase is certainly complex, and its understanding necessitates a better knowledge of surface topology.

ROLE OF MOBILE PHASE COMPOSITION

Manipulation of mobile phase composition is undoubtedly the most powerful means for adjusting both absolute and relative retentions in chromatography, particularly in RPLC. Indeed, not only the water content can be adjusted (and thus the extent of solvophobic effect), but the nature of the organic modifier — such as methanol (MeOH), acetonitrile (ACN), and tetrahydrofuran (THF) — can also be changed. Moreover, it is generally possible to control the pH and ionic strength of the eluent, which has a large effect on the chromatographic behavior of ionic compounds. It is also possible to obtain particular solvation effects by adding adequate organic solvents, or to modify solute behavior by means of various agents for ion-pairing, ligand-exchange, and charge-transfer complexes or stereochemical recognition. The following section will examine only the effects of water content, the nature of the organic solvent(s), pH, and ionic strength. Readers are referred to other sources for discussion of secondary equilibria (31).

Role of the Water Content

There are basically two steps for optimizing mobile phase composition. Determining the water content in order to elute the peaks in a convenient range of capacity ratios is the first step. The second is to find the organic solvent(s) that will produce the required selectivity. The first step is generally the simpler, and, most often, two preliminary experiments are sufficient to determine the correct water concentration because $\log k'$ is generally linearly related to the mobile phase water content (expressed in volume fraction), at least in the practical range of k' values. In the following discussion x will represent the volume fraction of organic modifier in the eluent.

According to the solvophobic theory (12,18,22), k' is related to water content via solvent surface tension:

$$\log k' = A + \frac{1}{2.3\,RT}\,[N\Delta A + 4.836\,N^{1/3}\,(\chi^e - 1)\,V^{2/3}]\gamma \qquad [15]$$

where V is the average solvent molecular volume and A is a constant including various parameters. The second term in Equation 15 is not a simple function of γ because both χ^e and V depend on solvent composition and thus on γ. Moreover, the surface tension of a hydroorganic mixture is not a linear function of x. The combination of the two effects, however, results in a quasilinear variation of $\log k'$ with x for the methanol/water mixtures and an almost linear variation for acetonitrile/water mixtures. A 10% increase of x results in about a threefold decrease in k' for MeOH/H_2O or ACN/H_2O mixtures. These findings have been experimentally confirmed; Karger and co-workers (37) were among the first to report a perfect linearity between $\log k'$ and x in the whole concentration range for MeOH/H_2O mixtures using n-hexanol and n-octanol as samples.

The linearity of $\log k'$ versus x can be conveniently described by the following equation:

$$\log k' = \log k'_w + Sx \qquad [16]$$

where k'_w is the solute hypothetical capacity ratio in pure water and S is a constant depending on the organic component(s) of the mobile phase and, possibly, on the solute nature. Karger and co-workers (37) have proposed the following equation:

$$\log k' = x_w \log k'_w + x_{org} \log k'_{org} \qquad [17]$$

where k'_{org} is the solute capacity ratio in pure organic solvent and x_i is the volume fraction of solvent i in the mixture. If x_w is replaced by $(1 - x_{org})$ and x_{org} by x, one obtains:

$$\log k' = \log k'_w - x(\log k'_w - \log k'_{org}) \qquad [18]$$

Because $\log k'_w$ is always larger than $\log k'_{org}$, Equation 18 indicates that $\log k'$ decreases with increasing values of x. The comparison of Equations 16 and 18 shows that S is $\log(k'_w/k'_{org})$.

According to Snyder and co-workers (57,58), S does not change considerably from solute to solute and can be considered as a constant (at least for gradient elution optimization). Schoenmakers and colleagues (51, 59–61) and others (53,54) have shown, however, that S is strongly related to the solute. It seems reasonable that S should depend on the sample because, according to the solvophobic and other theories, the extent of change in retention upon a change of solvent composition depends on solute molecular size.

A quadratic relationship between $\log k'$ and x has also been proposed (51,53–55,59–61). Schoenmakers et al. (59–61) have used the relationship between the solubility parameter of a mixture, δ_m, and those of its components, δ_i, to derive the dependence of $\log k'$ on x:

$$\delta_m = \sum_p x_i \delta_i \qquad [19]$$

where x_i is the volume fraction of component i. For a binary mixture, combination of Equations 11 and 19 gives:

$$\log k' = Ax^2 + Bx + C \qquad [20]$$

The complete expression derived from the solubility-parameter model shows that B is strongly negative, whereas A is weakly but significantly positive. The $\log k'$ versus x plots thus have a convex shape, which has indeed been observed very frequently. An equation similar to Equation 20 has also been proposed by Colin et al. (53–55, 63), who used the empirical model of the interaction indices (Equation 12). Using either the solubility-parameter approach or that based on interaction indices, it can be calculated that, in most cases, the quadratic term plays a negligible role with $MeOH/H_2O$ mixtures. Its importance increases when using ACN and THF, as has been shown in many experimental results. At least for practical purposes, it is generally justified to make linear regression plots of $\log k'$ versus x plots over a limited range of $\log k'$ values (1–1.5). This is extremely important for optimization of mobile-phase composition (see below). Attention must be paid, however, to the fact that the coefficients of the regression line may have no physical significance.

The elution strength of the mobile phase can be conveniently expressed in terms of eluotropic strength, $\varepsilon°$. The solvent strength results from three types of intermolecular interactions: dispersion, orientation, and hydrogen bonding. By analogy to Snyder's treatment of adsorption on polar surfaces (62), the change of solvent strength when going from solvent 1 to solvent 2 can be defined according to:

$$\log (k'_1/k'_2) = S(\varepsilon°_2 - \varepsilon°_1) \qquad [21]$$

where S is a solute-dependent parameter. It can be tempting to identify S as the solute molecular surface area or volume in connection with the solvophobic treatment of RPLC. This approach fails to describe retention accurately because it gives solute-dependent $\varepsilon°$ values. This is basically a result of the fact that, contrary to the situation in NPLC, solvent/solute interactions play a very critical role in RPLC. A simple analysis of the role of water in RPLC can be made as follows. Increasing the water content enhances polar interactions in the mobile phase, which tends to decrease retention. The result of these two opposite effects is, in most cases, a larger sample retardation with increasing water content, the rate of increase being solute-dependent (hydrocarbonaceous backbone and polar functional groups). This indicates that it is not possible to characterize the role of the water content simply by considering changes in k' value. It may be interesting to express the solvent strength in terms of two contributions, one related to the solvophobic effect and the second to specific solvation effects:

$$\varepsilon° = \varepsilon°_{solv} + \varepsilon°_{spec} \qquad [22]$$

Using Equation 22 it is then possible to compare the properties of solvent mixtures having the same solvophobic strength but containing different organic solvents. This was suggested several years ago by Karger and co-workers (37) for normalization of solvent condition.

Using this approach, $\varepsilon°_{solv}$ is easily measurable from the retention of homologous series. Indeed, the change in α_{CH_2} (selectivity in homologous series) with solvent composition gives the change in $\varepsilon°_{solv}$ according to Equation 23:

$$\alpha_{CH_2}^{(1)} - \alpha_{CH_2}^{(2)} = S (\varepsilon°_{solv}{}^{(1)} - \varepsilon°_{solv}{}^{(2)}) \qquad [23]$$

The concept has been described by Colin and associates (63) and Yun (64), who calculated $\varepsilon°_{solv}$ for various solvent mixtures using water as a reference ($\varepsilon°_{solv} = 0$) and considering S as the molecular volume of a CH_2 group ($S = 16.8$ mL). Because $\log k'$ versus carbon number plots are parallel lines, the $\varepsilon°_{solv}$ values are independent of the series chosen.

It has been often reported that changing the water content of the mobile phase changes absolute retentions and also affects selectivity. It is clear from Equation 16 that if S is solute-dependent, then α is too. Although this is not an absolute rule, in many cases the larger the solute molecular volume, the faster the rate of increase of k' with increasing water content. Depending on the pair of solutes investigated, increasing water content may either induce an increase or a decrease of selectivity. In some cases the selectivity remains constant and in others the elution order is reversed.

Before discussing the role of the organic solvent(s) it must be indicated that, under certain conditions, retention may decrease with increasing water content. Such anomalies have been reported with dipeptides (65,66), iodoamino acids (67), and other polar compounds (68,69). Horváth and co-workers (68,70) provided a theoretical treatment of this treatment that is predominantly a stationary-phase effect that results from a mixed retention mechanism (normal reversed-phase behavior + abnormal behavior because of the free silanol groups at the surface of the stationary phase). Increasing the water content of the solvent thus has two opposite effects: the reversed-phase contribution tends to increase retention, and the normal-phase effect gives a decreasing contribution because of increasing solvent strength. It is very likely in that respect that in many cases the nonlinear plots of log k' versus solvent composition may simply be explained in terms of residual silanol contributions without involving the quadratic mobile phase effect. The effects of residual silanols can be minimized by addition of small quantities of an amine to the mobile phase.

Role of the Organic Modifier

The main role of the organic solvent (or organic modifier) is to provide the necessary selectivity once the water content has been adjusted to give acceptable capacity ratios. Because the selectivity is the result of specific solvation effects — in addition to the solvophobic contribution discussed previously — it is difficult to characterize in an absolute manner the role of the organic modifier. Moreover, the role of the organic solvent depends also on the stationary phase (71,72).

The comparison of different solvent mixtures can be made in several ways. Binary aqueous mixtures can be compared on the basis of their water contents. This procedure, however, is not very informative because too many chromatographic parameters are varied at the same time: the capacity ratio, solvophovic effect, and specific solvation effects. A second possibility is to compare eluents (aqueous or nonaqueous) that give the same solvophobic effect, that is, the same $\varepsilon^{\circ}_{solv}$ value (37,63,73). A third possibility consists of normalizing the capacity ratio for a given compound, generally a "neutral" solute, such as benzene (37,74). Finally, it is also possible to compare solvents having the same polarities (59,75). A classification of solvents based on the $\varepsilon^{\circ}_{solv}$ value is presented in Table I. Also given in Table I are the values of the polarity indices (P'). The agreement between values presented by Snyder and co-workers (57) and by Colin et al. (63) is remarkable, Snyder's values being approximately 10% higher. It is interesting to observe that $\varepsilon^{\circ}_{solv}$ values are apparently correlated with P' (Figure 2). The linear regression of $\varepsilon^{\circ}_{solv}$ versus P' gives the equation:

$$\varepsilon^{\circ}_{solv} = -0.56\,P' + 5.86 \qquad [24]$$

TABLE I

Comparison of Solvent Eluotropic Strength ($\varepsilon\,^\circ$) Calculated by Different Researchers*

Solvent	Colin et al. (63)	Snyder et al. (57)	Karch et al. (76)	Polarity index (P')
Water	0.00	0.00	0.00	10.2
Methanol	2.95	3.0	1.0	5.1
Ethanol	3.14	3.6	3.1	4.3
Acetonitrile	2.87	3.1	3.1	5.8
Ethylacetate	3.48	—	—	9.4
Acetone	3.19	3.4	8.8	5.1
Tetrahydrofuran	3.52	4.4	—	4.0

* Reprinted, with permission, from Reference 63.

with a value of $r = 0.998$ for $n = 10$ points. Using $P' = 10.2$ for water, it is possible to calculate $\varepsilon\,^\circ_{solv,w} = 0.15$, which is very close to the theoretical value of 0. The polarity indices thus appear to give a reasonable estimate of solvent solvophobicity. The correlation between $\varepsilon\,^\circ_{solv}$ and the total solubility parameter is, however, much less satisfactory ($r = 0.860$).

Karger et al. have studied the role of the organic modifier (MeOH, ACN, and THF) on selectivity for various substituted benzene derivatives under normalized

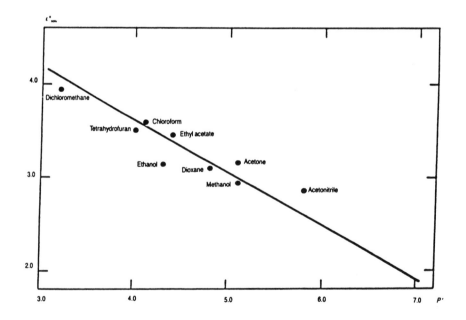

FIGURE 2.   Relationship between the solvophobic contribution to solvent strength ($\varepsilon\,^\circ_{solv}$) and the polarity index (P').

α_{CH_2} values (73). These data were obtained by adjusting the water content in order to keep $k'_{toluene}/k'_{benzene}$ constant ($\alpha \cong 2.0$). It must be noted that this does not correspond exactly to constant $\varepsilon°_{solv}$, because the effect of the CH_2 group addition directly on the benzene ring is not the same as in an alkyl chain. Under the normalization conditions chosen, the capacity ratio for benzene varies largely from solvent to solvent (Table II). There also are very important selectivity changes in going from one modifier to the other. The effects observed are in agreement with the results reported by Soczewinski et al. (77), who proposed the following sequence of increasing solvent strength (Me_2CO, Me_2SO, and AcOH are acetone, dimethylsulfoxide, and acetic acid, respectively):

(for anilines) $MeOH < Me_2CO < dioxane < ACN < Me_2SO < AcOH$ [25]

(for phenols) $MeOH < Me_2CO < dioxane < AcOH < ACN < Me_2SO$ [26]

(for quinolines) $MeOH < Me_2SO < Me_2CO < dioxane < ACN < AcOH$ [27]

It must be remembered, however, that although such results are very useful for selection of the proper mobile phase, the selectivity and the elution order for a series of compounds are strongly related to the water content.

TABLE II

Influence of the Organic Solvent on the Group Contribution*

	Mobile phase composition					
	MeOH/H_2O (50:50)		ACN/H_2O (30:70)		THF/H_2O (25:75)	
Compound (R)	k'	α^x	k'	α^x	k'	α^x
-H	3.19	1.00	6.78	1.00	8.28	1.00
-CH_3	6.38	2.00	13.62	2.00	16.12	1.95
-CH_2CH_3	12.35	3.87	27.20	4.00	30.20	3.65
-$CONH_2$	0.35	0.11	0.46	0.07	0.50	0.06
-CH_2OH	0.96	0.30	1.06	0.16	1.43	0.17
-OH	0.96	0.30	1.59	0.23	3.70	0.45
-CHO	1.28	0.40	2.77	0.41	2.53	0.31
-CN	1.47	0.46	3.62	0.53	3.48	0.42
-CH_2CH_2OH	1.47	0.46	1.58	0.23	2.10	0.25
-$COCH_3$	1.60	0.50	3.11	0.46	2.58	0.31
-NO_2	2.04	0.64	5.42	0.80	6.61	0.80
-OCH_3	3.00	0.94	6.67	0.98	7.26	0.88
CO_2CH_3	3.29	1.03	6.19	0.91	5.32	0.64
-Cl	6.54	2.05	14.44	2.13	18.64	2.25
-$CO_2CH_2CH_3$	6.48	2.03	12.47	1.84	10.15	1.23
-$CO_2CH(CH_3)_2$	11.93	3.74	24.67	3.64	18.31	2.21

* $\alpha^x = k'_{\phi-x}/k'_{\phi}$; temperature = 30°C; ϕ = benzene ring; data taken from Reference 73.

The chromatographic data are more difficult to interpret when the normalization of retention is made for a given compound. This anomaly arises because both nonspecific and specific selectivities are varied simultaneously.

Besides the chemical aspect, the organic solvent influences the physical properties of the mobile phase. Critical parameters are viscosity, detector compatibility, toxicity and convenience of use (price and so forth), as well as compatibility with the column.

MeOH and ACN can be used at low wavelengths provided they do not contain highly absorbing impurities, but THF and acetone can hardly be used even isocratically below 240 nm. It is sometimes considered that the stability of chemically bonded phases is adversely affected by THF. The latter effect, in addition to its toxicity and high viscosity, limits the use of THF despite its established ability to provide remarkable selectivity changes.

Ternary and More-Complex Solvent Systems

In some difficult separation problems, it is not always possible to obtain the desired selectivity using a binary mobile phase. Changing the organic modifier often improves the selectivity between two given compounds but worsens the selectivity for another pair. Using ternary (or more complex) solvent mixtures may dramatically help to solve difficult separation problems; however, two points are worth mentioning. First, it is better for the sake of reproducibility and simplicity to avoid using multicomponent systems. This is particularly true when one of the organic solvents is at a low concentration (< 5% v/v). Second, if complex solvent mixtures must be used a well-defined optimization procedure must be followed; otherwise a trial-and-error approach, which is always time-consuming, may actually give worse results than those obtained with a binary mixture.

Basically, three approaches have been proposed to describe retention in complex (mainly ternary) solvent mixtures. The first is more mathematic than chromatographic (78), the second is based on solubility parameters (59,79), and the third is based on the interaction indices model (54,80–82). For a detailed review of the various optimization procedures, the reader is referred to the references cited above.

Adsorption of Mobile Phase Components by the Stationary Phase

The selective extraction of organic component(s) of the mobile phase by the stationary phase has been the subject of several studies (9,19,20,83–88). This phenomenon is typical of liquid-solid systems and reflects the larger affinity of the stationary phase for certain components of the mobile phase. In RPLC, this effect is of utmost importance because it may affect (and possibly determine) the retention mechanism. The layer of extracted material is sometimes responsible for the observed selectivity (73,89). Moreover, it determines the volume of the moving mobile phase in the column and thus influences the V_o value and the capacity ratios.

The situation is further complicated by the dependence of the thickness of the extracted layer on the nature of the organic solvent(s), the concentration of the solvent(s), the nature of the bonded silica, and the temperature. The thermodynamic treatment of this phenomenon is complex, and only some conclusions drawn from the fundamental papers of Kováts and co-workers (19,20) will be outlined.

Although the adsorption of mobile phase component(s) on the stationary phase is more of a thermodynamic than chromatographic concern, it nevertheless raises the very critical problem of the void volume determination. From the work of Kováts et al. (19,20), it is clear that it is not possible to define one dead volume. The value of the dead volume actually depends on the convention adopted for the adsorption process. A direct consequence of this dependence is that the concept of capacity ratio is certainly not as straightforward as it appears to be. This means, in turn, that many chromatographic theories, particularly those involving extrathermodynamic relationships, must be considered mainly as empirical or semiempirical.

The problem associated with the determination of column hold-up volume has been discussed many times. Several methods have been proposed to measure this volume with, apparently, a high degree of disagreement among them. Most of these contradictions can be explained with the theory of Kováts et al. (19,20), from which it appears that, most often, chromatographers do not measure the *same* volume in the sense that they use (consciously or not) a different adsorption convention. Before reviewing the various approaches for V_o (or t_o) determination, it must be noted that different unretained solutes (with all the uncertainty associated with the word *unretained*) may explore different column volumes, depending on their molecular size. This problem has been discussed by Horváth and Lin (90), who defined that the dead volume as:

$$V_o = V_e + \psi V_i \qquad [28]$$

where V_e and V_i are the extraparticular and intraparticular volumes, respectively. ψ is a constant ($0 < \psi < 1$) that describes the ability of the solute to penetrate the porous structure of the phase. ψ depends on the solute, but it may also depend on the solvent and certainly on the stationary phase, the conformation of which is related to solvent composition (extent of chain collapsing). Second, depending on its surface tension and viscosity, the mobile phase may or may not completely penetrate the porous structure. The extent of penetration should also depend on the solvent pressure; higher pressures favor solvent penetration into the pore. This hypothesis has been confirmed by McCormick and Karger (83), who have shown that the retention volume of deuterated water (D_2O) with a mobile phase of 100% H_2O depends on the column pressure drop (and, thus, the flow rate).

It is convenient to express the void volume in terms of porosity, ε. The maximum ε value (ε_{max}) is given by the following equation:

$$\varepsilon_{max} = \frac{V_o^{max}}{V_G} \qquad [29]$$

where V_o^{max} is the volume of empty space in the column, defined as:

$$V_o^{max} = V_G - V_S \qquad [30]$$

where
V_S = the volume of the support (silica plus bonded moieties).

It is possible to measure V_o^{max} using the method proposed by Van de Venne (91) and Slaats et al. (86). The method consists of weighing the column filled with solvent of widely different densities. V_o^{max} is then given by:

$$V_o^{max} = \frac{M_1 - M_2}{d_1 - d_2} \qquad [31]$$

where M_i and d_i are the mass of the column filled with solvent i and the density of that solvent, respectively. This method assumes that there is no change in stationary phase conformation from one solvent to another. This assumption is reasonable with many solvents, but it may be questionable for such solvents as water.

Various methods have been proposed to measure V_o and have been discussed by several authors (23,92–96). The most commonly used are:
• injection of unretained solutes
• minor disturbance of the composition of the mobile phase
• injection of isotopically labeled components of the mobile phase
• linearization of plots of log k' versus carbon number for homologous series.

UV-active salts, such as sodium nitrate, benzene sulfonate, potassium dichromate, and so forth, should be used with caution because the retention time may depend on the amount injected and on the solvent dielectric constant if some precautions are not taken (92,93,97). Berendsen et al. (92) have suggested that, at low electrolyte concentration in unbuffered eluents, salts are excluded from the pores of the packing probably because of electrical charges at the phase surface (Donnan effect). It is therefore important to ensure that the retention volume of the salt does not depend on the amount injected. Other "unretained" solutes — uracil (37), cytosine (98), tartrazine (93), acetone (95), and *N,N*-dimethylformamide (99) among others — have been reported. Vit et al. (100) have investigated the possibility of determining the dead volume by extrapolating the retention volumes of hydroxybenzenes containing five or six hydroxy groups in a molecule. In the case of fluorescence detection, Street (101) has reported that injection of water using Raman scattering or universal fluorescence detection conditions gave a V_o value similar to that obtained by injection of nonretained salts or by linearization of homologous series.

There are two additional points of importance to mention concerning the injection of labeled compounds. With D_2O, isotopic exchange of deuterium with hydrogen atoms of water (or methanol, if present) and even residual silanols may cause a problem, as discussed by Slaats (23). Second, the retention volume depends on the degree of deuteration (20). Therefore, the use of a radioactive carbon label as proposed by Knox (102) would be preferable, although much less convenient.

From previous discussion of the determination of the dead volume, three important conclusions can be drawn. First, it is not possible to define a single dead volume. The dead volume results from conventions concerning the adsorption process. Among these conventions, some are more realistic and/or are associated with a simpler thermodynamic treatment than others. In any case, determination of dead volume requires the injection of labeled components of the mobile phase. Second, because there is not one dead volume but several, a convention can be adopted to give convenient chromatographic behavior (such as linear log k' versus n_c plots). The thermodynamic value of such a convention is, however, not at all clear. The third conclusion is probably the most important. The necessity to define an adsorption convention must be kept in mind each time the retention must be dealt with in terms of the log k' values. It must not be forgotten that certain conventions are associated with negative k' values for certain samples and mobile phases (94).

Effects of Mobile Phase Additives

As was indicated earlier, this chapter will not be concerned with ion-pair chromatography and techniques based on ligand-exchange or charge-transfer complexes. The points that will be discussed are the effects of pH and ionic strength, the use of exotic modifiers and special additives to obviate the adverse effect of unreacted silanol groups, and, finally, the use of nonaqueous reversed-phase chromatography.

The eluent pH has no effect on the retention of neutral solutes; it only affects the chromatographic behavior of ionized or ionizable (in the range of pH investigated) compounds. The effect of pH on retention has been studied by many authors (33,103–109). The technique of adjusting pH to control ionization of solutes is generally called ionization-control (IC) chromatography. It offers two advantages: improvement in peak shape and adjustment of absolute and relative retentions.

On a quantitative basis, it is simple to predict the effect of pH. As long as the solute is not retained, pH has no effect. When the solute is fully ionized, further pH change also has no effect, but solute retention is decreased because ionized molecules interact more strongly than neutral ones with the polar mobile phase. In the pH range corresponding to $pK_a \pm 2$ units, a smooth transition should occur between the two extreme k' values (ionized and nonionized forms). Although it is possible to work in the intermediate pH range in which the solute is partially ionized, this is generally not recommended because excessive band broadening may occur because of the simultaneous presence of the solute under two forms, neutral and ionized (129).

The ionization of a monoprotic acid, HA, is governed by the following equilibrium:

$$HA \rightleftharpoons H^+ + A^-$$ [32]

The equilibrium constant, K_A, is given by:

$$K_A = \frac{[H^+]_m[A^-]_m}{[AH]_m}$$ [33]

where $[X]_m$ represents the concentration of species X in the mobile phase. The capacity ratio k' can be written:

$$k' = \phi \frac{[AH]_s + [A^-]_s}{[AH]_m + [A^-]_m}$$ [34]

The subscript s denotes the concentration in the stationary phase, and ϕ is the phase ratio. If k'_o and k'_- are the capacity ratios of the nonionized and deprotonated species, respectively, then Equation 34 can be rewritten:

$$k' = \frac{k'_o + k'_-/[H^+]_m}{1 + K_A/[H^+]_m}$$ [35]

Equation 36 was derived by Horváth et al. (33), who also treated monoprotic bases as well as diprotic compounds and zwitter ions. For a monoprotic base, Equation 34 must be rewritten:

$$k' = \frac{k' + k'_+[H^+]_mK_A}{1 + [H^+]_m/K_A}$$ [36]

where k'_+ is the capacity ratio of the protonated form. The expression for zwitterions is similar to Equations 35 and 36 (36); the dependence of k' on pH is sigmoidal.

It is important to mention that attention must be paid to the effect of ionic strength and pH on pK_a values (109). When organic solvents are added to an aqueous buffered eluent, corrected pH (pH*) and pK_a (pK_a*) values should be introduced. The difference between the true pH (pH*) of a hydroorganic mixture and the apparent pH (pH^{app} = the observed pH value) have been studied (references contained in 109) for various organic/water mixtures.

Eluent pH can be controlled either by use of a buffer or by addition of suitable compounds such as trimethylamine. It is important to take some precautions in order to avoid (partial) destruction of the stationary phase and to obtain high enough buffer capacity (it is preferable to use buffer concentrations greater than $10^{-3}M$). The buffer counterion may also affect retention in certain cases (107); although it is likely that buffer ions will not interact significantly with the nonpolar stationary phase, they do interact with unreacted silanol groups and mask their effect (68,70,106,109). The most commonly used buffers are phosphates, tartrates, acetates, and citrates. It has been reported that acetate buffers frequently result in relatively low column efficiencies, possibly because of the formation of nonpolar complexes between the acetate ions and cationic solutes and the relatively low kinetics of this process (18). Ammonium acetate, however, has a high buffering capacity in the range of pH 3–6. Phosphate buffers are usable in a wide range of applications. Sodium phosphate has a high buffering capacity at pH values below 3 and near 7, and citrate buffer is useful in pH range 3–4 because it has a much higher buffer capacity than phosphate at that pH.

If the separation is to be carried out in a basic medium, attention must be paid to the stability of the silica matrix at pH values > 8. The most popular buffers are of the trialkylammonium phosphate type. Triethylammonium phosphate has been shown to give remarkable results for the analysis of peptides and proteins (110); the effects of amine modifiers have also been studied (111–115). Quaternary ammonium ions, however, hydrolyze the silica matrix, and the use of highly hydrophobic tertiary amines may give a mixed retention mechanism because of the buildup of a thick extracted layer.

A consequence of the addition of salts to the mobile phase is an increase of the ionic strength. The surface tension of the eluent (γ) is a function of its ionic strength, I, according to:

$$\gamma = \gamma_o + \sigma'I \qquad [37]$$

where

γ_o = the surface tension at zero ionic strength

σ' = a constant depending on the nature of the salt (36).

In the case of neutral solutes, $\log k'$ is a linear function of I (12,33). With ionized solutes, the situation is more complex because the electrostatic term's contribution to the free energy of transfer (Equation 6) is a complex function of the ionic strength. In general the ionic strength of the eluent used in chromatography is too high for the Debye-Hückel theory to be applicable (12). Under the conditions of high salt concentration, the capacity ratio of an ionized solute can be written as:

$$\log k' = \log k'_o + \varrho\,(B\,I^{1/3} + C\,I) + \varrho'I \qquad [38]$$

where

$k'_o = k'$ at $I = 0$

ϱ and ϱ' = constants for a given solute, salt, and column.

The values of the different parameters in Equation 38 are such that the $\log k'$ versus I curve has a minimum corresponding to $I = 0.3$ independent of the solute.

A drawback of Horváth's treatment of the effect of ionic strength on retention (22), at least for ionizable compounds, is that it does not take into account the effect of I on pK_a values (91), which is an important point considering the key role of pK_a on solute retention. The change in the ionization constant, ΔpK_a, depends on the stage of ionization but not on the sample. The results obtained by Van de Venne (91) are apparently in disagreement with the solvophobic theory because no change in k' value is observed for undissociated compounds, and a steady increase is observed for dissociated compounds with no appearance of an initial decrease. Pietrzyk and Chu also found no effect of I on retention of neutral solutes (116).

Besides special additives dedicated to techniques such as ion pairing, charge-transfer complexes, and ligand exchange, various organic compounds have been evaluated for selectivity control in RPLC. These compounds are used at small concentration (typically less than 10%). In many cases they are hydrophobic enough to be strongly extracted from the mobile phase. In addition to specific solvation effects in the mobile phase, their use results in a particularly high selectivity for certain types of samples.

Solvents other than MeOH, ACN, or THF have been used, and higher alcohols — ethanol, propanol, and isopropanol — have been investigated (75,117,118). These alcohols do not seem to offer significant advantages over MeOH: although they have higher solvent strengths, they have the main drawback of generating very large pressure drops. Dioxane has also been used (53); it gives selectivities similar to ACN — although in some cases it brings about interesting changes — but it creates higher pressure drops and has a lower UV transparency.

Billiet and colleagues (117) have reported the use of 2,2,2-trifluoroethanol (TFE). TFE gives very large selectivity changes not only for fluorine-containing solutes but for many other samples, particularly phenols. Complete reversals in the elution order may occur upon replacing MeOH with TFE. Bakalayar et al. (74) have evaluated ternary mixtures of MeOH/H_2O/X in which X is diethyl ether, dichloromethane, dimethyl formide (DMF), or dimethylsufoxide. The systematic use of DMF has also been suggested for the chromatography of polar basic compounds (118). Much better peak shapes and plate heights are obtained with 10% DMF in the solvent.

The use of micellar mobile phases has been described by Armstrong and co-workers (119,120) and Yarmchuk et al. (121). Highly selective partitioning of many solutes into micelles gives unique properties to the mobile phase. Micelles are spherical or ellipsoidal association colloids that form when the critical micelle concentration (CMC) of a given surfactant is exceeded. There are two basic types of interactions with a solute molecule and a micelle (119): electrostatic and hydrophobic. The first occur between the surface charge of the micelle (Stern layer) and the solute charges (in the case of ionized solutes). The second interactions are attributable to the nonpolar character of the micelle: nonpolar solutes have a much higher solubility in the nonpolar micelle than in the polar eluent.

The efficiency of micellar chromatography is somewhat lower than that of conventional RPLC (121), primarily because of slow mass-transfer kinetics associated with micelles. It is also possible to chromatograph hydrophobic and hydrophilic solutes simultaneously with aqueous micellar solutions, whereas traditional RPLC would require the use of gradient elution to accomplish the same task.

The addition of crown ethers to the mobile phase has also been used to control retention and selectivity (122–127). Crown ethers have been used for the optical resolution of amino acids and their ester salts (126), biogenic amines (127), catecholamines (127), and other compounds (122,123). This technique does not seem to offer significant advantages when compared to ion-pair formation.

Nonaqueous Reversed-Phase Chromatography

Nonaqueous reversed-phase chromatography (NARP) is characterized by the absence of water in the eluent, and it is thus associated with very low solvophobic selectivity. It is primarily used for the separation of compounds of low to medium polarity having a molecular weight larger than 250–300 (128). NARP has been used for the separation of polystyrene (129,130), olefins (131), and carotenoids (132,133) as well as fatty acids and lipids (134). The biggest advantages of NARP are enhanced sample solubility, high chromatographic efficiency and long column life, and high sample capacity and recovery. The low solvophobic selectivity, however, does not mean that overall selectivity is small.

The retention mechanism of NARP is not clear. It must be noted that stationary phases with high surface coverage are required, and mobile phases are typically binary mixtures consisting of a polar solvent such as (MeOH or ACN) and a medium polarity modifier (dichloromethane, acetone, THF, or ethylacetate). Occasionally, a small amount of a third solvent is added to optimize selectivity (132). It seems that the solvent polarity calculated for normal-phase systems cannot be used to predict solvent strength in NARP (132).

A general drawback of the NARP technique is that photometric detection cannot be carried out at low wavelengths because of solvent absorption. It is sometimes possible to use refractive index detection, but this then excludes the use of gradient elution. Light scattering detection seems to be an interesting alternative.

ROLE OF SOLUTE STRUCTURE

The role of sample structure on retention in RPLC has been studied by many authors, and it is impossible to make an exhaustive survey of the very abundant literature available on this subject. Most of the original work was performed using TLC, and studies by Soczewinski and co-workers are very important contributions to understanding the relationship between molecular structure and chromatographic parameters (134 and references therein). It must also be mentioned that the general problem of chromatography in quantitative structure-activity relationships (QSAR) has been the subject of intensive research (46,135,136 and references cited therein).

The quantitative prediction of retention is very difficult because not only the solute structure but also the mobile and stationary phases as well as the temperature have large effects on both absolute and relative (selectivity) retentions. Before discussing the role of the sample, it might be useful to recall the following qualitative rules for RPLC:

- retention is increased when the size of the hydrocarbonaceous skeleton is increased
- polar substituents generally decrease retention
- retention increases with increasing mobile phase water content and is larger for solutes with a large hydrocarbonaceous skeleton
- specific solvation effects in mobile phase decrease retention.

The following discussion will be divided into two parts: the contribution of the hydrocarbonaceous backbone and the effect of polar group substitution.

Solute Hydrocarbonaceous Skeleton

Enhancement of solute retention with increasing size of its hydrocarbonaceous backbone is a result of decreased solubility of the solute in the mobile phase and of increased interactions with the nonpolar stationary phase.

Using the solvophobic approach, Horváth (12) suggested that the reduction of surface cavity inside the solvent upon formation of the solute/stationary phase complex (term A in Equation 7) is proportional to the solute hydrocarbon surface area (HSA). It is thus possible to write $\log k'$ as (12):

$$\log k' = a + b(\text{HSA}) \qquad [39]$$

where a and b are parameters related to the type of solute, the solvent, and the temperature. Equation 39 suggests that, all other factors being constant, $\log k'$ is linearly related to HSA.

A particularly simple case as far as change in HSA is concerned is that of homologous series based on a $-CH_2-$ group increment. The behavior of such series in RPLC has been studied by many authors (30,37,38,42,44,56,76,137–141) who have reported a linear relationship between $\log k'$ and the carbon number, n_C, in the alkyl chain according to:

$$\log k' = \alpha_{CH_2} n_C + \beta \qquad [40]$$

where

α_{CH_2} = the methylene group selectivity

β = a constant depending on the residue of the series $[R-(CH_2)_{m-1}-CH_3]$.

It was shown in the previous section that α_{CH_2} can be used to measure solvent strength expressed in terms of solvophobicity. Different homologous series thus appear as parallel lines in $\log k'$ versus n_C plots. The intercept, β, characterizes the polarity of the residue and its solvation by the solvent (137,139). Homologous series can be used to evaluate the contribution of functional groups to retention. The linearity between $\log k'$ and n_C can be justified in solvophobic terms by assuming that the HSA of the solute is linearly related to the carbon number. It is also possible to use Martin's approach of free energy linear relationship (142):

$$\Delta G = \sum_i \Delta g_i \qquad [41]$$

where the total free energies of transfer (ΔG) of a sample from one phase to another one is expressed as the sum of the free energies of transfer of the different groups (Δg_i) constituting the molecule. Equation 41 is equivalent to Equation 40 if it is assumed that Δg_{CH_2} is a constant. It has recently been reported, however, that departures from linearity may be observed in certain conditions (30). This behavior may result from a contribution of either the mobile or the stationary phase. Very often, nonlinear log k' versus n_c plots are observed at small (< 3–5) n_c values because, for small chains, the behavior of the series residue (β, in Equation 40) may be affected when the chain length is changed. A second type of nonlinearity is observed when the length of the alkyl chain becomes larger than a certain critical value that depends on the length of the alkyl chains bonded on the silica (30).

Apart from homologous series, the role of the solute hydrocarbonaceous backbone also has received much attention. Branching and ring closure decrease retention relative to the linear chain, although some exceptions may occur (37). In order to characterize more quantitatively the effect of the hydrocarbonaceous structure, Karger et al. (37) have suggested use of the molecular connectivity.

It has been shown (143–146) that this index is related to the size of the cavity to be created in a solvent to accommodate the sample molecule, as well as to the partition coefficient in n-octanol water system and to the water solubility. This index is simply calculated using the following equation:

$$\chi = \sum_{k=1}^{k} 1/(\delta_i \delta_j)^{1/2} \qquad [42]$$

where the value of δ (1, 2, 3, or 4) represents the number of atoms attached to atoms i and j, and k is the number of bonds in the solute molecule or its nonpolar segment. In calculating χ values, only the hydrocarbonaceous skeleton is considered; hydrogen atoms are neglected. All of the bond values in the molecule are summed to give the $^1\chi$ index. When the nature of the atom is not taken into account, the index is referred to as the connectivity level (the number 1 refers to a summation of all subgraphs of one bond length). The concept of connectivity was expanded (145) to allow for differences in the identity of atoms and differences in bond order by introducing a valence level index ($^1\chi^v$). Connectivity indices have also been introduced to include indices of different orders (the order, i, being the number of bonds involved in a subgraph of the molecule, χ^i) as well as subgraphs composed of paths, clusters, and others (subscripts p, c, pc, and so forth). The details of the calculation of the various indices are given in Reference 146. The use of connectivity indices to correlate molecular structure with retention in RPLC has received much attention (37,42–45,147–153). It must be pointed out that the choice of the pertinent connectivity indices does not seem to be straightforward. It apparently depends not only on what is to be correlated but also on the stationary and mobile phases.

Various other predictions of retention have also been suggested, such as the hydrocarbon surface area (137) and the van der Waals volume (158,161,162). The chromatographic behavior of polyaromatic hydrocarbons has received considerable attention. Schabron et al. (154) introduced the correlation factor, F, expressed as: F = number of double bonds + number of primary and secondary carbons − 0.5 (for a nonaromatic ring). In their comparative study of the correlation between retention and F or χ, Hurtubise and co-workers (152) concluded that F gives higher correlation degrees than χ. It must be indicated, however, that these authors considered only molecular connectivity and did not try to include other connectivity indices, such as those previously mentioned. Radecki et al. (155) have defined the L/B parameter (length to breadth) based on the rectangle with minimum area that could envelop the molecule.

Various studies have investigated and compared different structure parameters (137,148,150,153). Depending on the type of compounds investigated, different results were obtained. Using pentanol isomers, Colin and Guiochon (137) have shown that HSA gave better correlation between volume and molecular connectivity as well as van der Waals volumes within a given class of compounds. It is clear, however, that the same correlation equation cannot be used for different families because these two descriptors do not take into account the chemical nature of the molecules. Rekker and colleagues (148,150) compared the connectivity approach with that based on log P values (the octanol/water partition constant) calculated using the Rekker fragment system (156) or the Leo-Hansch method (157). Their results demonstrate that, in the case of alkylbenzenes, both Rekker's system and the molecular connectivity give a good description of relative and absolute retentions, the Leo-Hansch approach being less reliable for correlative purposes. The situation is more complex for other chemical families such as benzophenones because of cross-conjugation and resonance effects. The achievement of high degrees of correlation (similar to those obtained with alkylbenzenes) necessitates special corrections, and the calculations of pertinent values of the predictors are much more difficult. It seems, however, that the Rekker system gives slightly better results than the connectivity indices. Jinno and Kawasaki (158) also compared molecular connectivity, van der Waals volume, and log P (calculated according to Hansch and Leo) and observed that connectivity gave better results on stationary phases with short aklyl chains, whereas van der Waals volume and the hydrophobic parameter (log P) were preferable for stationary phases with long alkyl chains.

Effect of Polar Group Substitution

As previously indicated, it is more difficult to quantify the effects of polar group substitution because these effects depend to a large extent on mobile phase composition and on the position of the functional group in the molecule. It has also been shown that in certain cases the transposition of results from one stationary phase to another is impossible.

The simplest approach to this problem is that based on Martin's extrathermo-dynamic rule, as described by Equation 41. According to this model, the contribution of a given group does not depend on the rest of the molecule. Chen and Horváth (159) have studied the behavior of catecholamines, and they proposed a matrix treatment for the evaluation of the group contributions based on the equation:

$$\log r_{i,p} = K_i - K_p = \sum_{j=1}^{m} \tau_{ji} \qquad [43]$$

where

K_i = the logarithm of the capacity ratio of compound i

$r_{i,p}$ = the selectivity

τ_{ji} = a substituent parameter that measures the change in chromatographic retention upon replacing a hydrogen atom by the substituent j

m = the maximum number of a substituent parameters for the set of congeners

p = the parent compound.

τ_{ji} actually can be written τ_j because it has been shown that the τ_{ji} values for a given substituent in a certain position are the same in different compounds i (160). By expressing the k_i' values as a function of k_p and the τ_j' values, a simplified matrix form is obtained:

$$[I_{ij}]_n^m - [\tau_j]_m' + [K_p]_n' = [K_j]_n' \qquad [44]$$

The matrix $[I_{ij}]_n^m$ contains the elements I_{ij}, which indicate whether ($I_{ij} = 1$) or not ($I_{ij} = 0$) the substituent j is present in the molecule i.

It must be kept in mind, however, that the I values are logarithmic increments and that small changes in values may generate very large changes in absolute retention times. From the data given in Reference 159, it is possible to calculate that the average difference between the experimental k' values and those calculated with the I values is about 17%–20%, depending on the stationary phase. This result clearly indicates that such an approach should be used with great care for qualitative analysis or solvent composition optimization.

CONCLUSION

This chapter has covered the theory of retention for reversed-phase liquid chromatography with the intention of elucidating its potential use in the isolation and quantitation of substances of pharmaceutical interest. It is clear that the nature of the sample and mobile phase composition pose ongoing challenges to the analyst; it also is clear that increasing volumes of work in the area will continue to build both an understanding of the mechanism of the technique and the method's utility.

REFERENCES

(1) R.J. Boscott, *Nature (London)* **159**, 342 (1947).

(2) G.A. Howard and A.J.P. Martin, *Biochem. J.* **46**, 532–538 (1950).

(3) J.J. Kirkland and J.J. De Stefano, *J. Chromatogr. Sci.* **8**, 309–314 (1970).

(4) D.C. Locke, *J. Chromatogr. Sci.* **11**, 120–128 (1973).

(5) J.H. Knox and A. Pryde, *J. Chromatogr.* **112**, 171–188 (1975).

(6) E.J. Kikta and E. Grushka, *Anal. Chem.* **48**, 1098–1104 (1976).

(7) D.E. Martire and R.E. Boehm, *J. Phys. Chem.* **87**, 1045–1062 (1983).

(8) M.C. Hennion, C. Picard, C. Combellas, M. Caude, and R. Rosset, *J. Chromatogr.* **210**, 211–228 (1981).

(9) R.P.W. Scott and P. Kucera, *J. Chromatogr.* **142**, 213–232 (1977).

(10) H. Hemetsberber, P. Behrensmeyer, J. Henning and H. Ricken, *Chromatographia* **12**, 71–76 (1979).

(11) D.C. Locke, *J. Chromatogr. Sci.* **12**, 433–437 (1974).

(12) Cs. Horváth, W. Melander, and I. Molnár, *J. Chromatogr.* **125**, 129–156 (1976).

(13) T. Hanai and K. Fujimura, *J. Chromatogr. Sci.* **14**, 140–143 (1976).

(14) H. Colin and G. Guiochon, *J. Chromatogr.* **158**, 183–203 (1978).

(15) M.J. Telepchak, *Chromatographia* **6**, 234–236 (1973).

(16) C.H. Lochmüller and D.R. Wilder, *J. Chromatogr. Sci.* **17**, 574–579 (1979).

(17) H. Colin and G. Guiochon, *J. Chromatogr. Sci.* **18**, 54–64 (1980).

(18) Cs. Horváth and W. Melander, *J. Chromatogr. Sci.* **15**, 393–404 (1977).

(19) N. Le Ha, J. Ungváral and E.sz. Kováts, *Anal. Chem.* **54**, 2410–2421 (1982).

(20) F. Riedo and E.sz. Kováts, *J. Chromatogr.* **239**, 1–28 (1982).

(21) H. Colin and G. Guiochon, *J. Chromatogr.* **141**, 289–312 (1977).

(22) W. Melander and Cs. Horváth, in *High Performance Liquid Chromatography, Advances and Perspectives,* Volume 2, Cs. Horváth, ed. (Academic Press, New York, 1980).

(23) E.H. Slaats, Thesis, University of Amsterdam, 1980.

(24) A. Pryde, *J. Chromatogr. Sci.* **12**, 486–498 (1974).

(25) V. Rehak and E. Smolkova, *Chromatographia* **9**, 219–229 (1976).

(26) D.C. Locke, J.J. Schermud, and B. Banner, *Anal. Chem.* **44**, 90–92 (1972).

(27) M.C. Spanjer and C.L. De Ligny, *J. Chromatogr.* **253**, 23–90 (1982).

(28) R.G. Bogar, J.C. Thomas, and J.B. Callis, *Anal. Chem.* **56**, 1080–1084 (1984).

(29) G.E. Berendsen and L. de Galan, *J. Chromatogr.* **196**, 21–37 (1980).

(30) A. Tchapla, H. Colin, and G. Guiochon, *Anal. Chem.* **56**, 621–625 (1984).

(31) A.M. Krstulović and P.R. Brown, in *Reversed Phase High-Performance Liquid Chromatography* (Wiley Interscience, New York, 1982).

(32) W.E. Hammers, G.J. Meurs, and C.L. De Ligny, *J. Chromatogr.* **246**, 169–189 (1982).

(33) Cs. Horváth, W. Melander, and I. Molnár, *Anal. Chem.* **49**, 142–154 (1977).

(34) T. Halicioğlu and O. Sinanoğlu, *Ann. N.Y. Acad. Sci.* **158**, 308–317 (1969).

(35) O. Sinanoğlu and S. Absulnur, *Fed. Proc.* **243** (part III, supplement 15), 12–23 (1965).

(36) H.S. Frank and M.W. Evans, *J. Chem. Phys.* **13**, 507–532 (1945).

(37) B.L. Karger, J.R. Gant, A. Hartkopf, and P.H. Wiener *J. Chromatogr.* **128**, 65–78 (1976).

(38) N. Tanaka and E.R. Thornton, *J. Am. Chem. Soc.* **99**, 7300–7306 (1977).

(39) C. Tanford, *The Hydrophobic Effect* (John Wiley & Sons, New York, 1973).

(40) R.D. Cramer III, *J. Am. Chem. Soc.* **99**, 5408-5412 (1977).

(41) M.J.M. Wells and C.R. Clark, *J. Chromatogr.* **235**, 31-41 (1982).

(42) M.J.M. Wells, C.R. Clark, and R.M. Patterson, *J. Chromatogr.* **235**, 43-59 (1982).

(43) M.J.M. Wells, C.R. Clark, and R.M. Patterson *J. Chromatogr.* **235**, 61-74 (1982).

(44) M.J.M. Wells, C.R. Clark, and R.M. Patterson, *J. Chromatogr.* **243**, 263-277 (1982).

(45) M.J.M. Wells and C.R. Clark, *J. Chromatogr.* **244**, 231-240 (1982).

(46) E. Tomlinson, *J. Chromatogr.* **113**, 1-45 (1975).

(47) C.M. Riley, E. Tomlinson, and T.M. Jefferies, *J. Chromatogr.* **185**, 197-224 (1979).

(48) C.M. Riley and E. Tomlinson, *Anal. Proc.* 528-533 (December 1980).

(49) W. Melander and Cs. Horváth, *J. Chromatogr.* **185**, 129-152 (1980).

(50) R. Tijssen, H.A.H. Billiet, and P.J. Schoenmakers, *J. Chromatogr.* **122**, 155-203 (1976).

(51) P.J. Schoenmakers, H.A.H. Billiet, and L. De Galan, *J. Chromatogr.* **185**, 179-185 (1979).

(52) T.L. Hafkenscheid and E. Tomlinson, *J. Chromatogr.* **264**, 47-62 (1983).

(53) P. Jandera, H. Colin, and G. Guiochon, *Anal. Chem.* **54**, 435-441 (1982).

(54) H. Colin, P. Jandera, and G. Guiochon, *Anal. Chem.* **55**, 442-446 (1983).

(55) H. Colin, P. Jandera, and G. Guiochon, *Chromatographia* **17**, 83-87 (1983).

(56) K.O. Killer, B. Masloch, and H.J. Möckel, *J. Anal. Chem.* **283**, 109-113 (1977).

(57) L.R. Snyder, J.W. Dolan, and J.R. Gant, *J. Chromatogr.* **165**, 3-30 (1979).

(58) J.W. Dolan, J.R. Gant, and L.R. Snyder, *J. Chromatogr.* **165**, 31-58 (1979).

(59) P.J. Schoenmakers, H.A.H. Billiet, and L. De Galan, *J. Chromatogr.* **218**, 261-284 (1981).

(60) P.J. Schoenmakers, H.A.H. Billiet, R. Tÿssen, and L. De Galan, *J. Chromatogr.* **149**, 519-537 (1978).

(61) P.J. Schoenmakers, H.A.H. Billiet, and L. De Galan, *J. Chromatogr.* **205**, 13-30 (1981).

(62) L.R. Snyder, in *Principles of Adsorption Chromatography* (M. Dekker, New York, 1968).

(63) H. Colin, G. Guiochon, Z. Yun, J.C. Diez-Masa, and P. Jandera, *J. Chromatogr. Sci.* **21**, 179-184 (1983).

(64) Z. Yun, Thesis, University Pierre and Marie Curie, Paris, 1982.

(65) M.T.W. Hearn and B. Greco, *J. Chromatogr.* **265**, 75-87 (1983).

(66) M.T.W. Hearn and B. Greco, *J. Chromatogr.* **225**, 125-176 (1983).

(67) M.T.W. Hearn and B. Greco, *J. Liq. Chromatogr.* **7**, 1079-1088 (1984).

(68) A. Nahum and Cs. Horváth, *J. Chromatogr.* **203**, 53-63 (1981).

(69) Z. Varga-Puchony and Gy. Vigh, *J. Chromatogr.* **257**, 380-383 (1983).

(70) K.E. Bij, Cs. Horváth, W.R. Melander, and A. Nahum, *J. Chromatogr.* **203**, 65-84 (1981).

(71) N. Tanaka, *J. Amer. Chem. Soc.* **98**, 1617-1619 (1976).

(72) I.D. Coilson, C.R. Biebly, and E.D. Morgan, *J. Chromatogr.* **238**, 97-102 (1982).

(73) N. Tanaka, H. Goodell, and B.L. Karger, *J. Chromatogr.* **158**, 233-248 (1978).

(74) S.R. Bakalayar, R. McIlwrick, and E. Roggendorf, *J. Chromatogr.* **142**, 353-365 (1977).

(75) H.J. Issaq, J.R. Klose, and W. Cutchin, *J. Liq. Chromatogr.* **5**, 625-641 (1982).

(76) K. Karch, I. Sebastian, I. Halász, and H. Engelhardt, *J. Chromatogr.* **122**, 171-184 (1976).

(77) E. Soczewinski and M. Waksmundzka-Hajnos, *J. Liq. Chromatogr.* **3**, 1625-1636 (1980).

(78) J.L. Glajch, J.J. Kirkland, K.M. Squire, and J.M. Minor *J. Chromatogr.* **199**, 57-79 (1980).

(79) J.W. Weyland, C.H.P. Bruins, and D.A. Doorbos, *J. Chromatogr. Sci.* **22**, 31-39 (1984).

(80) P. Jandera, H. Colin, and G. Guiochon, *Chromatographia* **16**, 132-137 (1982).

(81) H. Colin, A. Krstulović, G. Guiochon, and J.P. Bounine, *Chromatographia* **17**, 209-214 (1983).

(82) P. Jandera, J. Churacek, and H. Colin, *J. Chromatogr.* **214**, 35-46 (1981).

(83) R.M. McCormick and B.L. Karger, *Anal. Chem.* **52**, 2249-2257 (1980).

(84) R.M. McCormick and B.L. Karger, *J. Chromatogr.* **199**, 259-273 (1980).

(85) C.R. Yonker, T.A. Zwier, and M.F. Burke, *J. Chromatogr.* **241**, 269-280 (1982).

(86) E.H. Slaats, J.C. Kraak, W.J.T. Brugman, and H. Poppe, *J. Chromatogr.* **149**, 255-270 (1978).

(87) E.H. Slaats, W. Markowski, J. Fekete, and H. Poppe, *J. Chromatogr.* **207**, 299–323 (1981).

(88) R.P.W. Scott and C.F. Simpson, *Faraday Symp.* **15**, 13–25 (1980).

(89) H. Colin, A. Krstulović, Z. Yun, and G. Guiochon, *J. Chromatogr.* **255**, 295–309 (1983).

(90) Cs. Horváth and H.J. Lin, *J. Chromatogr.* **126**, 401–420 (1976).

(91) J.L.M. Van de Venne, PhD Thesis, University of Eindhoven, Eindhoven, 1979.

(92) G.E. Berendsen, P.J. Schoenmakers, L. De Galan, G. Vigh, Z. Varga-Puchony, and J. Inczedy, *J. Liq. Chromatogr.* **3**, 1669–1686 (1980).

(93) M.J.M. Wells and C.R. Clark, *Anal. Chem.* **53**, 1341–1345 (1981).

(94) A.M. Krstulović, H. Colin, and G. Guiochon, *Anal. Chem.* **54**, 2438–2443 (1982).

(95) K. Jinno, N. Ozaki, and T. Sato, *Chromatographia* **17**, 341–344 (1983).

(96) H.J. Möckel and T. Freyholdt, *Chromatographia* **17**, 215–220 (1983).

(97) O.A.G.J. van der Houwen, J.A.A. van der Linden, and A.W.M. Indemans, *J. Liq. Chromatogr.* **5**, 2321–2341 (1982).

(98) Sj. van der Wal and J.F.K. Huber, *J. Chromatogr.* **149**, 431–453 (1978).

(99) S.H. Unger and T.F. Feuerman, *J. Chromatogr.* **176**, 426–429 (1979).

(100) I. Vit, M. Popl, and J. Fahrnich *J. Chromatogr.* **281**, 293–298 (1983).

(101) K.W. Street, Jr., *J. Chromatogr. Sci.* **22**, 225–230 (1984).

(102) J.H. Knox, R. Kaliszan, and G.J. Kennedy, *Symp. Faraday Soc.* **15**, 113–125 (1980).

(103) I. Molnár and Cs. Horváth, *Clin. Chem.* **22**, 1497–1502 (1976).

(104) W.E. Rudzinski, D. Bennett, and B. Garcia *J. Liq. Chromatogr.* **5**, 1295–1312 (1982).

(105) W.E. Rudzinski, D. Bennett, V. Garcia, and M. Seymour, *J. Chromatogr. Sci.* **21**, 57–61 (1983).

(106) M. Otto and W. Wegscheider, *J. Liq. Chromatogr.* **6**, 685–704 (1983).

(107) E. Papp and Gy. Vigh, *J. Chromatogr.* **259**, 49–58 (1983).

(108) J.L.M. van de Venne, J.L.H.M. Hendrikx, and R.S. Deelder, *J. Chromatogr.* **167**, 1–16 (1978).

(109) B.L. Karger, J.N. Le Page, and N. Tanaka, in *High Performance Liquid Chromatography, Advances and Perspectives,* Volume 1, Cs. Horváth, ed. (Academic Press, New York, 1980).

(110) J.E. Rivier, *J. Liq. Chromatogr.* **1**, 343–366 (1978).

(111) N.J. Eggers and G.M. Saint-Joly, *J. Liq. Chromatogr.* **6**, 1955–1967 (1983).

(112) R. Gill, S.P. Alexander, and A.C. Moffat, *J. Chromatogr.* **247**, 39–45 (1982).

(113) A. Sokolowski and K.-G. Wahlund, *J. Chromatogr.* **189**, 299–316 (1980).

(114) D. Westerlund and E. Erixson, *J. Chromatogr.* **185**, 593–603 (1979).

(115) K.-G. Wahlund and A. Sokolowski, *J. Chromatogr.* **151**, 299–310 (1978).

(116) D.J. Pietrzyk and C.-H. Chu, *Anal. Chem.* **49**, 757–764 (1977).

(117) H.A. Billiet, P.J. Schoenmakers, and L. De Galan, *J. Chromatogr.* **218**, 443–454 (1981).

(118) M. Ryba, *Chromatographia* **15**, 227–230 (1982).

(119) D.W. Armstrong and R.Q. Terrill, *Anal. Chem.* **51**, 2160–2163 (1979).

(120) D.W. Armstrong and S.J. Henry, *J. Liq. Chromatogr.* **3**, 657–662 (1980).

(121) P. Yarmchuk, R. Weinberger, R.F. Hirsch, and L.J. Cline-Love, *Anal. Chem.* **54**, 2233–2238 (1982).

(122) T. Nakagawa, A. Shibukawa, and T. Uno, *J. Chromatogr.* **239**, 695–706 (1982).

(123) T. Nakagawa, A. Shibukawa, and H. Murata, *J. Chromatogr.* **280**, 31–42 (1983).

(124) T. Nakagawa, A. Shibukawa, and T. Uno, *J. Chromatogr.* **254**, 27–34 (1983).

(125) T. Nakagawa, H. Mizunuma, A. Shibukawa, and T. Uno, *J. Chromatogr.* **211**, 1–13 (1981).

(126) L.R. Sousa, D.H. Hoffman, L. Kaplan, and D.J. Cram, *J. Amer. Chem. Soc.* **96**, 7100–7101 (1974).

(127) M. Wiechmann, *J. Chromatogr.* **235**, 129–137 (1982).

(128) N.A. Parris, *J. Chromatogr.* **149**, 615–624 (1978).

(129) D.W. Armstrong and K.H. Bul, *Anal. Chem.* **54**, 706–708 (1982).

(130) J.J. Lewis, L.B. Rogers, and R.E. Pauls, *J. Chromatogr.* **264**, 339–356 (1983).

(131) R.W. McCoy and R.E. Pauls, *J. Chromatogr. Sci.* **22**, 493–496 (1984).

(132) H.J.C.F. Nelis and A.P. De Leenheer, *Anal. Chem.* **55**, 270–275 (1983).

(133) R.J. Bushway and A.M. Wilson, *J. Can. Inst. Food. Sci. Technol. J.* **15**, 165–169 (1982).

(134) M. Ciszewska and E. Soczewinski, *J. Chromatogr.* **111**, 21–27 (1975).

(135) R. Kaliszan, *J. Chromatogr.* **220**, 71–83 (1981).

(136) K.K. Bhutani, *Pharmacos* **20**, 12–16 (1975).

(137) H. Colin and G. Guiochon, *J. Chromatogr. Sci.* **18**, 54–63 (1980).

(138) H. Colin, G. Guiochon, and J.C. Diez-Masa, *Anal. Chem.* **53**, 146–155 (1981).

(139) H. Colin, A.M. Krstulović, M.F. Gonnord, G. Guiochon, Z. Yun, and P. Jandera, *Chromatographia* **17**, 9–15 (1983).

(140) W. Melander and Cs. Horváth, *Chromatographia* **15**, 86–90 (1982).

(141) G.E. Berendsen and L. De Galan, *J. Chromatogr.* **196**, 21–37 (1980).

(142) A.J.P. Martin, *Biochem. Soc. Symp.* **3**, 4–10 (1949).

(143) L.H. Hall, L.B. Kier, and W.J. Murray, *J. Pharm. Sci.* **64**, 1974–1977 (1975).

(144) W.J. Murray, L.H. Hall, and L.B. Kier, *J. Pharm. Sci.* **64**, 1977–1980 (1975).

(145) L.B. Kier and L.H. Hall, *J. Pharm. Sci.* **65**, 1806–1809 (1976).

(146) L.B. Kier and L.H. Hall *Molecular Connectivity in Chemistry and Drug Research* (Academic Press, New York, 1976).

(147) M. Randic, *J. Amer. Chem. Soc.* **97**, 6609–6615 (1975).

(148) A. Kakoulidou and R.F. Rekker, *J. Chromatogr.* **295**, 341–353 (1984).

(149) Gy. Szasz, O. Papp, J. Vamos, K. Hanko-Novak, and L.B. Kier, *J. Chromatogr.* **269**, 91–95 (1983).

(150) R.E. Koopmans and R.F. Rekker, *J. Chromatogr.* **285**, 267–279 (1984).

(151) K. Jinno and K. Kawasaki, *Chromatographia* **17**, 445–449 (1983).

(152) R.J. Hurtubise, T.W. Allen, and H.F. Silver, *J. Chromatogr.* **235**, 517–522 (1982).

(153) T. Hanai and J. Hubert, *J. Chromatogr.* **290**, 197–206 (1984).

(154) J.F. Schabron, R.J. Hurtubise, and H.F. Silver, *Anal. Chem.* **49**, 2253–2260 (1977).

(155) A. Radecki, H. Lamparczyk, and R. Kaliszan, *Chromatographia* **12**, 595–599 (1979).

(156) R.F. Rekker, in *The Hydrophobic Fragmental Constant. Its Derivatives and Applications. A Means Of Characterizing Membrane Systems,* W.Th. Nanta and R.F. Rekker, eds. (Elsevier, Amsterdam, 1977).

(157) C. Hansch and L. Leo, in *Substituent Constants for Correlation Analysis in Chemistry and Biology* (John Wiley & Sons, New York, 1979).

(158) K. Jinno and K. Kawasaki, *Chromatographia* **17**, 337–340 (1983).

(159) B.K. Chen and Cs. Horváth, *J. Chromatogr.* **171**, 15–28 (1979).

(160) I. Molnár and Cs. Horváth, *J. Chromatogr.* **145**, 371–381 (1978).

(161) K. Jinno and K. Kawasaki, *Chromatographia* **17**, 445–449 (1983).

(162) T. Hanai and J. Hubert, *J. Chromatogr.* **290**, 197–206 (1984).

COMPUTER-ASSISTED OPTIMIZATION OF ISOCRATIC MOBILE PHASES

Mark Canales

Nelson Analytical
Cupertino, CA 95014

In the past decade, high performance liquid chromatography (HPLC) has flourished as a nondestructive technique for the analysis of soluble, nonvolatile materials. As analyses moved from the one-time-only to the routine, the process of method development and support has become more difficult. The analyst faced with a less than optimal separation must ask whether the chromatogram can be improved and, if so, how. In the past, only trial-and-error experiments or serendipity could answer such questions. The advent of the low-cost microcomputer in the laboratory has allowed researchers to use a more systematic approach for method optimization, and the recent literature contains a number of schemes for measuring the "goodness" of a separation as a tool in method development.

The qualities most frequently sought in a chromatogram are good peak separation and short run time. Other goals may include improved detector sensitivities, conversion of a gradient method to an isocratic one, reduction of band broadening or peak splitting, manipulation of elution order, and method ruggedness. For routine analyses, the last goal may be the most important of all.

Although establishing initial conditions is often straightforward, achieving any of the above goals may tax the patience and ingenuity of even the most experienced chromatographer. Computer-aided optimization cannot supplant the user's knowledge, but it can speed the evaluation of experimental results and, if properly used, simplify the method development process.

WHAT IS AN OPTIMIZATION?

Any optimization scheme can be reduced to three steps. The starting point is always an initial set of conditions provided by the user. After a chromatogram is obtained using these conditions, some quality of separation is measured. This separation factor can be a simple or complex function that is termed a response. The chromatographic process is repeated until enough responses are collected to make an evaluation of the response surface — that is, how the response changes with experimental conditions. The next step depends on the optimization algorithm.

If a searching strategy were used, the algorithm would predict the next set of experimental conditions — which would be run — and a new prediction would be made. The search would repeat until a criterion is met that says an optimum has occurred. In a mapping strategy, the experimental points are predetermined. Typically, no evaluation is made until all the data are collected. Then a map is generated and one or more optima are predicted.

The last step again depends on strategy. A search strategy tries to maximize a single optimum. If it proves unsuccessful or the user believes that other optima existed and were not found, then a new set of experiments must be set up and the search rerun. A mapping strategy only predicts optima; although it should find all possible optima with a response surface — that is, the best separation parameters given the initial conditions — mapping does not always put them in the right place. As with a searching strategy, a great deal depends on the definition of the response function. If an optimum is found, the chromatographer must validate the fit of the mapping model by testing the predicted run conditions. If more than one optimum is found, each should be checked. If none are found, then new conditions must be established and the experiment rerun.

TOOLS IN OPTIMIZATION

Given this general outline, the first question to ask is, "What can be manipulated?" It is possible with the aid of some elementary statistics to optimize any set of experimental parameters. In practice, variables such as stationary phase, particle size, column length, buffer, and flow rate are best chosen through preliminary experiments.

A point to remember about any optimization is that the chromatographer's choice of initial conditions determines the response surface to be optimized. Changing any of the initial parameters will influence the optimum to a greater or lesser degree. Changing the column length, for example, may have little effect on the optimum, but changing the buffer or run temperature may have a profound effect on the optimum. Careful selection of the initial conditions is an important part of any optimization. The reader is referred to a paper by Otto and Wegscheider (1) as an example of the effect of buffer on peak resolution.

The three most easily modified parameters in reversed-phase chromatography (RPC) are temperature, pH and ionic strength, and organic mobile phase (that is, solvent selectivity). The solvophobic interaction theory proposed by Nahum and Horváth (2) and Bij et al. (3) suggests that solvent selectivity would be the most influential parameter in RPC retention mechanisms. It is not surprising, therefore, that most optimization schemes developed to date involve mobile phase manipulation. For ionizing solutes, silanophilic interactions can play a significant role in the separation (1); the effects of temperature, pH, and ionic strength become correspondingly more important. The strategies for optimizing these different parameters will be discussed in this chapter. Secondary chemical equilibria can also be used to minimize silanophilic interactions (4). Although mobile phase additives may play a critical role in the final separation, their application is straightforward and will not be considered as part of an overall optimization strategy.

THE RESPONSE FUNCTION

As mentioned previously, the two most commonly desired qualities in a chromatogram are good peak separation and short run time. In order to know how an optimization is proceeding, some function is needed to describe the "goodness" of the separation at any point — that is, a response function. A number of response-function definitions have been presented in the literature. Most have been reviewed by Weyland et al. (5), Debets et al. (6), or Wegscheider et al. (7). The reader is referred to these accounts for a critical analysis of the merits of each response-function definition; it is worth noting that the reviewers come to different conclusions.

To illustrate how response functions are derived, the simple case of separating two peaks will be considered. Three general properties are used to measure separation: resolution (R_s), valley-to-peak ratio (V), and peak separation (P). The first parameter has its basis in chromatographic theory, and the latter two are empirically derived properties of the chromatographic output.

Resolution is a function of relative retention times and peak widths. It requires a measurement of how far apart the peaks are and how much the components have been spread by the column. Retention is properly measured in units of volume, although it is often expressed in units of time or as a capacity factor (6). The capacity factor, k', is a measure of the column's ability to retain the solute. In time units, it is calculated using Equation 1:

$$k' = \frac{t_r - t_0}{t_0} \qquad [1]$$

where
   t_r = the retention time
   t_0 = the retention time of an unretained solute or a measure of the column void
        volume.
The relationship is identical in volume units. If symmetrical or Gaussian peak shapes are assumed, then R_s can be calculated as (8,9):

$$R_s = \frac{t_2 - t_1}{2(s_1 + s_2)} \qquad [2]$$

or

$$R_s = \frac{2(t_2 - t_1)}{w_1 + w_2} \qquad [3]$$

where

t_i = the retention time

s_i = the standard deviation

w_i = the width at $4s$ of the ith peak.

If peaks are assumed to be asymmetric, then t_i becomes the peak maximum and s_i becomes the square root of the variance or second moment of the peak (10). If the column's separation efficiency is defined as an average plate number, N, then R_s can be written as Equation 4 (11):

$$R_s = \frac{(t_2 - t_1)N^{1/2}}{2(t_2 + t_1)} \qquad [4]$$

If retention is formulated in terms of k', R_s can be calculated by using Equation 5 (11):

$$R_s = \frac{\left(\dfrac{N^{1/2}}{2}\right)\left(k'_2 - k'_1\right)}{k'_1 + k'_2 + 2} \qquad [5]$$

An alternative approach to calculating R_s was derived by Snyder and Kirkland (8) using Equation 6:

$$R_s = \left(\frac{N^{1/2}}{4}\right)\left(\frac{\alpha - 1}{\alpha}\right)\left(\frac{k'}{1 + k'}\right) \qquad [6]$$

where

α = the selectivity (k'_2/k'_1)

k' = the average capacity factor for the two peaks.

Selectivity alone has been used as the optimization response. As will be seen below, Equation 6 is not equivalent to Equations 2–5.

The valley-to-peak ratio, V, is the ratio of the valley height between the peaks, a, and the height of the smaller peak, b (12):

$$V = a/b \qquad [7]$$

To derive the peak separation, P, a straight line is projected between adjacent peak maxima (13). The valley depth, f, is the distance from the intervening valley to a straight line at the valley's minimum. The height, g, is the distance from the baseline to the straight line at the same point. P is then calculated from Equation 8:

$$P = f/g \qquad\qquad [8]$$

As peaks begin to separate, resolution will provide a continuous response starting from zero at total overlap; as $t_2 \gg t_1$, R_s will approach the limiting value of $N^{1/2}/2$. The valley-to-peak ratio will return a value of one from total overlap to the limit of valley detection (about 2σ); from there it will proceed to zero at baseline separation of the peaks. Peak separation will be zero until a valley can be detected, and it will rise to one at baseline separation. Neither V nor P will provide useful information about the quality of the separation beyond baseline separation. The importance of this limitation will be seen in later discussions in which more complex response functions are derived.

The principal drawback of the resolution equations is that they all require some estimate of the separation efficiency of the column. If the chromatographer does not have separate standards of all the components of interest, calculating each peak width may be an impossible undertaking. In the absence of deconvolution techniques, the analyst is forced to make a simplifying assumption. In theoretical studies, it is often stated that the plate number, N, may be considered to be constant for a given column (8), but the chromatographer's experience suggests that this assumption is rarely correct. Mechanistically, N is a function of a number of physical parameters including sample loading and separation mechanisms (8). If components are not present in similar concentrations, if the mobile phase is modified, or if one component is retained by a hydrophobic mechanism while another is silanophilic (2–4), N will not be a constant. All three conditions usually occur during the optimization of biologically important materials.

A better approach would be to presume that every change in mobile phase will affect the plate number. If only one standard is available, then one should use its N as the average plate number. If more than one is available, one should determine how N varies across the chromatogram and make an estimate for those components for which no standard is available. In the case of shoulder peaks, if the user is confident that N for these peaks varies little with mobile phase, the N for one chromatogram may used in another in cases in which the peaks are severely overlapped. In general, N in the resolution equation is a scaling factor for the response. Assuming constant N throughout an optimization removes an important parameter in separation quality. Narrower assumptions bring the analyst closer to measuring the physical phenomenon represented by the chromatogram.

Because the response function is supposed to measure the relative separation of two peaks, it is important that it return the same numeric value for that separation at any point in time. Obviously, P and V are independent of elution time; resolution, however, is not independent of time.

If one sets up an experiment in which two peaks are separated by $P = 0.99$ and their retention times are allowed to vary from $k' = 1$ to $k' = 10$, one would find that the R_s calculated by Equation 4 is different from the R_s calculated by Equation 6 (Figure 1). Equation 4 returns a constant value at any point in the chromatogram if the plate number and the separation are held constant. As calculated by Equation 5, R_s decreases asymptotically to:

$$R_s = \frac{N^{1/2}(t_2 - t_1)}{4t_1}$$ [9]

A variant of Equation 4 has been derived in which t_1 replaces t_2 in the denominator. Regardless of its other merits as a definition of resolution, Equation 6 would not be satisfactory as part of an optimization response factor because it provides the computer with too much room for error in discriminating between good and bad resolution.

The results in Figure 1 confirm another point already discussed. The value of R_s is clearly a function of the time spread between peaks, and it has been shown that a given peak separation can return the same value at any point in the chromatograms. One may ask why resolution tables show different values for resolution as a function of peak size (14). The answer rests in their relative widths. If N were varied and the percent peak separation were held constant in the experiment, it would be expected that R_s would not change, but that the relative retention required to obtain the same separation would. Conversely, if the relative retention were held constant while N was varied, the R_s would match the variance in N. If N were presumed to be constant and was not, R_s would vary in a fashion that did not reflect the actual peak separation.

Attempts to create more complex response functions can be categorized in one of three classes:
• worst-pair analysis
• weighted summation or product functions
• functions with additional penalty terms.

Worst-pair analysis involves measuring the response, P_v, V, or R_s, for all peak pairs and constructing a phase diagram for each. These phase diagrams are overlapped, and a minimum at each point is determined. A phase diagram of the minima completes the analysis. This technique was developed by Laub and Purnell (15–17) and applied by Deming and Turoff (18) to optimize pH versus resolution. The overlapping resolution maps (ORM) of Glajch et al. (19) are a multidimensional form of the analysis. In this example, a response function measuring the percent of desired retention and the desired analysis time was used to produce the phase diagrams of each peak pair. These were overlaid to find the minimal responses in the system.

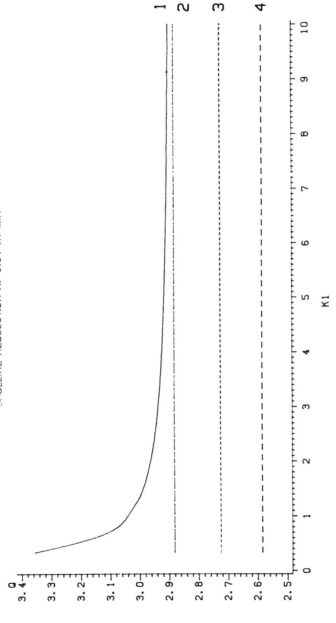

$K'_1 = 1-10$

BASELINE RESOLUTION AT 0.01 HT MIN

K1

FIGURE 1. A comparative plot of $2R_s$ versus k' of the first of two peaks. The second peak retention time was adjusted to $P = 0.99$. Curve 1 is the plot of R_s as calculated by Equation 6. Curves 2–4 are plots of R_s as calculated by Equation 4 with the denominator set to $4t_2$ for Curve 2 and $4t_1$ for Curve 4.

The second general type of response function is the product or summation functions, which are constructed from a response — for example, R_s or P — or from a response ratio — for instance, R_s versus the desired R_s. The response for each pair of peaks or adjacent pair of peaks may be multiplied by a weighting factor before it is summed or multiplied with the responses from other peak pairs. For purposes of this discussion, two functions will be considered: a product function based on R_s and a summation/product function based on P. The reader is again referred to Wegscheider et al. (7), Debets et al. (6), and Weyland et al. (5) for analyses of other functions.

Schoenmakers et al. (11) described a product function, ΠR_s, as the product of all resolutions between adjacent peak pairs (R_s was defined in Equation 4):

$$\Pi R_s = \prod_{i=1}^{n-1} R_{s_{i+1,i}} \qquad [10]$$

where

$(n - 1)$ = the number of adjacent pairs.

Based on further experiments, the authors reported that the function could increase while the separation was deteriorating, so the optimum was not accurately predicted (20). The function was redefined as Equation 11:

$$r = \frac{\displaystyle\prod_{i=1}^{n-1} R_{s_{i+1,i}}}{\left[\left(\displaystyle\sum_{i=1}^{n-1} R_{s_{i+1,i}}\right)(n-1)\right]^{n-1}} \qquad [11]$$

The denominator is the maximum of ΠR_s possible in a given chromatogram. The function varied from zero at the point of overlap to one at the point of equal resolution of all pairs. A value of one, however, did not ensure good separation because all pair separations might have been equally poor. The authors were also concerned about the analysis, believing that the optimum separation might require an unreasonable run time. A new function, r^*, was defined by Equation 12:

$$r^* = \frac{\displaystyle\prod_{i=0}^{n-1} R_{s_{i+1,i}}}{\left[\displaystyle\sum_{i=0}^{n-1} R_{s_{i+1,i}} (n-1)\right]^{n-1}} \qquad [12]$$

where the zeroth component is a designated minimum retention time with an imaginary peak width equal to that of the first component. The tendency of the function will be to optimize on evenly resolved peaks early in the chromatogram.

A chromatographic response function (CRF) based on peak separation was proposed by Morgan and Deming (21) and defined as:

$$CRF = \sum_{i=1}^{n-1} \ln P_i \qquad [13]$$

where

$(n - 1)$ = the number of adjacent pairs.

The natural logarithm of P varies from negative infinity when all peaks overlap to zero when all peaks are separated at the baseline. This relation weights the sensitivity of the function toward small changes in poorly separated peak pairs and away from small changes in adequately separated pairs. Wegscheider et al. (7) reported that this CRF is not able to distinguish between a number of well-resolved peaks and a few poorly resolved peaks and in another study (22), redefined the CRF as:

$$CRF = (1/t)\prod_{i=1}^{m-1} \left(\frac{f_i}{g_i + 2n_i}\right) = (1/t)\ SEP \qquad [14]$$

where

$(m - 1)$ = the number of adjacent pairs

f_i and g_i = the same definition as P in Equation 8

n_i = the noise at the valley between the peaks

t = time required to elute 95% of the most retained component.

When accuracy is the limiting factor — that is $g_i \gg n_i$ — then SEP has the same value as P. When the noise approaches g_i, as in trace analysis, then precision is the limiting factor and SEP < P. This CRF is one of the few response functions that considers the signal-to-noise ratio in the optimization criterion.

The CRF will approach a minimum of zero when any pair coincides, when any pair has a poor signal-to-noise ratio, or when t approaches a large number. The upper limit will be $1/t$ when all peaks are baseline separated. As was the case in Equation 12, the CRF will tend to optimize on equal separation with a minimum analysis time. Wegscheider et al. (7) provide an interesting comparison of this and other CRFs in which a number of optima were generated.

Because of its limited dynamic range, the CRF in Equation 14 will not distinguish between the case in which only one pair is poorly resolved and the case in which all pairs are adequately resolved in the same analysis time. In a six-component system, for example, one chromatogram may have peak separations of 0.26, 1.0, 1.0, 1.0, and 1.0, while another may contain equal separations of 0.8. The resultant SEP will be 0.26 in both cases, and the CRFs will be equivalent if the analysis times are equivalent.

The last general group of response functions are those in which various penalty terms have been added to product or summation functions. Nickel and Deming (23) modified Equation 13 as follows:

$$\text{CRF} = \sum_{i=1}^{n-1} \ln P_i - 100(M - N) \qquad [15]$$

where
 M = the expected number of peaks
 N = the observed number of peaks.
Remembering that Equation 13 varied from negative infinity at complete overlap of all peaks to zero at complete separation of all peaks, it can be calculated that each missing peak will add a penalty of -100 to the summation. This procedure weights the CRF toward finding the maximum number of peaks and requires the user to make a rough estimate of the number of peaks in the chromatogram, which may be difficult to do in a sample as complex as a biological fluid.
 Berridge (24) defined a CRF as:

$$\text{CRF} = \sum_{i=1}^{L} R_i + L^x - a|T_M - T_L| + b(T_1 - T_0) \qquad [16]$$

where
 R_i = defined by Equation 3 [w in Equation 3 is defined as 2(area/height) to avoid questions of asymmetry]
 L = the total number of peaks detected
 T_M = the maximum allowed retention time
 T_L = the retention time of the last peak
 T_1 = the retention time of the first peak
 T_0 = the minimum allowed retention time
 x, a, b = user-selected weighting factors in the range of 0.5–2.0.
This function will tend to optimize resolution, number of peaks, and retention time of the first component while minimizing the run time. The need to select so many factors (that is, T_M, x, a, and b) precludes any general assessment of the utility of this function. The chromatographer would clearly need to know each variable's relative importance in the separation before beginning the optimization.
 This examination of response functions should indicate that the ideal function has not yet been defined. Each of the functions that have been examined requires that certain assumptions about the separation must be true in order for the function to be successful. The disadvantage of resolution is that it incorporates some measure of the separation efficiency for each peak, which is difficult to determine for overlapping peaks without individual standards. A constant plate number is a simplifying assumption that will, most likely, introduce an unwanted variance into the

function. Peak separation and valley-to-peak ratio are empirical measures of separation quality, but they lack the dynamic range needed to be completely useful.

Worst-case analysis will correctly indicate minimum separations, but it may not provide enough information about the rest of the chromatogram. Product and summation functions suffer from a tendency not to discriminate between certain mixtures of good and bad separations. The addition of penalty terms typically requires the user to assess each penalty's relative importance in the separation before optimization can take place.

Before choosing any response function as an optimization criterion, the chromatographer should be aware of the function's implicit constraints. Because most of the literature concentrates upon model systems, the best approach may be to choose a response function whose model best matches the analyst's perception of the task at hand. Although this philosophy is neither foolproof nor universally true, it will at least provide a starting point for optimization.

THE RESPONSE SURFACE

A response surface may be defined as a contour map of the way in which a response varies as a function of the optimization variables. If one imagines a chromatographic space defined by the experimental variables, the response function will measure a surface wherein the optimum lies. If there is more than one high point in the surface, then the highest response is called the *global* optimum. The remaining high points are termed the *local* optima.

Earlier in this discussion, temperature, pH and ionic strength, and percent of organic modifier were chosen as experimental variables; each could be optimized separately. If the process were repeated in a different order, however, it would be likely that a different result would be reached. This consequence is particularly true if any of the components are protic or hydrogen-bonding species. One reason for this behavior is that the experimental variables are capable of interacting with one another in a significant way; that is, they are interdependent rather than independent.

There are therefore two choices in defining the response surface. Either all but one set of variables are held constant and the remainder are manipulated or one tries to manipulate as many as is feasible. Depending upon the optimizing variable, even a single-variable analysis may return a complex response surface. On the one hand, in optimizing pH versus resolution, Deming and Turoff (18) reported that, although the response surface of any peak pair might be described as a simple curve, the composite response was a complex surface not easily described by a single function. On the other hand, every chromatographer has done a simple two-solvent optimization, such as methanol/water. If an isocratic separation exists, it occurs within a very narrow region of the gradient and determination of the response surface is very straightforward.

In general, optimizations of more than one type of experimental variable produce complex response surface contours that look much like mountain ranges with multiple maxima and minima. The dependence of these contours — for example, of the capacity factor — on pH, ionic strength, and temperature is often nonlinear and species-specific (1). The response surface of an organic modifier optimization is a function of the elution order and tends to behave in a more linear fashion.

Regardless of how many solvents are used in the actual experiment, optimization of the organic phase is actually an experiment in solvent selectivity. Any solvent will interact with the stationary phase and a solute through a combination of four interactions: dielectric, dipole, dispersion, and hydrogen bonding. The response surface then describes a mixing model of these various interactions. One can measure solvent strength in terms of the Hildebrand solubility parameter, Δ (25), or the polarity index, P' (26–28). In either case, the relationship between these interactions and a particular solvent can be defined in terms of Snyder's solvent selectivity triangle concept (8). The usefulness of this model was confirmed by Schoenmakers et al. (11). These authors carried out an extensive survey of ternary mobile phases. By measuring analysis times, a series of phase diagrams was generated that showed the apparent constant solvent strength or isoeluotropic behavior of these mobile phase mixtures. Based on averages of a large number of solutes, the authors were able to predict the composition of isoeluotropic mixtures from solubility parameters. They observed that the behavior of individual solutes may deviate from the norm by twofold, but this behavior was the exception and not the rule. Given this information, a much better *a priori* analysis of the response surface for solvent selectivity can be made than can be made of other optimizing parameters.

In the solvent selectivity triangle concept, the mobile phase consists of a base solvent and one or more organic modifiers (8). For RPC, the base solvent is water; for normal-phase chromatography (NPC), the base solvent might be an alkane or fluorocarbon. The mobile phase response surface will be defined by a mixture of base solvent and organic modifier (which was the basis for the earlier definition of this as a mixing model).

In a binary mobile phase — water/methanol, for example — the response surface of a single pair of peaks will be a curve related to the gradient. If, for example, the best separation occurs at 50:50 water/methanol, then the peaks will overlap at higher methanol concentrations and may not elute at lower concentrations. Thus, the response surface contour will begin at infinite separation and analysis time. Within a narrow region, the response will begin to decrease rapidly to zero, and at some point it will pass through an optimum combination of separation and analysis time. For multiple peaks, the surface contours will be a combination of these elution curves. In theory, the optimum will now be the best compromise between run time and separation.

In a ternary mobile phase system, the response surface of a single pair becomes a curved plane, but the upper and lower bounds remain the same. The contour of the plane is a function of solvent strength — that is, how rapidly the peaks elute — and of equivalent separation power. Note that the surface was not described in terms of its isoeluotropic behavior because the surface must tell something about the separation characteristics of the mobile phase, not just its eluting power. This point will become important in the discussion of quaternary mobile phases.

For multiple peaks, the response surface can take one of several shapes. If resolution is limited by a single pair of peaks, the surface may be a smooth plane with a single optimum; if there is more than one pair, the user will encounter multiple optima (1,5). Each local optimum is the result of limiting resolution for an individual peak pair. Ideally, there will be a set of conditions that maximizes resolution of all pairs and minimizes analysis time; this set is the global optimum and will clearly depend on what the response function defines as optimal.

Before discussing the shape of the quaternary mobile phase response surface, how it is defined should be examined. The ternary case was a combination of two binary systems — water/methanol and water/acetonitrile, for example — with different solvent selectivity parameters. The ternary model must first address the question of where resolution occurs before seeking a best resolution. The quaternary system, as a combination of three binaries, can assume resolution occurs and therefore search immediately for the optimal response.

In Snyder's selectivity triangle [as applied by Glajch et al. (19)], the strategy of choosing the binaries entails selecting modifiers whose selectivity parameters are as different as possible. In this way, the chromatographer can take advantage of the widest range of selectivity behavior to manipulate resolution. Glajch proposed that binaries be chosen by optimizing the response in one binary and then choosing the two other isoeluotropic binaries (19). The response surface represents an intersecting plane in which the solvent strength is approximately constant as a function of the capacity factor of the last eluting peak. In terms of optimizing resolution, this limitation is unnecessary. Schoenmakers et al. (11) and Drouen et al. (20) have shown that the behavior of $\ln k'$ is not necessarily linear in mixing isoeluotropic binaries. Requiring a constant k' for each binary does not guarantee the best response from each binary or any combination of binaries. If the intent is to optimize the best response from the resolution or separation function, the binary mixtures or the vertices of the response surface should themselves be chosen for their best response. Then the response surface can be described as the combination of the individual binary best selectivities and the interaction terms of their mixing. If analysis times become an important consideration, then that can either be built into the response function as seen above or can be achieved by censoring the data (23).

If a quaternary mobile phase response surface is defined as a map of the available solvent selectivity interactions, then any optima must exist within the plane. The shape of the plane is a function of where it intersects the binaries. Depending on the choice of binaries and other column conditions, one can imagine an infinite number of intersecting planes. The ternary mobile phase response surface is an edgewise cross-section of these planes showing the response behavior along two of the three coordinates. The binaries are the individual axes of the planes, and their response surface is the behavior of the response function along that axis.

In theory, the quaternary response surface will contain more information about the global optimum and any local optima than either of the other two systems. The author's experience suggests that if every pair of peaks can be separated in one of the binaries — not necessarily the same one — then an optimal combination of resolutions will exist within the response surface. Conversely, it appears that a pair of peaks that will not resolve in any of the binaries does not appear to do substantially better in any combination. The other experimental conditions should be reevaluated before continuing with the optimization.

In practice, the global optimum may have an acceptable counterpart in a binary or ternary response surface (29). The chromatographer will need to rely on experience to determine which response surface should be optimized. The experiment may ultimately be determined by the available time and equipment rather than any theoretical considerations.

OPTIMIZATION STRATEGIES

Current strategies available for optimization can be divided into two classes: searching and mapping. In a *searching* strategy, the user runs an initial set of experiments and, after evaluating the results, chooses a new set of experimental conditions according to some criterion and tests them. The process is repeated until either an optimum is found or the testing criteria are satisfied. In a *mapping* strategy, a predetermined set of experiments are run and the results are used to map the response surface. The response surface is checked for the presence of an optimum. If none is found, the chromatographer chooses a new set of initial conditions and the new response surface is mapped.

Two types of searching algorithms will be considered: sequential simplex and predictive searches with confidence limit testing; three types of mapping algorithms will also be evaluated: window diagrams, factorial designs, and constrained factorial designs or mixing models. Each strategy has been successful in the examples cited by the appropriate authors. These four questions should be answered:

• What does the strategy require as prior knowledge?
• What will the strategy provide as new information?
• What are its strengths as a strategy?
• What are its weaknesses?

A simplex is a simple geometrical figure that consists of one more side than the number of variables to be analyzed (30). For a two-dimensional analysis — optimization of two binary solvent mixtures, for example — the simplex is a triangle. For three variables, it is a tetrahedron, and so forth. A sequential simplex optimization begins with an evaluation of the initial simplex; the worst result is rejected and a new point is selected and tested. Based on the relationship between the new result and previous results, a new simplex is defined. As long as there is movement toward an optimum, the process is repeated. The experiment is concluded when an optimum is indicated or a predetermined number of experiments have been performed (31). Berridge has applied the technique to optimization of flow rate and a binary isocratic mobile phase, a binary gradient elution, and a ternary isocratic mobile phase (24). Nickel and Deming used the same strategy with a different response factor to optimize pH and percent acetonitrile in the separation of 19 PTH-amino acids (32).

The rules for determining the initial simplex were reviewed by Yarbro and Deming (33). The initial size may be large or small; a large simplex will collapse on itself, have a faster approach to the optimum, and be more sensitive to noise in the response than will a small simplex. As many variables as are convenient may be chosen, and they need not be ranked in any way. The following points should be considered when choosing a variable. Is the variable a limiting reagent? If so, then changing the concentration will skew the response. If two variables, solvent binaries, for example, are related, they should be treated as one. The units for the variable should be correct. If a modifier is being added to the mobile phase, its units should reflect the effective concentration, not the amount added. Effective boundaries should be determined, a range of 0–14, for example, for pH.

Once the initial simplex is determined and the responses measured, the first step will be to reflect the worst point and measure its response. Deming and Morgan (31) have proposed rules for the acceleration or deceleration of the simplex as modified by Nelder and Mead (34). Following Figure 2, the initial simplex was BNW, the best point, the next best point, and the worst point, with R as the reflection. R is determined by reflecting the distance between W and the midpoint P about P. If the response of R is greater than that of B, then a new point, S, an expansion, is determined. If the response of S is greater than B, then the new simplex is BNS. Otherwise it is BNR. If the response of R is equal to or between the responses of B and N, then the new simplex is BNR. If the response of R is less than that of N, then one of two contractions is tested. If R is worse than W, then the response of contraction T is measured. If R is greater than or equal to W, then the contraction will be U. If T or U is still the worst point in the new simplex, the vertices should be corrected so that the simplex will not become stranded.

If a vertex of an n-dimensional simplex appears as the best point in $n+1$ simplexes, then it should be rechecked. If not in error and the simplexes have circled the point, then it is an optimum. If a new vertex is outside the boundary conditions, the data should be censored to produce an undesirable response. Berridge (24) used Equation 16 in a simplex optimization. If no peaks were detected, the CRF was assigned a value of -50. If one variable was outside its boundary condition, the CRF was assigned a value of -100. The process of determining new vertices continues until an optimum is reached or the allowed number of experiments have been performed. Berridge proposed that the optimum was found when there was

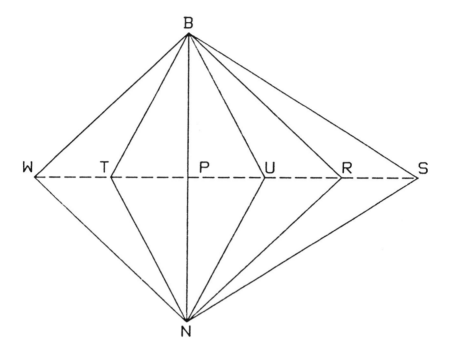

FIGURE 2.   A schematic diagram of the simplex searching strategy. Points B, N, and W constitute the original simplex. Point P is the midpoint between B and N. Points R, S, T, and U are options for the next simplex point as outlined in the text.

less than 3% change in n variables during $n + 1$ experiments and suggested limiting the number of experiments to 30 (35). If the simplex does not approach an optimum by that time, the experiment should be considered a failure.

Other than defining the variables and their boundaries, the simplex strategy requires little prior knowledge of the system. As a result, the simplex strategy seeks the nearest optimum of the response function without regard to the rest of the response surface. It is a well-defined method with clear procedures, and there is no limit on which response function or surface it optimizes. Conversely, its results will only be as good as the response function's description of the resolution surface. If the response function has weighting factors, such as in Equation 16, these factors can skew the response surface away from a true optimum. This result is a limitation of the response function, not of the strategy.

Sequential simplex optimization is a rugged method, and its calculations are simple enough to be done by hand. Because it is a search and not a map, this technique will provide little insight into how the response surface's contours vary in the experimental region. As a consequence, it will not find more than one optimum if more than one exists. Berridge (36) has recommended that, for unknown samples, the procedure be repeated at least once from a different set of starting points to increase confidence in the chosen optimum. If the response is noisy and the response function does not correct for the noise, the simplex can be easily deflected from the true optimum by a falsely high CRF. Because of the number of experiments involved, if the analysis time is long it would be faster to map the response with a factorial design than to search for one or more optima with a sequential simplex.

Drouen et al. (20) have proposed an alternate strategy for selecting the optimal ternary mobile phase mixture. This strategy is based on plots of the variance in ln k' with changes in the mobile phase for each solute. An optimal response is predicted from the ln k' data and tested; the process repeats until an optimum is found or the response meets certain confidence limit criteria.

The procedure consists of four basic steps:
- A water/methanol gradient is run to determine whether an isocratic separation is possible.
- Chromatograms are developed in the isocratic binary found in step one, as are the corresponding isoeluotropic binaries of water/acetonitrile and water/tetrahydrofuran. Each solute peak is identified, and the change in its k' between mobile phases is recorded.
- A linear plot of ln k' versus solvent composition is constructed for each solute between pairs of binaries. Values from the plot are used in an appropriate response function to predict an optimal ternary composition.
- A chromatogram is developed in the predicted optimal mobile phase. The calculations are repeated until the "best" optimum is found or the predicted change falls within the confidence limits of the previous prediction.

Jandera et al. (37) have suggested that the linear relationship between ln k' with the volume fraction of organic modifier provides a good prediction of retention behavior in ternary solvents. Although the strategy initially assumes linearity, deviations are predicted in theory and observed in practice. The authors developed Equations 10–12 as resolution criteria for the optimization. Predicted values of R_s were derived from the ln k' plots. The response curves were calculated, and the predicted maximum response at fractional composition x was determined. To accelerate the search and to compensate for the behavior of ln k', the predicted optimum is shifted according to Equation 17:

$$x' = x + 2f\,[0.5 - (x - x_1)/(x_2 - x_1)] \qquad [17]$$

where

f = a value between zero and one
x = the predicted optimum
x_1 and x_2 = previously measured compositions on either side of x.

The authors indicated that a value of $f = 0.2$ provided a rapid search with adequate precision.

Each new data point defines a confidence range (CR) as:

$$CR = \Delta(x)/2 - (1/2)[\Delta(x)^2 - (4\,\delta/|A|)] \qquad [18]$$

where

$\Delta(x)$ = the unsearched solvent range on either side of x

δ = the allowed uncertainty in $\ln k'$ (usually 0.01)

A = the curvature coefficient for $\ln k'$.

With each new data point, the confidence limits increase until they cover the entire solvent range; the authors indicate, however, that an optimum is normally reached before this occurs.

After each new point is examined, the new $\ln k'$ data are added to the plot and a new response surface for the optimizing criterion is calculated. The new optimum is found and the cycle repeats until the confidence limits are satisfied or no further change is indicated. The authors suggest that this will occur within 7–10 analyses. Figure 3 illustrates an example of the corrected phase diagram.

This strategy has two initial requirements: the chromatographer must be able to determine two isoeluotropic mixtures from solvent selectivity parameters; more important, each solute peak must be tracked during the optimization. In the examples cited, the strategy provided information about each of the optima resulting from changes in elution. The authors' use of several resolution functions provides a useful insight into how response functions can fail to correlate with the observed peak separation in a chromatogram.

By definition, this predictive strategy is limited to optimizing ternary solvent mixtures. This method does not have the general applicability of the simplex search, but it takes advantage of solvent selectivity parameters to search the response surface rapidly and is not limited to determining a single optimum. This approach will work with any response function that can be calculated from k', suggesting that CRFs based on peak separation would not be appropriate.

Because the strategy is based on solvent selectivity, it should be noted that it only searches the edge planes of the solvent selectivity triangle. The authors believed that enough information could be obtained from the ternary mixtures that further investigation of the quaternary response surface was unnecessary. This conclusion is in disagreement with the results of Glajch et al. (19). As for the critical requirement of identifying each solute peak, the authors indicated that multiwavelength detection was a possible alternative to having a standard for each peak. Although this method was successful in their examples, it presumed that the UV/Vis spectrum of each peak contains enough information to identify it unambiguously. It is questionable whether this assumption would be true for compounds lacking a strong chromophore, sharing a common chromophore, or existing in a complex matrix.

The window diagram technique was developed by Laub and Purnell (15) for the optimization of mixed stationary phases in gas chromatography and was later applied by Deming and Turoff (18) to liquid chromatography. The strategy measures

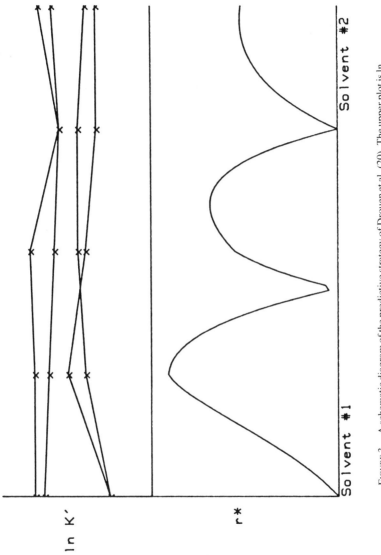

FIGURE 3. A schematic diagram of the predictive strategy of Drouen et al. (20). The upper plot is ln k' versus solvent composition for each of four components. The bottom curve is a plot of the corresponding r^* (Equation 12) for these compounds.

the retention time of each solute at several combinations of the optimization variables. A response function is used to determine the resolution curves of all possible pairs. These are overlapped to map the response surface of the worst separated pair. Depending on the particular variables, this surface may involve more than one limiting pair. Figure 4 is an example of a typical window diagram.

Deming and Turoff (18) have applied the technique to the optimization of pH in the separation of several substituted benzoic acids. In this example, the retention times of all the components were measured at four pH values. A model relating pH and retention time was fitted to the data, and values of k' were calculated for each solute across the experimental region. Resolution curves were calculated from a variant of Equation 6, from which N was deleted; that is, constant plates were assumed and plotted as a function of pH. The lowest points in the overlapping curves corresponded to the response of the worst separated pair. Treating these minima as a unique response surface allowed the authors to determine the best resolution of the worst resolved pair. The region around the optimum was considered a good starting point for further optimization with a more critical strategy, such as a sequential simplex optimization. In later reports (38,39), the strategy was extended to the two-dimensional optimization of pH versus concentration of an ion interaction reagent and percent organic modifier versus concentration of an ion interac-

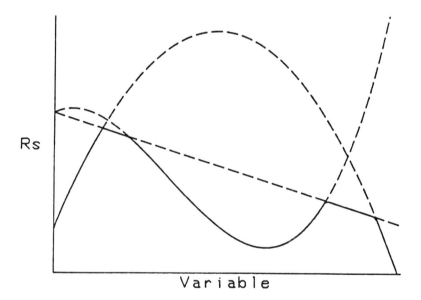

FIGURE 4.   A window diagram of the response of three components to an experimental variable. Each component is represented by a dashed curve and the resultant window is represented by the solid window.

tion reagent. This technique requires that each experimental variable be independent and that a model be available for its interaction with the dependent variables of the response function (36). The strategy also requires that each solute peak be identified and that the shift in its retention time can be fit with the interaction model. Unfortunately, many of the variables one might consider in an optimization are interdependent rather than independent.

Window diagrams are an elegant way of rapidly providing the analyst with a global set of optimal elution conditions for a well-defined sample. The technique would be of little use when an interaction model is not available or the sample's composition is unknown. In the multidimensional case, the difficulties associated with visualizing and interpreting the response surface limit the utility of the technique. The three-dimensional plots reported above might have better been presented as topographical maps, particularly in the case of the minimum alpha plots, because much of the information is below the plane of the upper bounds and therefore is not seen.

Factorial designs may be viewed as a mapping analogue of the sequential simplex. Whereas the simplex design will search the response surface of any set of variables, the factorial design attempts to map the surface with preselected settings of the experimental variables. Factorial designs are used extensively in the design and analysis of experiments outside of chromatography (21).

The factorial design strategy usually consists of four steps. First, data are acquired in a predetermined set of experiments. For example, a two-factor, three-level full factorial design (Figure 5) would consist of three settings for each of the two variables (1, 2, 3 and a, b, c). The response for each combination (nine in all: $a1$, $a2$, $a3$, $b1$, $b2$, $b3$, $c1$, $c2$, and $c3$) is measured in the initial step. A mathematical model is next fit to the data using statistical curve fitting techniques. The model may be linear or nonlinear and usually is a polynomial. The response surface contours are then mapped to determine where a maximum response might occur. Finally, if the predicted optimum lies outside the initial design, a new set of conditions can be set up with the design and analyzed. If the predicted optimum lies within the design, the design might be rotated or otherwise modified and reanalyzed to confirm the maximum. [The reader is referred to Box et al. (40) for the general rules governing factorial design; Morgan and Deming (21) provide examples of several simple factorial designs for chromatographic optimization.]

Otto and Wegscheider (1) have applied the factorial design strategy to the optimization of pH, volume percent of methanol, and ionic strength with the goal in mind of defining from first principles a species-specific response surface whose maximum indicated the chromatographic conditions giving the best selectivity. The factorial design was a three-factor $6 \times 3 \times 2$ design (Figure 6) — that is, 6 pH settings, 3 percent-methanol settings, and 2 ionic-strength settings for 36 analyses in all. The run order of the experiments was randomized to reduce bias and to preserve column integrity. Six additional mobile phases were tested to provide an added one-dimensional fit of pH dependence. The retention times for each solute

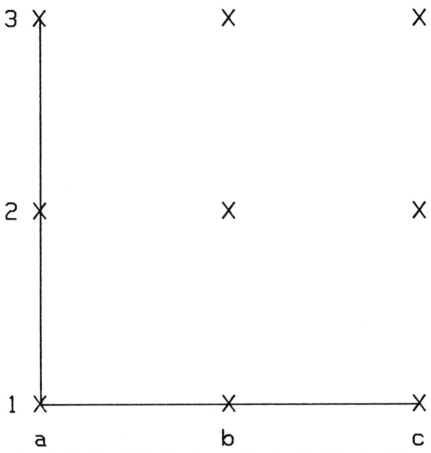

FIGURE 5.   A 3×3 factorial design. Experiments would be performed at all nine settings of the two variables.

were tracked individually and converted to k', and the capacity factor data were fit by multiple nonlinear least-squares estimates.

Rather than fit the data with a purely empirical model, terms were included for the dependence of k' on pH (41,42), ionic strength (43), and percent methanol (38). The three-factor model was formulated with six linear parameters and seven nonlinear parameters as:

$$k' = [C_0(F_1 + F_2e^{-K_3(\%M)})/S]$$
$$+ [C_1(F_3 + F_4e^{-K_4(\%M)})([H^+]/K_{a1}P_1)/S]$$
$$+ [C_{-1}(F_5 + F_6e^{-K_5(\%M)})(K_{a2}P_2/[H^+])/S] \qquad [19]$$

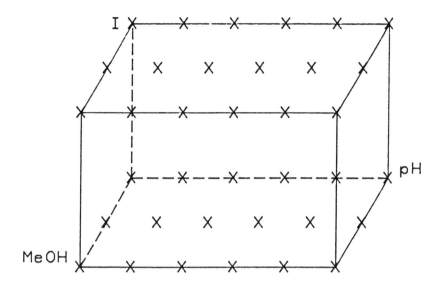

FIGURE 6.   The $6 \times 3 \times 2$ factorial design used by Otto and Wegsheider (1) (details are discussed in the text).

where

$$S = 1 + ([H^+]/K_{a1}P_1) + (K_{a2}P_2/[H^+]).$$

F_1, F_3, and F_5 are intercept terms; F_2, F_4, F_6 are the capacity factors for *HS*, *H₂S*, and *S* in the absence of methanol; K_{a1} and K_{a2} are the dissociation constants; K_3, K_4, and K_5 are solvent strength constants; C_0, C_1, and C_{-1} are corrections for ionic strength from the Davies equation (44); P_1 and P_2 are correction factors for ionic strength and nonlinear solvent effects; and $\%M$ is the volume percent of methanol. The size of Equation 19 is indicative of the complexity associated with fitting a factorial model.

Some parameters were found to be meaningless either as a result of high correlation of some fit coefficients (that is, near dependence on one another) or of some ionic forms not contributing to the performance of k'. The intercept terms of k' versus solvent strength were usually omitted. Because the maximum was not readily apparent, a computerized grid search (45) was performed in all three dimensions in step widths of pH = 0.1, $\%M = 2\%$, and ionic strength (I) = 0.01 M. Once the maximum was determined, a chromatogram developed under the predicted optimal conditions was analyzed. The differences between the expected and the observed retention times were within experimental error.

Factorial design has been shown to be a powerful tool for tracking the interactions of experimental variables. The method is general enough to require no prior knowledge of the relationships between optimizing parameters. It is also flexible enough to allow a response surface description from first principles of acid-base and solvent interactions. Unlike the simplex search, factorial design can account for experimental noise as well as the presence of multiple optima. In the example reported above, sample tracking was an integral part of the analysis, which is a function of the particular model and not the strategy. It is not clear what the effect of a purely empirical model would have been in this case, although, as a general rule, the better the model is designed, the better the design will perform. A drawback is that, unlike the simplex search, factorial design is computation intensive and requires that the user have knowledge of statistical design to develop the model and interpret the results. Depending on the design, a large number of experimental points can be required; nonetheless, factorial design remains the best approach for mapping (as opposed to searching) a response surface about which little is known.

The mixing model used in solvent optimization was first proposed by Snee (46). The model is a factorial design similar to those discussed above with two constraints: the proportion of any component in the mixture must lie between 0 and 1, and the sum of the proportions of all components must equal one. The design for three variables consisted of ten points (Figure 7) with relative proportions of 1 at

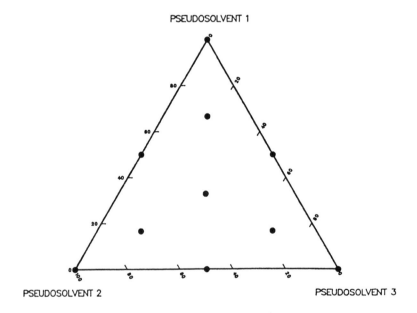

FIGURE 7.　The ten-point mixing model proposed by Snee (46). In a four-solvent system, each binary composition is treated as a pseudosolvent. The ten points are linear combinations of the three pseudosolvents as outlined in the text.

each vertex, 1:1 at the midpoints, 1:1:1 at the center of the triangle, and 4:1:1 at the centers of the triangles formed by the midpoints. Figure 7 illustrates its usual representation as a phase diagram with trilinear coordinates. The response is a fourth dimension above and below the plane of the diagram.

Glajch et al. (19) applied the model to Snyder's solvent selectivity triangle to map the effect of mobile phase mixing upon resolution; their design later consisted of the first seven points of Snee's model (47). Six points were used to fit the data with a quadratic model and the seventh, the centroid, was used to test the model's fit. The response of an eighth point, the predicted optimum, was also used to check the fit.

The basic optimization strategy consisted of six steps. First, the sample was eluted in water/methanol and an isocratic binary was established. Next, the compositions of two more isocratic binaries, water/acetonitrile and water/tetrahydrofuran, were calculated from their respective solvent polarity indices. These compositions were adjusted to produce equivalent k' ranges for the sample. The three binaries formed the vertices of an approximately isoelutropic response surface. Step three entailed running the remaining four compositions based on linear combinations of the three binaries. In each of seven analyses, the identities of the solute peaks were tracked with standards. In step four, the data were fitted to the model and the resolution maps of all limiting pairs were obtained. In the authors' initial work, the response function was a chromatographic optimization function (COF) consisting of the weighted sums of the natural logarithm of the ratio of the observed resolution to the desired resolution with a penalty term for analysis time (19); this function was later supplanted by Equation 5 with the N term removed (29). Once the maps were completed, the regions where a resolution of 1.5 or greater could be expected were determined. In the final step, these regions were overlaid in an overlapping resolution map (ORM), and the mobile phase composition or region with the best resolution of the worst separated pair was determined. This result is similar to that obtained from the window diagram strategy described above.

The ORM strategy maps the widest range of available solvent selectivities and takes advantage of the interdependence of solvent compositions to simplify both the experimental design and the mathematical model. To keep track of elution-order changes, as with most of the models discussed here, each solute peak of interest must be identified in the test chromatograms. Although burdensome, this effort does allow the method to compensate for peak crossovers and the resultant multiple optima.

The Snee mixing model has a tendency to weight edge effects — that is, the model measures more points around the edges than within the region. Using only seven points rather than ten only emphasizes this tendency. All of the four solvent interactions are calculated from a single response, the seventh point. Requiring that the initial binaries be isoeluotropic may further limit the dynamic range of the resolution available in the ORM. The visualization of the results is, to date, limited to manually constructed ORMs or a three-dimensional plot (36). A topographical map of the response contours would enhance the predictive power of the mixing model strategy.

An alternate implementation of Snee's mixing model has been reported (48). All ten points of the design were tested and the response was resolution as measured by Equation 5. The N term was set to the average plates of the adjacent pair in each mobile phase; the plates were derived from chromatograms of the individual components. If the peaks were unlikely to overlap, the standards were analyzed in groups to shorten the method development time. The data were then fit with a special cubic model (46):

$$R_s = b_1 f_1 + b_2 f_2 + b_3 f_3 + b_{12} f_1 f_2 + b_{13} f_1 f_3 + b_{23} f_2 f_3 + b_{123} f_1 f_2 f_3 \quad [20]$$

where

R_s = the resolution at any point for a given peak pair

b_i and f_i = the fit coefficients and volume fractions of the first seven points.

A least-squares regression was performed with all ten data points on the response model, and response contours for all pairs were calculated. If a given pair exhibited any limiting resolution behavior, it could be omitted at this point. Each of the vertex binaries was treated as a pseudocomponent (49); the remaining contours were mapped in 2.5% increments across the solvent composition range. The lowest value at each point was sought and placed in a 41 × 41 matrix, the various optima were determined and ranked, and the points were then plotted. The resultant response surface represents the best predicted resolution of the worst resolved pair.

Figure 8 illustrates a plot of retention time versus test run number for a set of four peaks (experimental details in Reference 49). Note that several peak crossovers were observed. Figure 9 shows the resolution map of these peaks. Three optima were predicted as a function of elution order with the global optimum in the lower right-hand corner; the exact composition varied between columns, but the region of the optimum remained the same. Figure 10 maps the response contours if constant plates are assumed. Note that a different optimum is predicted. Subsequent experiments verified that the optimum predicted in Figure 9 was the true global optimum.

The analytical samples were substituted aromatic acids. Esterification minimized any silanophilic interaction and simplified the response contours (Figure 11). One rather than three elution orders was now observed with a single optimum. The response can be constrained by the analysis time. Figure 12 shows the resolution map of a normal-phase separation in which one component did not elute in cyclohexane/methyl-t-butyl ether. The map is characteristic of one in which the resolution is limited by the response of a single adjacent pair. The compression at the lower left is attributable to censoring the data for analysis time.

This variation of the ORM strategy has many of the same strengths as its predecessor, and it has the additional features of making full use of Snee's mixing design, of better precision in the estimation of the interaction terms, and of a topo-

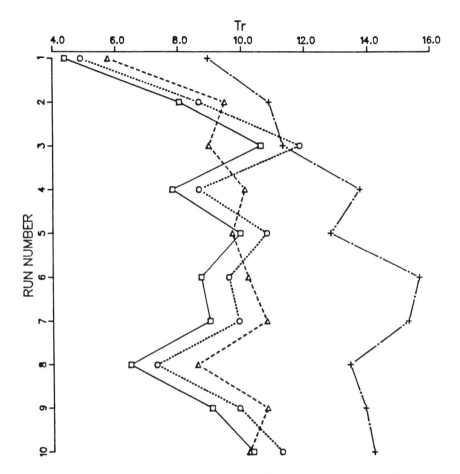

FIGURE 8.   A plot of retention time versus run number for a ten-point optimization of four substituted aromatic acids. Note that there are three distinct regions in which peaks cross over.

graphical presentation of the response contours. The method also is more computation intensive; specialized software is required to do both the model fit and the plotting. Finally, it suffers from the same limitation of requiring sample tracking, which clearly limits the method's applicability to unknown samples.

SUMMARY

Optimizations of chromatographic conditions have in the past been largely trial-and-error endeavors. Relative success was a function of the chromatographer's experience and serendipity. The advent of the computer in the laboratory has allowed the tools of statistical design to be applied to the task.

35.0% WATER 65.0% METHANOL

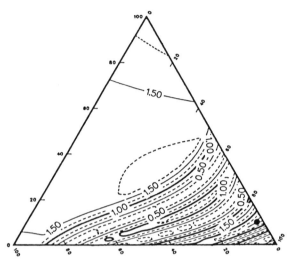

52.0% WATER 48.0% ACETONITRILE   55.0% WATER 45.0% THF

FIGURE 9.   A topographical map of the limiting resolution of the four components in Figure 8. Resolution was calculated by Equation 4 with each peak's width determined from standards. The position of the global optimum is indicated by the filled circle.

35.0% WATER 65.0% METHANOL

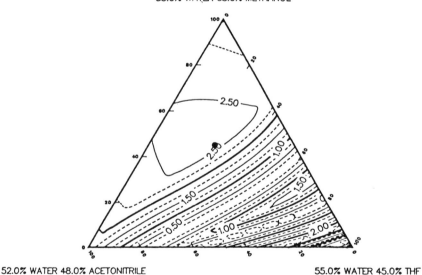

52.0% WATER 48.0% ACETONITRILE   55.0% WATER 45.0% THF

FIGURE 10.   A topographical map of the same experiment as plotted in Figure 9. In this case, constant plates were assumed. Note the shift in the apparent global optimum.

34.0% WATER 66.0% MEOH

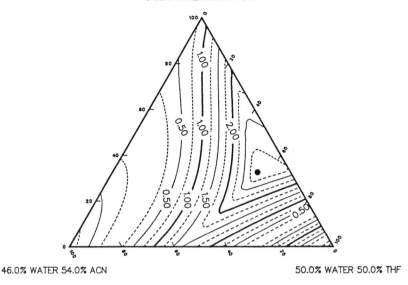

46.0% WATER 54.0% ACN   50.0% WATER 50.0% THF

FIGURE 11. A topographical map of the esters of the four components optimized in Figures 9 and 10. Note the simplification of the resolution surface as a result of a decrease in silanophilic interactions.

93.0% CYCLOHEXANE 7.0% ISOPROPANOL

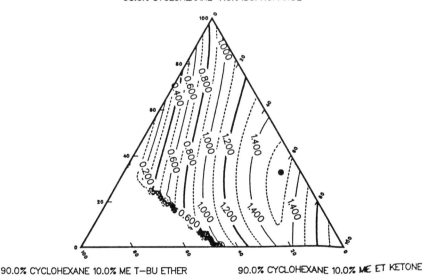

90.0% CYCLOHEXANE 10.0% ME T-BU ETHER   90.0% CYCLOHEXANE 10.0% ME ET KETONE

FIGURE 12. A topographical map of the limiting resolution surface for a family of near homologues. Two of the six compounds analyzed had additional functional groups that retarded their elution. The resolution surface on the right is indicative of a single pair controlling the limiting resolution. The blank region on the left is the result of censoring in the function to compensate for undesirable retention times.

The optimization process has been examined and it was observed that the process may be broken down into several key elements: the response, the response surface, and the strategy for measuring the response. The choice of optimization criteria will clearly affect any predictions of chromatographic performance. An analysis of the response surface allows one to take advantage of simplifying assumptions and prior knowledge to describe the interactions within the experimental region. This chapter has reviewed the major strategies available through the current literature; each one was designed to address a particular set of concerns. As yet, none provides a *universal* solution to the problems associated with chromatographic optimization.

If one were to design an ideal workstation for optimizing chromatographic separations — one that might present a universal solution to optimization problems — it would offer the user a choice of response functions and optimization strategies. The chromatograph and data acquisition would be automated under the computer's control. The analyst would also have the freedom to reanalyze data and to try new models after all experiments are finished. The recent interest of various instrument vendors offers hope that such a workstation may be developed in the near future; until then, chromatographers face a complex (but interesting) challenge.

REFERENCES

(1) M. Otto and W. Wegscheider, *J. Chromatogr.* **258**, 11 (1983).

(2) A. Nahum and Cs. Horváth, *J. Chromatogr.* **203**, 53 (1981).

(3) K.E. Bij, Cs. Horváth, W.R. Melander, and A. Nahum, *J. Chromatogr.* **203, 65** (1983).

(4) Z. Varga-Puchony and Gy. Vigh, *J. Chromatogr.* **257**, 380 (1983).

(5) J.W. Weyland, C.H.P. Bruins, H.J.G. Debets, B.L. Bajema, and D.A. Doornbos, *Anal. Chim. Acta* **151,** 93 (1983).

(6) H.J.G. Debets, B.L. Bajema, and D.A. Doornbos, *Anal. Chim. Acta* **151**, 131 (1983).

(7) W. Wegscheider, E.P. Lankmayer, and M. Otto, *Anal. Chim. Acta* **150**, 87 (1983).

(8) L.R. Snyder and J.J. Kirkland, *Introduction to Modern Liquid Chromatography* (John Wiley & Sons, New York, 2nd Ed., 1979).

(9) H. Engelhardt, *High Performance Liquid Chromatography* (Springer-Verlag, Berlin, 1979).

(10) J.J. Kirkland, W.W. Yau, H.J. Stokolosa, and C.H. Dilks, *J. Chromatogr. Sci.* **15**, 303 (1977).

(11) P.J. Schoenmakers, A.C.J.H. Drouen, H.A.H. Billiet, and L. de Galen, *Chromatographia* **15,** 688 (1982).

(12) A.B. Christophe, *Chromatographia* **4,** 455 (1971).

(13) R.E. Kaiser, *Gas Chromatographie* (Geest und Portig, Leipzig, 1960).

(14) L.R. Snyder, *J. Chromatogr. Sci.* **10**, 200 (1972).

(15) R.J. Laub and J.H. Purnell, *Anal. Chem.* **48**, 799 (1976).

(16) R.J. Laub and J.H. Purnell, *Anal. Chem.* **48**, 1720 (1976).

(17) R.J. Laub, A. Peller, and J.H. Purnell, *Anal. Chem.* **51**, 1878 (1979).

(18) S.N. Deming and M.L.H. Turoff, *Anal. Chem.* **50**, 546 (1978).

(19) J.L. Glajch, J.J. Kirkland, K.M. Squire, and J.M. Minor, *J. Chromatogr.* **199**, 57 (1980).

(20) A.C.J.H. Drouen, H.A.H. Billiet, P.J. Schoenmakers, and L. de Galen, *Chromatographia* **16,** 48 (1983).

(21) S.L. Morgan and S.N. Deming, *Sep. Purif. Methods* **5,** 333 (1976).

(22) W. Wegscheider, E.P. Lankmayer, and K.W. Budna, *Chromatographia* **15**, 498 (1982).

(23) J.H. Nickel and S.N. Deming, *LC, Liq. Chromatogr. and HPLC Mag.* **1**, 414 (1983).

(24) J.C. Berridge, *J. Chromatogr.* **244**, 1 (1982).

(25) P.J. Schoenmakers, H.A.H. Billiet, and L. de Galen, *Chromatographia* **15**, 205 (1982).

(26) L. Rohrschneider, *Anal. Chem.* **45**, 1241 (1973).

(27) L.R. Snyder, *J. Chromatogr.* **92**, 223 (1974).

(28) L.R. Snyder, *J. Chromatogr. Sci.* **16**, 223 (1978).

(29) J.L. Glajch and J.J. Kirkland, *Anal. Chem.* **55**, 319A (1983).

(30) W. Spendley, G.R. Hext, and F.R. Himsworth, *Technometrics* **4**, 441 (1962).

(31) S.N. Deming and S.L. Morgan, *Anal. Chem.* **45**, 278A (1973).

(32) J.H. Nickel and S.N. Deming, *Amer. Laboratory* **16**, 69 (1984).

(33) L.A. Yarbro and S.N. Deming, *Anal. Chim. Acta* **73**, 391 (1974).

(34) J.A. Nelder and R. Mead, *Computer J.* **7**, 308 (1965).

(35) J.C. Berridge, *Anal. Proc.* **19**, 472 (1982).

(36) J.C. Berridge, *Trend Anal. Chem.* **3**, 5 (1984).

(37) P. Jandera, J. Churacek, and H. Colin, *J. Chromatogr.* **214**, 35 (1981).

(38) B. Sachok, R.C. Kong, and S.N. Deming, *J. Chromatogr.* **199**, 317 (1980).

(39) B. Sachok, J.J. Stranahan, and S.N. Deming, *Anal. Chem.* **53**, 70 (1981).

(40) G.E.P. Box, W.G. Hunter, and J.S. Hunter, *Statistics For Experimenters* (John Wiley & Sons, New York, 1978).

(41) W.R. Melander and Cs. Horváth, in *High-Performance Liquid Chromatography — Advances and Perspectives*, Vol. 2, Cs. Horváth, ed. (Academic Press, New York, 1980), p. 113.

(42) E.P. Kroeff and D.J. Pietrzyk, *Anal. Chem.* **50**, 497 (1978).

(43) J.L.M. van de Venne, J.L.H.M. Hendrix, and R.S. Deelder, *J. Chromatogr.* **167**, 1 (1978).

(44) C.W. Davies, *J. Chem. Soc. (London)* 2093 (1938).

(45) P.R. Bevington, *Data Reduction and Error Analysis for the Physical Sciences* (McGraw-Hill, New York, 1969).

(46) R.D. Snee, *Chemtech* **9**, 702 (1979).

(47) J.L. Glajch, J.J. Kirkland, and L.R. Snyder, *J. Chromatogr.* **238**, 269 (1982).

(48) M. Canales, paper presented at Eighth International Symposium on Column Liquid Chromatography (New York, New York, 1984, Paper 201).

(49) B.A. Hohne and M. Canales, in preparation.

C: ADVANCES IN DETECTION AND IDENTIFICATION

NEW DEVELOPMENTS IN
LC DETECTORS

Edward S. Yeung

Department of Chemistry, Iowa State University
and Ames Laboratory, USDOE
Ames, IA 50011

Inherent in any application of liquid chromatography (LC) is the problem of detection. Except in the case of preparative separations, the LC detector is the key to obtaining qualitative and quantitative information about the analytes separated. Even though significant advances in chromatographic instrumentation and applications have been made in recent years, most of the systems in use today are still based (in decreasing order of popularity) upon the absorption detector, the refractive index (RI) detector, or the fluorescence detector.

For the development of pharmaceuticals, very often there are special problems of detection in LC. High sensitivity is needed to screen for impurities or to follow metabolic products in serum or in urine. Because good chromophores must be present to use an absorption or fluorescence detector, derivatization may be necessary if chromophores are lacking. In addition to the tedious sample preparation that is involved in derivatization, the sample of interest may be altered during the process. It is thus of interest to develop universal detectors that are more sensitive than standard RI detectors for LC.

Universal detectors can also aid in the quality control of pharmaceuticals because all impurities will be detected. An important need is to quantitate the impurities without standards and without identification. This way, one limits the scope of the analysis by considering further only those impurities above a certain concentration level. Nonetheless, because of the complexity of the samples in drug analysis in general, interferences must be minimized to obtain reliable information, which calls for highly selective LC detectors. Even if the chromatographic separation does not clearly isolate the component of interest, a highly selective detector can provide essentially the same information about the component as when interferences are absent.

It is beyond the scope of this chapter to review the many uses of LC detectors in the development of pharmaceuticals. Instead, emphasis will be placed on three distinct detection schemes that address some of the special needs described above.

REFRACTIVE INDEX METHODS

   While absorption detection is the most frequently used detection method in LC, the technique does require that the analytes absorb in a convenient wavelength region. The development of the variable-wavelength absorption detector has extended the usable spectral range, but only to the extent that the chromatographic mobile phase does not absorb light. Various derivatization methods have been developed to attach desirable chromophores to molecules so that the absorption detector can be used. An important example is the derivatization of amino acids (1). There are, however, situations in which derivatization is not possible. In the survey of impurities, one is faced with analytes with unknown chemical properties. The universal response of an RI detector makes it ideal for the analysis of truly unknown samples. Commercial RI detectors, unfortunately, only provide limited detectability, typically 3×10^{-8} RI units, which means that the analytes must be present in the 1 μg range for quantitation.

   Some new instrumental concepts have been demonstrated to improve detectability in RI detection. One commercially available interferometric RI detector, for instance, uses white light interference in a two-beam arrangement (2). The detectability depends on the contrast in the interference fringes. For this reason, a Fabry-Perot interferometer is inherently superior (3). The Fabry-Perot interferometer consists of two end mirrors that can be translated relative to each other by a piezoelectric crystal. As the mirror separation, d, is changed, constructive and de-

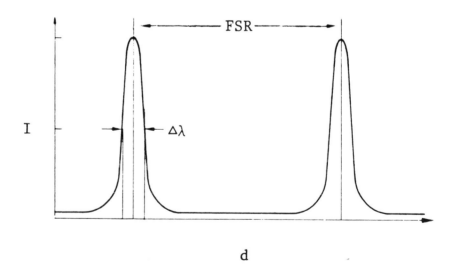

FIGURE 1.   Transmission properties of a Fabry-Perot interferometer as a function of mirror separation, d. $\Delta\lambda$ = width of interference peak; and FSR = free spectral range.

structive interferences occur for monochromatic light (as shown in Figure 1), and the peaks are given by the expression:

$$m\lambda = 2dn \qquad [1]$$

where
  λ = wavelength of light
  n = refractive index
  m = any integer.
The change in location of the interference peak rather than the transmitted intensity is monitored. Flicker noise in the light source thus only contributes indirectly to the measured signal. To compensate for temperature changes and resultant baseline drifts, a double-beam arrangement such as that illustrated in Figure 2 can be used (4). In this case, the reference flow cell can be used to follow the frequency instabilities of the laser, and an ordinary helium-neon (HeNe) laser can be used. Figure 3 illustrates a chromatogram obtained with such an instrument, providing a limit of detection (LOD) of 4×10^{-9} RI units (S/N = 3).

Three additional improvements can be made in interferometric RI detectors. The original design relies on a computer to locate the interference peak during the scanning of the interferometer, which limits the scanning rate to 3 Hz (3,4). An analog system to perform the same function has recently been built by the author and

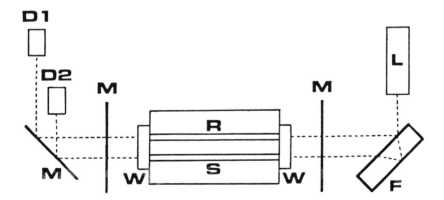

FIGURE 2. RI detector based on a dual-beam Fabry-Perot geometry. M = mirrors; W = antireflection coated cell windows; R = reference flow cell; S = sample flow cell; L = laser; F = optical flat; and D1, D2 = photodetectors.

colleagues, which increases the scanning rate to 100 Hz and also increases the S/N ratio. It is also possible to use confocal geometry, that is, concave end mirrors rather than plane-parallel mirrors, for the interferometer. A natural beam waist of very small volume exists in this arrangement, so that coupling of this detector to microbore LC systems is easy. Confocal interferometers can achieve a resolution (FSR/$\Delta\lambda$ in Figure 1) of 200, so that a 1-cm cavity with a 1-μL volume can measure an RI change of 8×10^{-8} units, and a limit of detection of 20 ng can thus be realized. A third improvement is to stabilize the interferometer at the half-intensity point on any constructive interference peak in Figure 1. RI changes are thus converted to intensity changes at the phototube. The commercial instrument is based

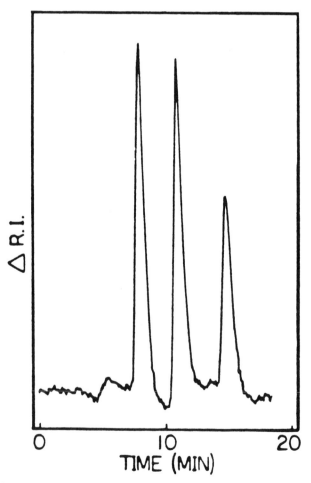

FIGURE 3.   RI chromatogram of glucose, sucrose, and raffinose for 0.72 μg of each injected. Column: 25 cm $\times$ 4.6 mm, 10-μm C$_{18}$; eluent: water; flow rate: 0.5 mL/min.

on essentially the same principle (2), but the LOD is improved because of the sharper peak.

One of the incentives for further development in RI detection is the possibility that quantitative information may be obtained without analyte identification and thus without standards (5). The analytical working curve is the traditional method used for quantitation, but implicit in the method is that the analyte must first be identified. When one is surveying for impurities in, for example, a pharmaceutical product, very often the identities of the contaminants are not known. It is important, however, to establish the concentration levels of these impurities in order to decide if isolation and identification of the impurities are needed. In synthetic chemistry, one is interested in reaction yields, particularly those of any side reactions. Very often it is not possible to isolate sufficient amounts of these minor products in pure form to establish an analytical working curve. Quantitation without standards, therefore, is important even if the identity of the analyte is known. In complex biological samples, it is difficult to produce complete separation of all the components. Yet in cases such as clinical diagnosis the overall distribution of components may be useful if the distribution can be quantitatively established without analyte identification.

The area measured for a chromatographic peak is determined by the refractive index of the eluent, the refractive index of the analyte, and the concentration of the analyte. Because the refractive index of the eluent can be measured separately, one has two unknown parameters contributing to the observed peak area. If one then injects the same sample into the chromatograph and elutes with an eluent with a different refractive index, a different peak area is observed. The two experimental peak areas can then be used to solve for the two unknown parameters. The concentration as well as the refractive index of the analyte can thus be determined.

The RI of a mixture consisting of an analyte and the chromatographic eluent can be predicted from the individual RIs, as shown in Figure 4 (5). The relationship is not linear, contrary to what is assumed by elementary discussions of the refractive index, which predict a horizontal line. Rather, quadratic behavior (a curved line) should be observed if the mixture is ideal. For all concentrations that are of chromatographic interest, the quadratic behavior can be approximated by the tangents to the curve (dashed lines). This greatly simplifies the calculations. For the analyte peak area S_1 when solvent 1 is used as the eluent,

$$S_1 K_1 = C_x(F_x - F_1)$$ [2]

where
  C = the volume fractional concentration
  x = the analyte
  1 = the eluent.
K is a constant that converts the area in arbitrary units to an actual ΔRI value and is a function of flow rate, detector calibration, RI of solvent 1, and the integration

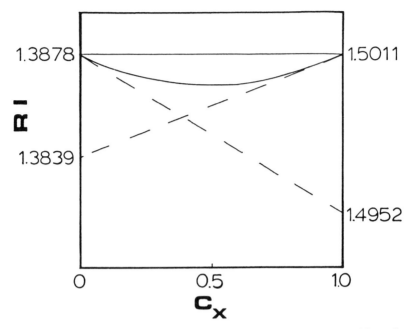

FIGURE 4. RI dependence on volume fraction for benzene in heptane. Solid horizontal line = linear interpolation, solid curve = true dependence, and dashed lines = limiting slopes.

interval used for determining the area. F_i is a function of the RI of the species, such that

$$F_i = \frac{n_i^2 - 1}{n_i^2 + 2} \qquad [3]$$

When the same sample is eluted with the second eluent, one obtains

$$S_2 K_2 = C_x(F_x - F_2) \qquad [4]$$

Because K_1, K_2, F_1, and F_2 can be measured separately, Equations 2 and 4 form a pair of simultaneous equations, and C_x can be obtained without using standards.

An even more elegant approach is to calibrate the four constants characteristic of the solvents under the same experimental conditions used for analysis. If the solvents can be eluted in each other in the chromatographic system, the areas S_a and S_b, respectively, can be obtained for concentrations C_2 and C_1 eluted in solvents 1 and 2, respectively. Equations 5 and 6 describe these conditions:

$$S_a K_1 = C_2(F_2 - F_1) \qquad [5]$$
$$S_b K_2 = C_1(F_1 - F_2) \qquad [6]$$

By combining Equations 2 and 4–6,

$$C_x = \left(\frac{S_1 C_2}{S_a} + \frac{S_2 C_1}{S_b} \right)$$ [7]

If the solvents do not elute in each other, one can arbitrarily choose two calibrating substances, 3 and 4, and determine their peak areas using the same two eluents.

$$S_3 K_1 = C_3 (F_3 - F_1)$$ [8]
$$S_4 K_2 = C_3 (F_3 - F_2)$$ [9]
$$S_5 K_1 = C_4 (F_4 - F_1)$$ [10]
$$S_6 K_2 = C_4 (F_4 - F_2)$$ [11]

Combining Equations 2, 4, and 8–11,

$$C_x = \left[\frac{S_1 - S_2 \left(\dfrac{S_3/C_3 - S_5/C_4}{S_4/C_3 - S_6/C_4} \right)}{S_3 - S_4 \left(\dfrac{S_3/C_3 - S_5/C_4}{S_4/C_3 - S_6/C_4} \right)} \right] C_3$$ [12]

Either Equation 7 or Equation 12 allows C_x to be determined without identification — and without knowing any of the physical properties of the analyte, the solvents, or the calibrating substances. If two of the refractive indices are known, one can also predict the RI of the analyte (6).

It is naturally possible that the equations relating the chromatographic peak areas are not independent of each other. Then, the solutions to Equations 7 or 12 will not be significant. One can avoid this situation by choosing eluents 1 and 2 to have very different RIs and by choosing the calibrating substances 3 and 4 to have very different RIs (but not necessarily different from those of the solvents). Another assumption is that the chromatographic peaks must be correlated with each other in the two eluents. This condition is possible in certain kinds of chromatography (for example, in gel-permeation chromatography or ion chromatography) in which the elution orders are preserved. Otherwise, a statistical procedure must be used to correlate the elution orders in the two solvents (7). In the case of overlapping chromatographic peaks or coelution, Equation 7 or Equation 12 predicts the total concentration of species eluted at that time, which is still useful quantitative information. The calculated RI, however, will not be meaningful as qualitative information.

An example of the application of this quantitation method for complex samples is shown in Figures 5 and 6. The same sample of crude oil is eluted from a gel-permeation column first using toluene and then using chloroform as the eluent. Because the elution order of components is expected to be constant, one can divide the individual chromatograms in Figure 5 into 1-s time segments. S_1 and S_2 in

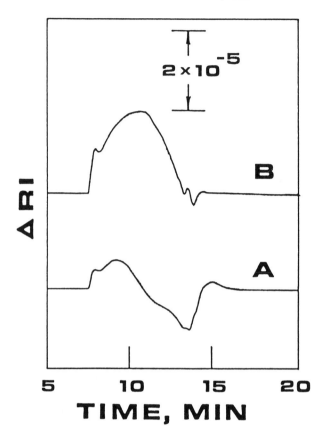

FIGURE 5. RI chromatograms of North Slope crude oil using gel-permeation chromatography. (a) Toluene eluent; (b) chloroform eluent.

Equations 2 and 4 then refer to the same time segment in the two chromatograms, and a concentration can be calculated for this segment. A concentration chromatogram can then be constructed from these calculations and is displayed in Figure 6a. Meaningful information for characterization of the crude oil is obviously contained in this chromatogram, unlike the original chromatograms. The absorbance chromatogram, Figure 6b, provides an incorrect picture of size distribution because most of the crude oil components do not absorb at 365 nm. As mentioned above, the RI for materials eluted in each time segment can also be calculated, as shown in Figure 6c. The extension of this example to the study of biological materials with high molecular weights using gel-permeation chromatography is straightforward.

A unique application of this quantitation scheme is for testing the purity of a chromatographic peak. Following the same procedure as above, the RI can be calculated at various points in a chromatographic peak. This procedure is similar to

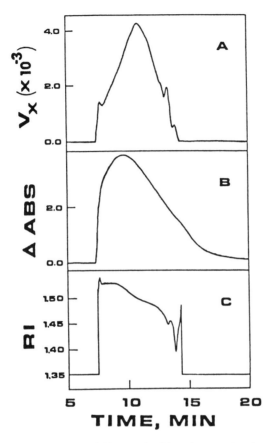

FIGURE 6. Characterization of North Slope crude oil by gel-permeation chromatography. (a) Volume fraction of materials eluted each second; (b) absorbance of materials at 365 nm; (c) RI of the materials; volume injected: 10 μL.

that in Figure 6c, but the time segments will be much shorter, down to the limit of the detector time response. If a chromatographic peak is pure, a constant RI will be obtained across the peak; RI variations during a peak will be indicative of incomplete resolution of components. A similar scheme based on the absorption ratio at two wavelengths has been suggested (8). The RI method, however, is more universal because it is applicable to analytes that do not absorb.

POLARIMETRY

Optical activity is an interesting property of chiral molecules because the rotation of the polarization direction of light is usually an indication of biological activity. In complex samples of clinical, geological, or biological importance, this type of selectivity is useful. Commercial polarimeters, however, do not have the sensitivity required for or cannot handle the small volumes required in liquid chromatography. A laser-based polarimeter has been used for detection in the study of sugars in urine (9), cholesterol in serum (10), and chiral components in coal extracts (11) after liquid chromatographic separation. A small, collimated laser beam that provides good rejection of stray light and the reduction of birefringence in the optical components is essential to the success of the laser-based polarimeter (12). For analytical scale LC applications, a 5-cm path-length cell with a volume of 100 μL seems to be the best compromise between sensitivity and volume in designing such a system. The most recent version has been interfaced to a microbore LC system that has a detector volume of 1 μL and an optical path length of 1 cm (13). The cell is illustrated in Figure 7. The actual optical volume is even smaller; the machining process limits the volume to 1 μL. The applicability of the miniaturized version of the detector was tested in a reversed-phase elution of fructose using water as the eluent. A detectability of 10 ng of injected material was found (S/N = 3), and the peak did not show any broadening beyond the specified efficiency of the column used (N = 25,000 plates) at a k' of 2.1 (13). Because flow rates are substantially lower in microbore LC, it is possible to use exotic solvents as mobile phase components.

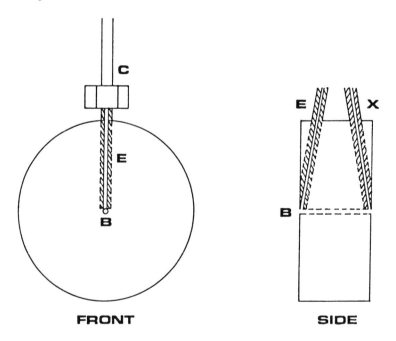

FRONT   **SIDE**

FIGURE 7.   Polarimetric 1 cm × 2.5 cm o.d. flow cell for microbore LC. C = column; E = entrance capillary; X = exit capillary; and B = 1-mm i.d. bore for laser beam.

For the initial survey and characterization of complex samples, however, the presence of detector selectivity is undesirable because the detectors may miss certain components. The only universal detector now routinely used in LC is the refractive index detector. The poor sensitivity of and the relatively large cell volumes required by commercial RI detectors make it worthwhile to seek alternatives for universal detection. The indirect approach, in which the detector is used to monitor some physical property of the eluent rather than that of the analyte, is attractive. When the analyte is eluted, a decrease in detector response is observed, and universal detection is accomplished. The critical point is that one must be able to detect a small change on top of the large signal produced by the eluent, so not all detectors can be used in this indirect mode. In polarimetry, the large rotation produced by an optically active eluent can be canceled out by the analyzer in the instrument, and a polarimeter is therefore ideal for operation in the indirect mode as a universal detector.

The concept of indirect polarimetry is illustrated in Figure 8. Optically active ($+$)-limonene ($[\alpha]_D = 106°$) is used as the eluent, and when each of the three optically inactive species elute from the column, a corresponding volume of the eluent in the detector cell is displaced and a lower amount of optical rotation is observed. Unless an eluting compound has a specific rotation identical to that of the eluent, a peak will be observed. This is then the basis of a universal detection scheme. The detectability is found to be the same as that using optically inactive eluents for species with $[\alpha] = 100°$, that is, 10 ng of injected material (S/N = 3) (13). Sensitivity is preserved despite the high optical rotation produced by the eluent because rotating the analyzer effectively suppresses the background. The additional noise caused by thermal fluctuations in the cell, which does not contribute to noise using optically inactive eluents, is apparently still below the 10^{-5} degree level. In this mode of operation, there appears a system peak attributable to the displacement of the eluent when the solute is retained on the column (Peak 1 in Figure 8). For weak eluents, this peak appears at the void volume and does not interfere with the retained peaks. Naturally, because of the cost of chiral solvents microbore LC is the method of choice for this detector. The polarimetric detector can thus replace the RI detector as the workhorse in the analytical laboratory.

The concept of quantitation without standards can be tested using the polarimeter as discussed in the earlier section. The only modification is that the function F_i in Equation 2 or Equation 4 now represents $[\alpha_i]\varrho_i$, where $[\alpha_i]$ is the specific rotation and ϱ_i is the eluent density. The two eluents used are ($-$)-2-methyl-1-butanol and ($\pm$)-2-methyl-1-butanol, each modified with an equal volume of acetonitrile. The chromatograms obtained are shown in Figure 9. Table I lists the results of the solution to the two simultaneous equations for each chromatographic peak given by Equation 12 compared to the true values. The results are within experimental error; in fact, if an eluent with a higher $[\alpha]$ is used, the precision should be even better. The specific rotation of the analyte can also be obtained from these chromatograms, thus allowing the determination of the optical purity of materials without

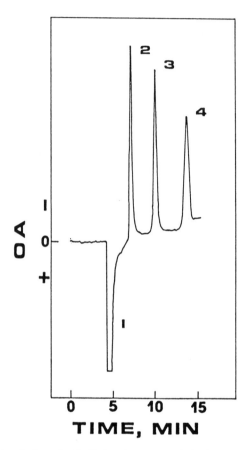

FIGURE 8.   Determination of optically inactive compounds by indirect polarimetry and normal-phase chromatography. Eluent: (+)-limonene, column: 25 cm × 1 mm, 10 μm silica; and flow rate: 20 μL/min; 1 = solvent peak, 2 = dioctyl phthlate, 3 = dibutyl phthlate, 4 = diethyl phthlate.

requiring the separation of the enantiomers. The cost of using an optically active eluent with microbore columns has been calculated, and it was found that it is in fact cheaper than using UV-grade LC eluents with standard analytical scale columns because, if one can sacrifice a little sensitivity, the eluent does not need to be optically pure.

Absorption detection for LC remains one of the most frequently used methods. Routine miniaturization of conventional absorption detectors sacrifices detectability because of the shorter path length and the reduced light throughput. The usual transmission mode of absorption detection is being supplanted by various calorimetric methods that probe the absorption indirectly (14). Principal among these has been thermal-lens calorimetry (15,16). The laser beam profile, which must be

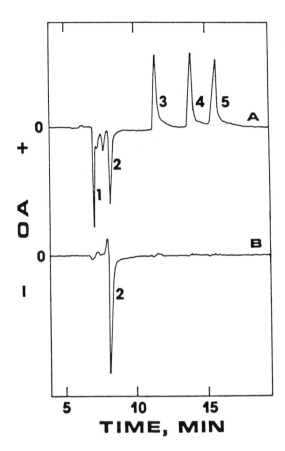

FIGURE 9. Optical activity chromatograms of a mixture of *l*-2-octanol (Peak 2), *N*-decane (Peak 3), *n*-tetradecane (Peak 4), and *n*-hexadecane (Peak 5); Peak 1 is void peak. (a) Optically active eluent; and (b) optically inactive eluent. Column: 25 cm × 1 mm, 5-μm C_{18}; flow rate: 20 μL/min.

TABLE I

Quantitation of Sample Components Using the Polarimeter

	(*l*)-2-Octanol	$C_{10}H_{22}$	$C_{14}H_{30}$	$C_{16}H_{34}$
True C_i ($\times 10^{-2}$)	4.00	8.00	8.00	8.00
Calc. C_i ($\times 10^{-2}$)	3.93 ± 0.38	7.74 ± 0.34	7.83 ± 0.36	8.36 ± 0.27
True $[\alpha_i]_{590}^{20}$	−9°	0	0	0
Calc. $[\alpha_i]_{488}^{27}$	−8.0° ± 2.8	0.07° ± 0.2	0.01° ± 0.2	0.01° ± 0.1

reproducible in its cross-sectional intensity distribution, is a critical consideration in this case. Lasers in the UV region and high-power lasers usually do not meet this criterion. A more general approach is to probe the bulk number density change in the optical region after heat is produced from the absorption process (3); therefore, if an optically active eluent is used, the heating effect produces an *expansion* in the liquid and a *decrease* in optical rotation is detected. This is the principle behind absorption detection via indirect polarimetry.

To determine the relationship between the amount of light absorbed and the observed rotation, two assumptions are necessary. First, the amount of light absorbed is equal to 2.303AI where A is the absorbance of the sample and I is the intensity of the light source in joules. This assumption is true only for small absorptions but will apply to most of the interesting LC cases. Second, it is assumed that all of the light absorbed eventually becomes heat. With these two conditions met, the temperature increase, ΔT(K), in the interaction region of cross-sectional area a (cm^2) and unit length is

$$\Delta T = \frac{2.303AI}{C_p \varrho a} \qquad [13]$$

where
   C_p = the specific heat of the solvent in Jg^{-1}K^{-1}
   ϱ = the density of the medium in g/cm^3.

The rotation α observed for a mixture is related to the individual specific rotations $[\alpha]_i$, such that

$$\alpha = \sum [\alpha]_i l \varrho_i V_i \qquad [14]$$

where
   l = the path length in decimeters
   V = the volume fraction of the material at the detector (13).

For low solute concentrations (which is the more interesting case) one can assume that the absorption-induced rotation is entirely attributable to the eluent, so that $\alpha = [\alpha] l \varrho$. If B is the coefficient of expansion of the eluent (K^{-1}), assuming small changes, there will be a change in the measured optical rotation given by

$$\Delta \alpha = [\alpha] l \varrho B \Delta T \qquad [15]$$

Finally, substituting into Equation 13, one obtains an explicit relation between the absorbance of the solution and the rotation observed:

$$A = \frac{\Delta \alpha \, a \, C_p}{2.303[\alpha] \, l \, IB} \qquad [16]$$

To assess the potential for measuring absorptions via indirect polarimetry, Equation 16 can be used to estimate the minimum detectable absorbance. The micropolarimeter of Reference 13 has a detectability of 1×10^{-5} degrees (S/N = 3). Using typical values for the other parameters — $a = 9 \times 10^{-4}\,cm^2$, $C_p = 1.8\,J/g\,K$, $[\alpha] = 192.0\,cm^3/g\,dm$, $l = 0.1\,dm$, $I = 8.0 \times 10^{-2}\,J$, and $B = 1.24 \times 10^{-3}\,K^{-1}$ — the minimum detectable absorbance calculated is 3×10^{-6} AU, which is superior to detection limits of conventional absorption detectors used in LC.

Figure 10 illustrates the use of the indirect polarimetric scheme as an absorption detector (17). The three chromatograms represent data taken at three different laser powers, and each have been normalized against the standard rotation produced by a DC solenoid to maintain constant sensitivity (12). Peak 3 is the indirect polarimetric signal resulting from the elution of dimethyl phthalate, and the peak height is approximately equal in the three chromatograms. Peak 2 is the signal for N-methyl-o-nitroaniline by the method of absorption via indirect polarimetry. The peak height increases linearly with increasing laser power as predicted by Equation 16. A linear calibration plot corresponding to laser powers from 31 mW to 55

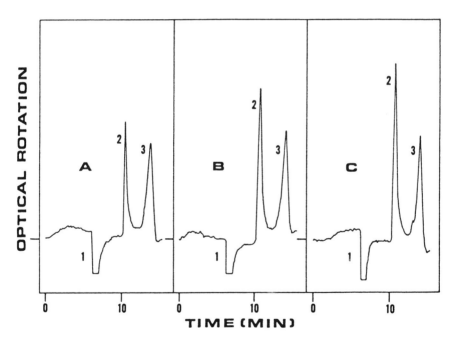

FIGURE 10.   Normal-phase separation of N-methyl-o-nitroaniline and dimethyl phthalate. Chromatographic conditions were identical to those in Figure 8, except that 100% limonene was used as eluent; $\lambda = 488.0$ nm. 1 = injection peak, 2 = absorption detection of 890 ng of N-methyl-o-nitroaniline, 3 = indirect polarimetric signal from 2 μg of dimethyl phthalate; (a) power = 31 mW, (b) power = 48 mW, (c) power = 55 mW.

mW results that has a correlation coefficient of 0.994. Figure 10 shows that one can use the same experimental setup for universal detection and for absorption detection by simply changing the laser power. Peaks that do not change with laser power correspond to indirect polarimetry, and peaks that vary with laser power are the result of absorption.

N-methyl-o-nitroaniline has a small absorption at 488 nm ($\varepsilon = 112$ L mol^{-1}cm^{-1}). Even so, the LOD with 32 mW of 488-nm radiation is 12 ng (S/N = 3) of injected material, which is identical to the LOD obtained with the indirect polarimetric detector. To take real advantage of the absorption process, one must use higher laser powers or rely on larger absorptivities. To improve the match with the absorption profile of N-methyl-o-nitroaniline, the experiment was repeated at 458 nm. At this wavelength, N-methyl-o-nitroaniline has a molar absorptivity of 1040. The LOD, using 73 mW of power, was found to be 36 pg (S/N = 3) of injected material. This corresponds to an LOD of 1.8×10^{-6} AU (peak volume 28 μL, S/N = 3). The improved LOD is also aided by a larger [α] for limonene measured at 458 nm ([α] =

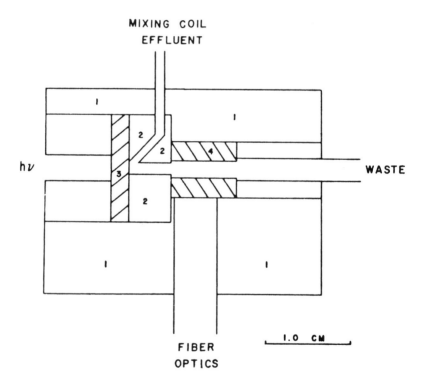

FIGURE 11. Flow cell for light-scattering detection in LC. 1 = stainless steel detector block; 2 = stainless steel entrance aperture; 3 = Pyrex window; 4 = Pyrex scattering cell. (Reprinted, with permission, from Reference 19.)

192 cm³/g dm) and a lower amount of solvent absorption at that wavelength. The observed detectability is in excellent agreement with that predicted by theory. α-Pinene was also investigated as an optically active eluent. The LOD for N-methyl-o-nitroaniline in this system was 17 ng of injected material (S/N = 3; 488 nm, 30 mW). This result, after correction for laser power and differences in eluent optical activity, is consistent with the limonene results.

The polarimeter is unique in its capabilities because it can provide very selective detection when used in the conventional mode and it can become a universal detector when used in the indirect mode. Moreover, the polarimeter can be used as an absorption detector with high sensitivity. Since the original instrument was developed (12), many improvements have been made in the system. The most recent experimental arrangement is based on a helium-neon laser that has about 7 mW of polarized output. The Faraday modulation coils have been wound directly onto the flow cell so that the chromatographic eluent acts as the Faraday medium. Only about 200 turns in a length of 2.5 cm are needed if a current of 0.1 A is used. This current can be derived directly from typical signal generators without additional switching. The calibration DC coil is also wound directly onto the liquid cell. The best result (S/N = 3) is a detectability of 4×10^{-6} degrees of rotation. The entire system costs about $10,000 and is thus within the price range of other LC instrumentation.

LIGHT SCATTERING

The scattering of light is related to many different types of molecular properties (18); therefore, a special type of selectivity can be obtained with light-scattering detection in liquid chromatography. The two properties that are of interest in LC are nephelometry and quasielastic light scattering.

The optical arrangement for nephelometry is not too different from that used for Raman or fluorescence spectroscopy, except that wavelength selection is not necessary to analyze the scattering. It is unlikely, however, that any solute in the LC effluent can directly contribute to a nephelometric signal beyond the Rayleigh scattering from the eluent. Because of this, postcolumn derivatization or precipitation is needed to produce particles or colloids that will scatter light. A flow cell with a 17-μL volume and fiber optics collection system (19) is illustrated in Figure 11. Only 0.5 mW of power from a laser at 633 nm is needed. It is interesting to note that in nephelometry the beam size of the laser should be substantially larger than the particle diameters, or refraction rather than scattering will be measured. For the test case of several nonpolar lipids, an ammonium sulfate solution that changes the solubility of those lipids is used as the precipitating reagent. The detector response is thus dependent on the ammonium sulfate concentration and its flow rate into the mixing chamber. The signal obtained is nonlinear relative to the injected quantity, approaching a square-root dependence for small samples. A

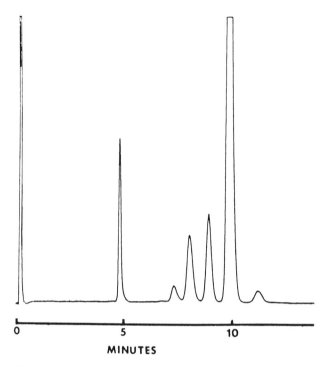

FIGURE 12.   Chromatogram of nonpolar lipids in human serum with light-scattering detection. Column: 10-μm C_{18}; eluent: 2:1 acetone/methanol with postcolumn addition of ammonium sulfate in water; sample loop: 50 μL; flow rate: 1.0 mL/min. (Reprinted, with permission, from Reference 19.)

chromatographic separation of nonpolar lipids in human serum is shown in Figure 12. Cholesterol, its esters, and the triglycerides all show a response. Detectability is on the order of 0.5 μg for lipids. It can be expected that any of the traditional precipitation reactions can be adapted to this scheme.

Instead of relying on the formation of precipitates, the alternative is to vaporize the solvent completely. The residual particles represent the analytes as they elute and can be monitored by light scattering (20). This technique is possible because most common LC solvents are highly volatile. The use of microcolumns also allows small amounts of solvent to be used. The main concern is solvent impurity, which, even at the 1-ppm level, can constitute a large background signal. A vaporization system for the LC effluent is shown in Figure 13. Carbon dioxide is used as the nebulization gas because of its high heat capacity and low cost. After nebulization, the droplets pass through a drift tube at a temperature of 40°–50°C. The solvent thus vaporizes and the only particles remaining are the nonvolatile impurities in the solvent and the analyte. A 1-mW helium-neon laser at 633 nm irradiates the particles, and the scattered light is collected by a glass rod and transmitted to a

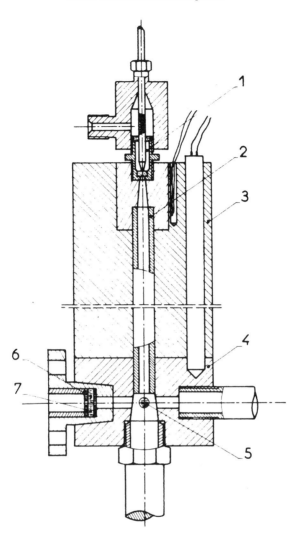

FIGURE 13.   Detector block for light-scattering detector. 1 = nebulizer; 2 = drift tube; 3 = heated copper block; 4 = light-scattering cell; 5 = glass rod; 6 = glass window; 7 = diaphragms. (Reprinted, with permission, from Reference 20.)

photomultiplier tube. For water-containing solvents, the temperature of the drift tube can be raised to 80°–85°C.

The intensity of the scattered light is a complex function of particle size, refractive index, wavelength, polarization, and observation angle. The system in Figure

13 has a wide distribution of particle sizes, and only a weighted average response is available. The signal is found to increase with the sample size to the 1.8th power. The detectability was found to be 0.55 µg injected, which corresponds to a concentration of 40 ppm in the mobile phase. An advantage of this detector is that the contribution to band broadening is extremely small and is on the order of 0.15 µL. Another advantage is that this detector is suitable for gradient elution because all of the solvent is vaporized. Figure 14 shows a separation of triglycerides obtained with gradient elution. The baseline is stable and column efficiency is preserved. A commercial version of this detector has recently become available (21).

Quasielastic light scattering, which is based on the slight Doppler shift in the scattered light as a result of Brownian motion, can be used to probe particles in the range of 50 Å to 2 µm in diameter. Highly monochromatic light is essential to these light-scattering measurements. The scattering signal depends on the angle of observation, but generally either very low angles or very high angles are used. The optical arrangements for the two configurations (22) are shown in Figure 15 and Figure 16, respectively, and it is easy to change from one mode to the other. Ade-

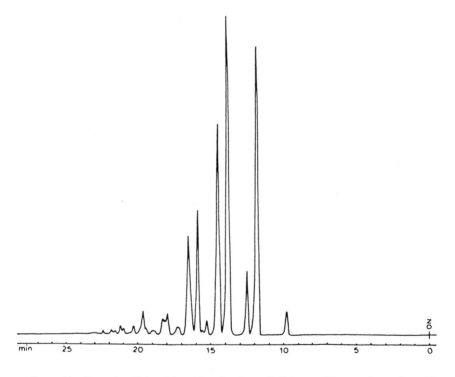

FIGURE 14.   Separation of the triglycerides of soybean oil. Column: 200 mm × 2 mm, 5-µm Li-Chrosorb RP-18; eluent: linear gradient from 33:67 acetone/acetonitrile to 99:1 acetone/acetonitrile in 25 min; flow rate: 0.3 mL/min. (Reprinted, with permission, from Reference 20.)

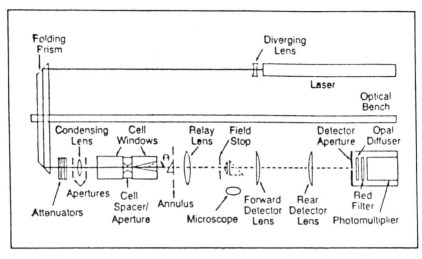

FIGURE 15.   Optical arrangement for quasielastic light scattering at low scattering angles. (Reprinted, with permission, from Reference 22.)

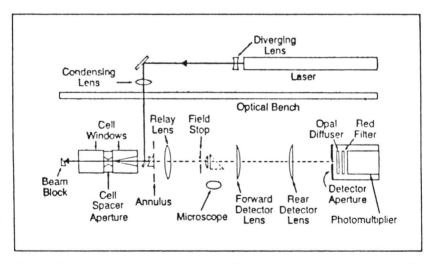

FIGURE 16.   Optical arrangement for quasielastic light scattering at high scattering angles. (Reprinted, with permission, from Reference 22.)

quate intensity is provided by a 2 mW laser. Because the important information is contained in the frequency shift, photon-counting autocorrelation must be performed after the signal leaves the photomultiplier tube. Latex spheres with diameters of 0.1 μm to 0.3 μm can be measured in the concentration range of 5 μg/mL to

100 µg/mL, which is still within the range of useful concentrations encountered in LC. The information that can be obtained from light-scattering techniques includes particle size, molecular conformation, molecular weight, molecular weight distributions, and molecular rotations. The main applications of quasielastic light-scattering detection are the studies of macromolecules and microemulsions in biochemistry and in polymer science, particularly in conjunction with gel-permeation chromatography.

SUMMARY

It is apparent that the absorption detector and the refractive index detector will dominate LC applications in the development and analysis of pharmaceuticals in the immediate future, but new detection schemes such as polarimetry provide unique advantages for studying complex samples. The instrumentation is reasonably straightforward and may eventually be available commercially. In the meantime, new information obtained with conventional detectors, such as quantitation without standards using polarimeters, can aid in monitoring quality control in pharmaceuticals.

Acknowledgments

The author thanks the many co-workers in his laboratory who have contributed to the work described here, particularly S.D. Woodruff, J.C. Kuo, D.R. Bobbitt, R.E. Synovec, and S.A. Wilson, and the U.S. Department of Energy, Office of Basic Energy Sciences, Division of Chemical Sciences, for partial research support through the Ames Laboratory, Iowa State University, under contract No. W-7405-eng-82.

REFERENCES

(1) H. Umagat, P. Kucera, and L.F. Wen, *J. Chromatogr.* **241**, 324 (1982).

(2) For example, Optilab 902 refractometer, Vallingby, Sweden.

(3) S.D. Woodruff and E.S. Yeung, *Anal. Chem.* **54**, 1175 (1982).

(4) S.D. Woodruff and E.S. Yeung, *Anal. Chem.* **54**, 2124 (1982).

(5) R.E. Synovec and E.S. Yeung, *Anal. Chem.* **55**, 1599 (1983).

(6) R.E. Synovec and E.S. Yeung, *J. Chromatogr.* **283**, 183 (1984).

(7) R.E. Synovec and E.S. Yeung, *Anal. Chem.* **56**, 1452 (1984).

(8) A.C.J.H. Drouen, H.A.H. Billiet, and L.D. Galan, *Anal. Chem.* **56**, 971 (1984).

(9) J.C. Kuo and E.S. Yeung, *J. Chromatogr.* **223**, 321 (1981).

(10) J.C. Kuo and E.S. Yeung, *J. Chromatogr.* **229**, 293 (1982).

(11) D.R. Bobbitt et al., *Fuel* **64**, 114 (1985).

(12) E.S. Yeung, L.E. Steenhoek, S.D. Woodruff, and J.C. Kuo, *Anal. Chem.* **52**, 1399 (1980).

(13) D.R. Bobbitt and E.S. Yeung, *Anal. Chem.* **56**, 1577 (1984).

(14) A.C. Boccara, D. Fournier, W. Jackson, and N.M. Amer, *Opt. Lett.* **5,** 377 (1980).

(15) M.J. Sepaniak, J.D. Vargo, C.N. Kettler, and M.P. Maskarinec, *Anal. Chem.* **56,** 1252 (1984).

(16) J.M. Harris and N.J. Dovichi, *Anal. Chem.* **52,** 695A (1980).

(17) D.R. Bobbitt and E.S. Yeung, *Anal. Chem.* **57,** 271 (1985).

(18) B. Chu, *Laser Light Scattering* (Academic Press, New York, 1974).

(19) J.W. Jorgenson, S.L. Smith, and M. Novotny, *J. Chromatogr.* **142,** 233 (1977).

(20) A. Stolyhwo, H. Colin, and G. Guiochon, *J. Chromatogr.* **265,** 1 (1983).

(21) Model F2110 mass detector (Anspec, Ann Arbor, Michigan).

(22) M.L. McConnell, *Anal. Chem.* **53,** 1007A (1981).

THE APPLICATION OF HIGH PERFORMANCE LIQUID CHROMATOGRAPHY/ MASS SPECTROMETRY TO SUBSTANCES OF PHARMACOLOGICAL INTEREST

David E. Games

Department of Chemistry
University College
Cardiff CF1 1XL
Wales, United Kingdom

This chapter will review the current status of combined high performance liquid chromatography/mass spectrometry (LC/MS) with respect to its ability to undertake qualitative and quantitative analysis of substances of pharmacological interest. Extensive use has already been made of gas chromatography/mass spectrometry (GC/MS) in this area. The advent of fused-silica capillary columns with bonded phases and the use of on-column injection techniques has considerably extended the range of compounds amenable to study. Also, chemical derivatization can be used to improve the gas chromatographic performance of compounds of interest. For gas chromatographic study, however, compounds require vaporization, so compounds that are thermally labile and/or are of low volatility cannot be studied. Although derivatization techniques assist in overcoming these problems, difficulties are encountered in GC as a result of reactions taking unexpected courses, of incomplete reaction, and of choice of suitable conditions if unknown compounds are being studied. LC can be used for the analysis of most substances of pharmacological interest and is also often the method of choice for the analysis of compounds that are amenable to GC because, by use of techniques such as precolumn concentration or column switching, minimal sample work-up is required. Thus, LC/MS should have considerable utility in the area because it will enable the approaches currently used in GC/MS to be applied in LC studies.

Extensive use has been made of off-line LC/MS in studies of a wide range of compounds of pharmaceutical and pharmacological interest. In this approach, fractions are collected from the liquid chromatograph and are then analyzed by mass spectrometry. It is not within the context of this chapter to review this approach, and the reader is referred to relevant reviews (1–3). For certain types of problems, however, particularly those in which molecules that are difficult to analyze with MS are being investigated, use of this approach with fast atom bombardment (FAB) or field desorption (FD) mass spectrometry may be the most effective solution.

Before proceeding to consider the combination of the liquid chromatograph with a mass spectrometer, it is appropriate to review those aspects of both mass spectrometry and liquid chromatography that are important to understand in terms of the combined technique.

MASS SPECTROMETRY

The mass spectrometer works under vacuum, and, on modern instruments, a variety of ionization methods are available. *Electron impact* (EI) is the most widely used technique and requires vaporization of the sample before its ionization by electrons. Source pressures must be on the order of 10^{-6} torr or lower for efficient use. The principles of EI are well understood, and the spectra obtained provide an abundance of structural information. In many cases the spectra can be matched with a library using the data system to provide identification, and programs such as STIRS provide a measure of structural interpretation as well. *Chemical ionization* (CI) also requires vaporization of the sample before ionization; in this case, an ionized gas is used to effect ionization. With this technique, source pressures up to 1 torr are used. The spectra obtained are simpler than EI spectra and usually have abundant protonated molecular ions that enable molecular weight to be readily determined. Fragment ions are often useful in providing an indication of the functional groups present in the molecule, and the extent of fragmentation observed is dependent on the proton affinity of the reagent gas used.

With both EI and CI, difficulties are experienced in obtaining spectra from compounds with low volatility and/or thermal instability because vaporization of the sample is necessary to obtain spectra. Use of rapid heating techniques, introduction of the sample — close to the electron beam — into the ion plasma, or use of inert sample holders or combinations of these techniques can enable useful spectra to be obtained from compounds that fail to provide such data from conventional direct insertion probes. This approach is exemplified in techniques such as desorption chemical ionization (DCI) (4). Of particular relevance to LC/MS was the discovery that a similar effect can be produced by introduction of the sample into a CI source in solution (5). Of further relevance is the fact that negative-ion CI (NICI) can be of considerable utility in two contexts: the use of the technique in the electron capture mode can result in enhanced sensitivity and selectivity for compounds with high electron affinity, and use of reagent-ion NICI can provide molecular weight information for compounds that fail to provide such data in the positive EI and CI modes. In addition, structurally useful fragment ions not observed in the positive mode are often present in the negative mode.

Because of the requirement of vaporization before ionization in EI and CI mass spectrometry, a large number of ionic and polar compounds fail to provide useful mass spectral data. This shortcoming has resulted in the development of a range of other mass spectral ionization techniques. This chapter will concentrate on those techniques that have utility for LC/MS; they can be divided into two broad classes — those in which a sample is introduced into the mass spectrometer's ion source in solution, and those in which ionization is effected from a surface.

Introduction in Solution

Atmospheric pressure ionization (API) can handle compounds introduced in solution (6). The solvent and sample are vaporized at atmospheric pressure, and the solvent molecules are ionized by either a radioactive- or a corona-discharge source. Ion-molecule reactions between the solvent ions and the sample molecules then result in ionization of the sample. Early systems could only handle relatively volatile compounds because vaporization at atmospheric pressure requires higher temperatures than vaporization conducted under reduced pressure. Recently it has been shown that nebulization of solutions into an API source enables mass spectra to be obtained from low-volatility compounds (7). An alternative approach that is particularly successful for ionic compounds is the use of liquid ion evaporation with an API system (8). With this technique, the solution is nebulized through a fine stainless-steel hypodermic needle into air at atmospheric pressure; a charge is induced on the sprayed droplets by means of a small electrode placed close to the sprayer.

Another technique that enables mass spectra to be obtained from difficult molecules is electrospray ionization, which is closely related to liquid ion evaporation. In early studies, a solution of a polymer was passed through a needle held at high voltage into a spray chamber flushed with nitrogen at atmospheric pressure; rapid desolvation was effected in an evaporation chamber, and ions were sampled into a time-of-flight mass analyzer using a nozzle and skimmer (9). Recent studies have modified this approach for use with a quadrupole mass analyzer and have shown that the technique is capable of ionizing mass spectrometrically difficult, biologically important molecules (10).

An electric field is also used to effect ionization in electrohydrodynamic ionization (11). The sample is dissolved in glycerol doped with a suitable electrolyte, which is usually an alkali metal halide. The solution is passed through a stainless-steel capillary, and a voltage of the order of 6 kV is applied to the end of the needle. Spectra containing abundant pseudomolecular ions of mass spectrometrically difficult molecules are produced, but these spectra are complicated by cluster ions formed from the glycerol and electrolyte.

Currently the most important ionization method in this class from the point of view of LC/MS is thermospray ionization (12–15). Sample, together with a volatile buffer, is introduced through an electrically heated capillary into a heated source, and a mechanical vacuum pump is connected opposite the vaporizer. This process results in the production of ions from a wide range of compounds including compounds not amenable to EI or CI study. Some of the ions produced are sampled through a sampling cone with a small diameter and pass into a mass spectrometer. Because the technique works best with aqueous solvent systems at flow rates of 1–2 mL min^{-1}, it is well suited to reversed-phase LC.

Ionization from a Surface

In surface ionization techniques, the compound of interest is desorbed and ionized from a solid or liquid layer deposited on a metal surface. Californium-252 plasma desorption (16) involves the use of ^{252}Cf fission fragments to collide with a sample that has been electrosprayed onto a thin aluminum holder. Each fission fragment producing ionization has a complementary fragment that travels in the opposite direction; this fragment is detected and used as a zero-time marker in a time-of-flight instrument and enables the ions generated from the sample to be mass analyzed. The technique has been used to obtain molecular weight from many large biomolecules.

Use of kiloelectron volt (keV) primary ions at low current densities can provide mass spectra from organic compounds adsorbed on a metal surface. The technique is called static secondary ion mass spectrometry (SIMS) and has been shown to be useful for handling a wide range of nonvolatile organic compounds (17). A technique closely related to SIMS is fast atom bombardment (FAB) mass spectrometry (18), which uses a neutral primary atom beam; in this case, the sample is coated onto a metal surface as a liquid matrix. The advent of this technique has had a major effect in that it is relatively easy to use and provides mass spectra from a wide range of involatile compounds.

Laser desorption (LD) mass spectrometry also enables mass spectra to be obtained from difficult compounds (19). In this technique, a short high-power laser pulse is used on a sample coated on a metal surface. Availability of commercial systems has resulted in an increase in its use for the analysis of organic molecules. For further discussion of all of the above and other ionization techniques, there are a number of excellent reviews (20–23).

The other instrumental factor to be considered that is relevant to LC/MS is the method of mass analysis used in the mass spectrometer. Most commercial systems use either quadrupole rods or a combination of magnetic and electric fields for this purpose. The major practical differences are that the former operate at low ion-source voltages, have limited mass ranges (currently 2000 daltons is the upper limit), and have only low resolution capability, whereas magnetic instruments operate at high ion-source voltages, have higher mass range capability, and can operate with high resolution. These are important considerations in choice both of instrument and interface.

HIGH PERFORMANCE LIQUID CHROMATOGRAPHY

In LC, separation is effected by differing distributions of components between a mobile liquid phase and a stationary solid phase. The composition of the mobile phase varies according to the type of chromatography being performed. Reversed-phase LC is the technique most widely used for studies of pharmacologically active compounds; the mobile phase used most often for reversed-phase LC usually consists of combinations of water with methanol or acetonitrile, often with modifiers added to improve chromatographic performance. These modifiers may be acids, bases, buffers, or ion-pair reagents and can be volatile or nonvolatile in nature. It should also be noted that, for studies of mixtures of compounds of varying polarities, gradient elution techniques in which the mobile phase composition is varied throughout the separation are used. Normal-phase LC is less widely used in this area but can be useful for studies of compounds that are of low polarity or are nonpolar. In this case, the mobile phases are organic solvents. It should also be noted that in LC the flow rates of mobile phases used can vary between 10 μL min^{-1} if small-bore columns are used and in excess of 5 mL min^{-1} if fast analysis is being undertaken.

COMBINED LIQUID CHROMATOGRAPHY/MASS SPECTROMETRY

When considering methods for the combination of liquid chromatography and mass spectrometry, there are three major considerations.
- A mass spectrometer conventionally works under vacuum in the gas phase, and normal instruments are not capable of handling the high gas volumes that would be generated from the mobile phase flow rates used for conventional LC.
- LC is used to study compounds ranging from volatile to nonvolatile, nonpolar to polar, and thermally stable to thermally labile. The ideal system should be capable of providing mass spectral information from the whole range of compounds, so considerable attention must be paid to the way in which ionization of the solute is effected.
- In any combined system, chromatographic performance should not be compromised. Thus, ideally, the system should be able to handle all of the various modifiers and solvent combinations used for LC without loss of chromatographic efficiency.

Currently there are two broad types of approach to effecting combined LC/MS.

One involves feeding the eluent from the liquid chromatograph (with or without enrichment of solute relative to solvent) into the mass spectrometer's ion source. The other involves removal of the solvent before the solute is ionized in the ion source of the mass spectrometer. The various approaches have been extensively reviewed (1,2,23–29); this section will concentrate on those systems that have been most commonly used and discuss progress that may result in improved systems.

Systems for LC/MS in Which Solvent Is Not Removed

There are at least three LC/MS systems in which solvent is not removed from the eluent before analysis: direct liquid introduction, monodisperse aerosol generation, and thermospray ionization. Each will be discussed here in turn.

Interfaces using *direct liquid introduction* (DLI) have been widely used for LC/MS. In these devices, a portion of the eluent from the liquid chromatograph (1%–5% for conventional columns) is fed into the ion source of a mass spectrometer configured for chemical ionization (CI). The solvent is ionized using a filament that, by ion-molecule reactions, ionizes the solute, resulting in the production of positive- or negative-ion CI-type spectra.

Early systems consisted of simple capillaries that could handle only volatile compounds (30–32). Considerable improvements in performance result from nebulization through a pinhole in a diaphragm (33,34), cooling of the probe, and incorporation of a desolvation chamber (33,35,36). Commercial systems incorporating these features are available from Hewlett-Packard (Palo Alto, California) and Nermag (Malmaison, France) and can analyze a wide range of organic compounds. One of the major limitations of the approach is that only a portion of the LC eluent can be handled, thus limiting detection levels. Incorporation of cryogenic pumping (31) results in an ability to accept higher solvent flow rates, and, if smaller-bore LC columns (1-mm i.d.) are used, all of the eluent can be handled, resulting in impressive sensitivity (37). One of the advantages of this approach is that homemade systems are inexpensive and relatively easy to construct; although they may not be capable of analyzing the full range of compounds that commercial systems are, they can still solve many problems (32,38–41).

Most of the work in this area has been performed on quadrupole instruments because the lower ion source potentials make interfacing easier. Systems for magnetic instruments have also been reported, and, in fact, the first developments in this area were on such instruments (30,31).

DLI systems have proved to be particularly effective for target-type analysis in which known compounds are sought and are amenable for use of a wide range of volatile modifiers in the mobile phase. Often spectra are relatively simple, consisting of mainly $[M+1]^+$ ions, so structural information may be limited with stable compounds. More labile molecules, however, often provide a wealth of structurally significant fragment ions, and there is a considerable volume of literature (see below) in which structural studies have been performed. When relatively simple spectra are produced, additional information can be obtained by use of tandem mass spectrometry (42). In this technique, ions of interest are separated in the first region of the instrument, collisional activation is performed to induce their decomposition in the second region, and the third region undertakes analysis of the ions formed.

An advantage of the DLI approach over moving-belt interfaces is that there is less thermal decomposition of labile compounds, enabling lower detection limits to be achieved with such compounds. Because of the presence of ions attributable to the mobile phase, however, scans are usually initiated above 100 u. It should also be noted that, because CI is the ionization process, detection of compounds will depend on the relative proton affinities of the mobile phases and of the compound; thus, compounds with lower proton affinities than the mobile phase will not be detected. The quality of spectra obtained is, as might be expected, also dependent on source pressure, temperature, and solvent composition (43). Finally, considerable care is necessary in attending to solvent quality, particularly in ensuring absence of particulate matter that, if it is not excluded by suitable filters, will cause blocking of the diaphragm's pinhole.

An alternative method for effecting nebulization is to induce a coaxial flow of gas. This approach originated from studies aimed at effecting enrichment of solutes through a jet separator (44). A number of systems have been described (38,45–47) and are most effective when used with small-bore LC columns because flow rates that can be handled are similar to those with the DLI systems described earlier. Given evidence available in the literature, this approach does not appear to have any particular advantage over conventional DLI systems.

By use of preconcentration of eluent via passage down a resistance-heated stationary wire, a system has been developed that enables much higher mobile phase flow rates to be accommodated (48). When normal-phase LC is performed, the concentrated eluent is sampled into the mass spectrometer's ion source through a capillary tube. For reversed-phase systems, ultrasonic vibration is used to effect nebulization of the eluent. The approach is called preconcentration DLI and is marketed by Extranuclear (Pittsburgh, Pennsylvania).

One of the earliest systems for LC/MS involved the direct introduction of all of the eluent from a conventional LC column into an atmospheric pressure ion (API) source (6,49). Early systems experienced difficulties in handling lower-volatility compounds. In the commercial system available from Sciex (Thornhill, Ontario, Canada), use of a nebulizer allows for handling of more difficult molecules (7). The system is available on a triple quadrupole instrument, which is advantageous because the spectra produced are usually very simple, being dominated by $[M+1]^+$ ions. The approach appears to be excellent for the identification and quantification of drugs in biological fluids. Because of the presence of cluster ions, however, poor total ion current traces are obtained that could restrict use of the approach for the identification of unknowns.

Liquid ion evaporation (8,50) is very effective for the ionization of ionic compounds and shows considerable promise for the LC/MS of such molecules. The technique has been incorporated into the API instrument available from Sciex. There are few examples of LC/MS with the system, and spectra have been reported from a range of mass spectrometrically difficult molecules.

An ionization method closely related to liquid ion evaporation, electrospray ionization (EI), has recently been adapted for LC/MS (51). In this method, the eluent from the liquid chromatograph is injected into a bath gas of nitrogen through a metal hypodermic tube at a potential of several kilovolts relative to the surrounding chamber walls. Charge is deposited on the surface of the emerging liquid, resulting in dispersal of liquid in a fine spray, and ions are desorbed into the ambient gas. Desorbed ions entrained in a dry bath gas pass through a glass capillary to a vacuum chamber into which they emerge in a supersonic jet of carrier gas. A portion of the free jet flow passes through the aperture of a skimmer into a second vacuum chamber equipped with a quadrupole mass analyzer. The approach has been demonstrated by the acquisition of spectra containing abundant ions in the molecular ion region from solutions of a range of mass spectrometrically difficult molecules, but as yet there are no on-line data. The technique appears to work best with mobile phase flow rates of 5–20 μL min^{-1}, which will restrict its use to small-bore LC columns for optimal detection limits.

Another approach using *monodisperse aerosol generation* is closely related to the DLI approach. It differs, however, in that by use of monodisperse aerosol generation the drop size of the aerosol is more uniform; second, evaporation of solvent occurs at atmospheric pressure to effect more rapid desolvation of the aerosol drops; finally, an aerosol beam technique is used to effect removal of solvent (52). The system shows good stability, sensitivity, and retention of chromatographic integrity. As yet, however, it has only been demonstrated with relatively easy compounds, and it will be interesting to see if it can handle the more-difficult polar, involatile compounds encountered in pharmaceutical research.

Many of the systems described above cannot handle the mobile phase flow rates generated from conventional LC and are limited in the range of compounds from which they can provide molecular weight information. *Thermospray ionization,* however, works best with flow rates of 1–2 mL min^{-1} for aqueous systems and has been shown to provide mass spectra from a wide range of mass spectrometrically difficult molecules (12–15). Thus it is well suited to LC/MS and is becoming widely used. Most studies have been undertaken on quadrupole instruments, and commercial systems of this type are available from Vestec (Houston, Texas), Hewlett-Packard, VG Masslab (Stamford, Connecticut), Finnigan MAT (San Jose, California), and Nermag. Recently, interfacing to a magnetic instrument was reported (53,54), and a commercial system is available from Kratos Analytical (Ramsey, New Jersey) (54).

The design of the Finnigan system differs from the others in that, instead of using a straight stainless-steel capillary heated through a copper block, a coiled directly heated capillary is used to provide rapid response. Systems available from Vestec, Hewlett-Packard, Kratos, and Nermag also have a filament that permits hot direct liquid introduction to be performed on the system. This approach enables the filament to be used to effect ionization and is very useful in that higher mobile phase flow rates can be handled than can be with conventional DLI; ionization can be effected without the addition of a buffer; more fragment ions are present in the spectra than with conventional thermospray ionization; and spectra are obtained from compounds not amenable to conventional thermospray ionization.

Thermospray ionization, although a major advance in LC/MS, is not a panacea. The optimal vaporizer and source temperatures are compound- and solvent-composition dependent. It has been found that best sensitivity is obtained with a high percentage water content (55,56). If low percentage, aqueous mobile phases are being used, or the systems do not have a buffer present, or gradient elution is being used, then conditions can be optimized by postcolumn addition of appropriate solvents through a coaxial tee (57). Programming of the vaporizer temperature can also be used for gradient elution studies (56).

A further problem is the wide variation of sensitivities of compounds to thermospray ionization (56,57). Because the ionization mechanism is not fully understood it is difficult to quantify, but this mechanism appears to be related to proton affinities. Availability of a filament and use of the negative mode can assist with this problem. It should be noted, however, that in the negative mode the enhanced sensitivities of the type reported with NICI have not yet been reported. In addition, spectra are solvent-composition dependent. Finally, many compounds provide very simple spectra, often almost exclusively $[M+1]^+$ or $[M+NH_4]^+$ ions. The simplicity of these spectra can cause problems for confirmation of identity, particularly if isomers are present; moreover, for characterization of unknowns one would like structurally significant fragmentations. This is an area in which tandem mass spectrometry should be of considerable assistance.

It is pertinent to make some observations on sensitivity with thermospray LC/MS. Good total ion current traces can only be obtained if acquired above 150 daltons because of solvent cluster ions. Sample amounts often have to be present at high nanogram levels to obtain good traces of this type. When target type analysis is being undertaken and mass chromatograms can be obtained, however, full-scan spectra are obtainable at low nanogram levels for many compounds. Selected ion monitoring can take detection limits down to low picogram levels. The wide variation in sensitivity observed in thermospray LC/MS even for very similar types of molecules should, however, be borne in mind.

Systems in Which Solvent Is Removed

The principle of operation of systems in which solvent is removed from the eluent is to feed the eluent from the liquid chromatograph onto a continuous moving belt (58). In early systems, solvent was removed with the aid of an infrared heater and in two vacuum locks. The residual solute is flash-vaporized into or in an EI/CI ion source of a mass spectrometer. Kapton is usually used as the belt material, and, because it is a relatively inert surface, the system uses flash-vaporization from such a surface to minimize thermal decomposition.

Early systems suffered from problems in handling high percentage aqueous mobile phases. Use of smaller-bore (1-mm or 2-mm i.d.) LC columns (54,60), of spray deposition (62–64), or of a combination of both overcomes these problems (61). Such systems provide EI and positive- and negative-ion CI data. In most cases the spectra are comparable with those obtained from a direct insertion probe, and library search systems can be used. With low sample amounts of thermally labile compounds, however, thermal decomposition can occur; moreover, in the selected ion monitoring mode one does not obtain the same measure of increase in detection limits as is observed with GC/MS. Low nanogram full-scan spectra can be obtained for many compounds, but because of the background, system sensitivity is not as good as with GC/MS. Such systems are, however, well suited to gradient elution studies, in which no variation in spectra is observed.

Three systems are commercially available: Finnigan MAT manufactures a system that interfaces to the ion source block for its 4000 series of quadrupole instruments (58) and also makes a system that is carefully positioned in the ion source for its 8200 and 8400 series of magnetic instruments (65). VG Analytical (Stamford, Connecticut) has a system that also enters the ion source for its magnetic range of instruments, and the same system is sold by VG Masslab for its quadrupole instruments (66). All systems are available with spray-deposition devices. The ability of these three systems to handle compounds that are more mass spectrometrically difficult has recently been compared (67). The older Finnigan MAT and VG systems show similar abilities in this direction and are equivalent to a good direct insertion probe, whereas the more recent Finnigan MAT system is equivalent to desorption chemical ionization in performance. It should be noted that the best systems in the EI/CI mode cannot provide molecular weight information from more difficult molecules amenable to thermospray ionization, such as quaternary ammonium compounds, tetrasaccharides, and underivatized peptides. Nonetheless, the large number of applications in the literature using these systems testifies to their ease of use and ability to assist in the solution of problems.

Moving-belt systems are amenable to use with surface ionization techniques; systems using laser desorption (68), secondary ion mass spectrometry (69), and fast atom bombardment (70) have been described. The latter system is available from VG Analytical. The potential of these systems has been demonstrated and the performance of SIMS and laser desorption systems compared (69). At present, all systems suffer from poor total ion current traces but have been shown to be capable of on-line studies with difficult molecules, as exemplified by recent FAB LC/MS studies of underivatized peptides (70).

Supercritical Fluid Chromatography/Mass Spectrometry (SFC/MS)

Although it is not strictly within the purview of this chapter, mention should be made of supercritical fluid chromatography/mass spectrometry (SFC/MS). SFC is currently undergoing a major renaissance. The technique has the advantage over gas chromatography in that the sample does not require vaporization and over LC in that higher efficiency can be obtained. Because the gas volumes generated are not as great as with LC, interfacing to the mass spectrometer is easier. Both capillary (71) and packed-column (72,73) SFC/MS systems have been described and impressive results reported. With capillary SFC/MS, both EI and CI data can be obtained. As yet the technique has only been applied to nonpolar or relatively low polarity compounds because carbon dioxide has been mainly used as the mobile phase. There would appear, however, to be no reason why mobile phases such as ammonia should not be used to handle more difficult molecules.

APPLICATIONS OF LC/MS TO PHARMACEUTICAL RESEARCH

A bibliography of LC/MS publications up to late 1982 (74) gives an indication of the wide-ranging applicability of LC/MS systems. In this section studies of compounds of pharmacological interest in which LC/MS has been used to assist in the solution of problems will be discussed. Areas of potential application in which evidence exists that LC/MS can handle the classes of compound involved will also be described.

Drugs and Their Metabolites

LC is extensively used for both qualitative and quantitative studies of drugs and their metabolites, and LC/MS has considerable potential in this area for identification of metabolites, for validation of assays, and for the performance of assays for which current LC detectors are not suitable or in which multicomponent peaks are present.

Ranitidine (Figure 1a) and its metabolites (Figures 1b, 1c, and 2) have been the subject of detailed LC/MS studies using both moving-belt (75,76) and DLI (77) interfaces; the studies serve to illustrate the relative merits of these systems in examination of thermally labile compounds. Use of a reversed-phase system for separation of the four compounds encountered problems with a moving-belt system because, during solvent removal with an infrared heater, thermal decomposition of ranitidine and its metabolites occurred. Use of a normal-phase system containing a mobile phase of methanol/2-propanol/5 M ammonium acetate (50:50:1), however, showed good chromatographic integrity on LC/MS with ammonia CI and produced characteristic spectra. Still, the N-oxide (Figure 1c) and S-oxide (Figure 2) failed to provide molecular weight information.

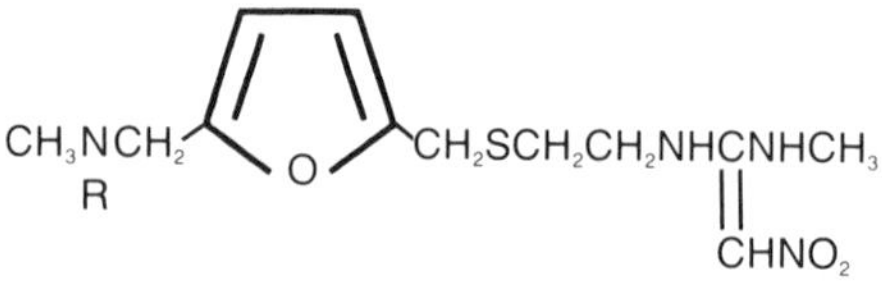

FIGURE 1. Structures of ranitidine and its metabolites.

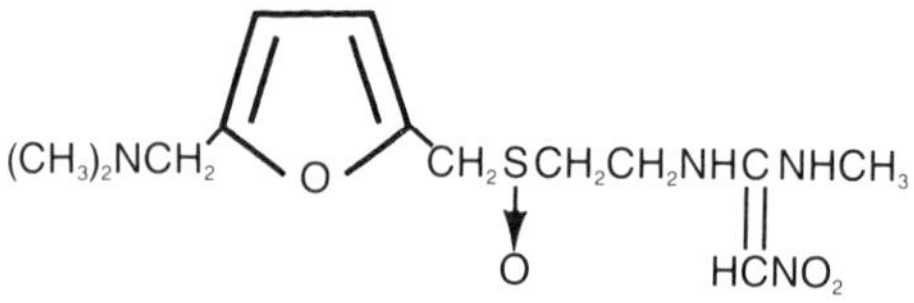

FIGURE 2. Structure of S-oxide of ranitidine.

Direct LC/MS analysis of rabbit and human urine samples enabled the drug and the three metabolites to be detected in each sample at the low microgram levels injected onto the column. Quantitative studies were performed for ranitidine in urine using selected ion monitoring (SIM) and [^{2}H$_3$]-ranitidine as the internal standard. A linear calibration was obtained for standards of ranitidine in urine over the range 0-21 μg mL^{-1}. For six replicate samples at the 1.05 μg mL^{-1} level, the standard deviation was 0.15 μg mL^{-1} and the coefficient of variation was 13.3%; better data were obtained at higher concentrations.

The technique was also used to quantify ranitidine in the urine of a patient. Use of a DLI system enabled reversed-phase LC to be used. All four compounds gave [M + 1]$^+$ ions in their spectra, and the lower limit of detection for ranitidine using SIM was 5 ng injected on-column. It should be noted that a 2-mm i.d. column was used, with 10% of the LC eluent entering the mass spectrometer's ion source. Evidence for peak tailing was noted, and there was a variation in the ratio of ion abundances across the peak, suggesting some thermal decomposition. Quantitative studies using SIM and [^{2}H$_3$]-ranitidine gave a linear calibration over the range from 5 ng to 100 ng of ranitidine injected. Ranitidine and its three metabolites were identified in the urine of a patient who received an oral dose of 1:1 mixture of ranitidine and [^{2}H$_3$]-ranitidine. Figure 3 shows the reconstructed total ion current trace obtained in this study; Figure 4 shows the spectra obtained (77).

These studies indicate that DLI rather than a moving-belt system is the method of choice for such studies when labile drugs and metabolites are involved. Unless a

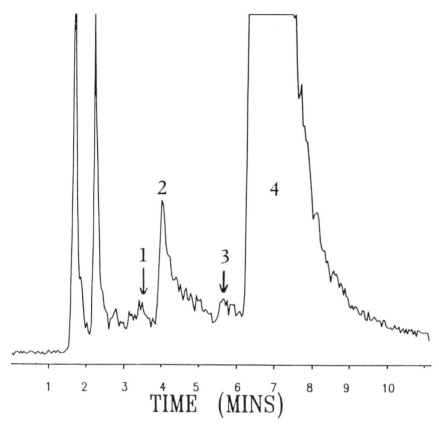

FIGURE 3.   Reconstructed TIC LC/MS from analysis of a 4–6 h urine from a patient who was given a 1:1 mixture of ranitidine and [^{2}H$_3$]-ranitidine. Peaks: 1 = ranitidine-S-oxide, 2 = ranitidine-N-oxide, 3 = desmethylranitidine, 4 = ranitidine. (Reproduced from Reference 77 with permission of Elsevier Science Publishers.)

microbore probe is used, however, detection limits are restricted in the DLI approach. Thermospray ionization can handle all of the eluent from the LC system, and recent studies (78) using a Finnigan MAT thermospray system enabled full-scan spectra of ranitidine-N-oxide to be obtained at the 50-ng injected level; using SIM the metabolite was detected to 5 pg.

Disopyramide (Figure 5a) and its metabolite (Figure 5b) have been studied (79) with moving-belt LC/MS, and a linear calibration was obtained in the 4 ng–1.9 μg range. The drug and its metabolite were identified in the urine of a monkey dosed with the ^{13}C-, ^{15}N-labeled drug. Using the same type of interface, phenobarbitone and its hydroxy metabolite were identified in the urine of a rat dosed with phenobarbitone (80); a hydroxy metabolite of nitrazepam was identified in rat bile (81); and paracetamol and acetanilide were identified in urine (66). Using medium reso-

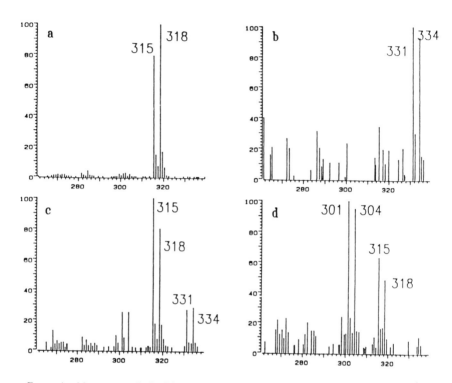

FIGURE 4.   Mass spectra obtained from reconstructed TIC shown in Figure 3. (a) Ranitidine/[^{2}H$_3$]-ranitidine, (b) ranitidine-S-oxide/[^{2}H$_3$]-ranitidine-S-oxide, (c) ranitidine-N-oxide/[^{2}H$_3$]-ranitidine-N-oxide, (d) desmethylranitidine/[^{2}H$_3$]-desmethylranitidine. (Reproduced from Reference 77 with permission of Elsevier Science Publishers.)

lution (4000), bromazepam can be detected to the 35-pg level using a moving-belt interface under CI conditions, and accurate mass measurements (better than 5 ppm) were obtained for low nanogram amounts of the drug in human serum extracts (82). Deactivation of the belt with a 50 ppm Carbowax 20M solution gave considerably improved spectra of this and other compounds and obviated recycling.

Ion-pair reagents are extensively used in LC studies of drugs and their metabolites, and the presence of nonvolatile modifiers can cause problems with all types of LC/MS interface. Use of a continuous liquid-extraction system with a moving-belt system has been shown to present a solution to the latter problem, and excellent data were obtained from a mixture of procainamide, N-acetylprocainamide, N-propionylprocainamide, and lidocaine (83). A similar system can be used on the micro scale and has recently been reported for use with DLI systems (84).

DLI LC/MS with flow switching has been used to detect bromocriptine in a tablet of Parlodel and the hydrogenated ergot alkaloid codergocrine mesylate in a Hy-

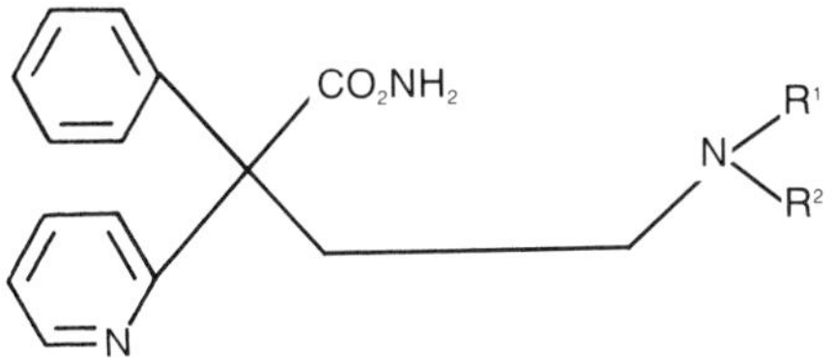

5a R¹ = R² = CH(CH₃)₂
 b R¹ = H, R² = CH(CH₃)₂

FIGURE 5.   Structures of disopyramide (a) and its metabolite (b).

dergine solution (85). Use of this approach enables diversion of excipients that elute without retention. The same technique was used to detect LSD in urine (85). Clobazam and its desmethyl metabolite have been shown to be amenable to study with a gas-nebulized DLI interface (47), and this approach has been used to analyze typical components in a cold medicine (86). Both of these studies used small-bore (1 mm i.d.) LC columns, and a similar approach has been used with DLI for low nanogram detection of 2-hydroxypromazine, acepromazine, chloropromazine (87) and other drugs (88), and for the detection of trichloromethiazide in horse urine (89). Using a Pirkle Type 1A stationary phase (Regis Chemical, Morton Grove, Illinois), DLI LC/MS was used to resolve optical isomers of derivatives of ibuprofen, amphetamine, fenoprofen, and benoxaprofen, and (S)-(+)-ibuprofen was shown to be the predominant enantiomer in an equine urine extract obtained after ibuprofen administration (90). In this study, the advantage of LC/MS over UV detection was shown in that an unknown component cochromatographed with the (R)-enantiomer.

Identification of conjugated metabolites presents a challenge for mass spectrometric techniques. Moving-belt systems have been shown to be capable of providing molecular weight and structurally useful fragment ions from glucuronides, but thermal decomposition occurs at low levels (91). Thermospray LC/MS, however, appears to be the method of choice for the analysis of such compounds because full-scan spectra that are characteristic of compounds of this class were obtained at the 10-ng injected-on-column level (94). The technique was used to characterize the diastereomeric glucuronides of propanolol formed by in vitro incubation of racemic propanolol with immobilized rabbit glucuronyltransferase. Recent studies indicate that negative-ion thermospray LC/MS has considerable potential for the analysis of sulfate conjugates of steroids; detection limits in the high picogram range are reported (95).

In a comparative study of candidate antimalarials using both DLI and thermospray LC/MS (96), thermospray LC/MS gave detection limits below 1 ng using SIM, whereas DLI gave detection limits of 30 ng. Thermospray LC/MS has also been used to analyze promazine N-oxide and N-oxide sulfoxide metabolites as well as the major urinary impramine metabolites isolated from alkalone extracts and enzyme hydrolysis extracts of equine urine (97). The potential and problems of using the technique for identification of by-products in the manufacture of drugs have also been described (98).

LC/MS using an API system has also proved to be useful in the analysis of drugs in biological fluids (99). Studies were performed on sulfamethazine and a mixture of sulfisoxazole, sulfadiazine, and sulfadimethoxine. The spectra obtained were very simple, consisting almost exclusively of $[M + 1]^+$ ions and hence lending little specificity for identification. Use of collisional activation in the second stage of a triple quadrupole provided more informative spectra, and the technique was used to identify sulfadimethoxine in both plasma and urine extracts from a racehorse.

Steroids

Both DLI and moving-belt interfaces have been used to identify corticosteroids in horse urine. In the latter case, ammonia negative-ion CI was found to be the ionization method of choice; dexamethasone, betamethasone, prednisolone, and hydrocortisone were found in the urine of horses administered therapeutic doses of veterinary preparations containing the steroids (100). None of the corticosteroids investigated in this study provided molecular weight information, and studies of the LC/MS of dexamethasone and betamethasone under methane positive-ion CI conditions show that the spectra obtained are concentration-dependent and that thermal decomposition occurs (101). The DLI studies also used negative CI and microbore LC, resulting in lower detection limits (41,102). Use of a desolvation chamber enabled high picogram detection to be obtained; the technique was used to identify betamethasone, 6-methylprednisolone, and their metabolites in horse urine (36).

One advantage of moving-belt interfaces is that they are available on magnetic instruments. As a result, high or medium resolution studies can be performed. Accurate mass data have been obtained at 7000 resolving power from a mixture of cholest-4-en-3-one, testosterone, and pregnenolone (81). Low-resolution LC/MS using a $[3,4-^{13}C_2]$ standard was used to quantify progesterone in human serum, and the values obtained were in good agreement with those obtained by radioimmunoassay (82).

An excellent illustration of the utility of LC/MS for the analysis of natural extracts is provided by the DLI LC/MS investigation of sterol peroxides in marine organisms (103). The technique provides spectra that offer molecular weight information and characteristic fragment ions. The same approach has been used to study cardiac glycosides (104). Although digitoxin gave only a very weak $[M + 1]^+$ ion in the positive mode, the $[\overline{M} - H]^-$ was the base peak in the negative mode. Similar data have been obtained using a moving-belt system (67,105).

As yet there are no reports of the use of thermospray ionization LC/MS in this area, other than in the study of steroid sulfates referred to earlier (95). Studies in the author's laboratory with a variety of steroids indicate that thermospray LC/MS should have considerable application in this area.

Amino Acids and Peptides

Amino acids and/or their derivatives have been shown to be amenable to study with DLI LC/MS (46,104), liquid ion evaporation (50), thermospray (56,106), and moving-belt systems (56). Thermospray LC/MS is the method of choice for underivatized amino acids, although in a study using the Finnigan thermospray interface (56), p-hydroxyphenylglycine was not detected. It appears to be a problem with the Finnigan interface, because the compound gave good response with other thermospray interfaces under the same conditions. This approach has been exploited as a method for obtaining sequence information from peptides (5,106). Underivatized peptide solutions are injected through a column containing immobilized carboxypeptidase Y; the amino acids are released, starting from the C-terminus, and are directly transported by a continuously flowing aqueous buffer into the thermospray source of a mass spectrometer, where they are detected and their $[M+1]^+$ ions are quantified.

Underivatized peptides present a major challenge to mass spectrometry. Although some simple underivatized peptides have been handled by a moving-belt system (80), extensive thermal decomposition occurs. FAB mass spectrometry has been particularly successful in providing both molecular weight and sequence information from underivatized peptides. Combination of this ionization method with a moving-belt system has been used to study peptides (70). Eluent from the chromatograph was fed onto the belt using a spray-deposition device, and the matrix necessary for generation of FAB spectra was applied in very low concentration (0.05% of the solvent employed) as part of the LC mobile phase. A mixture of seven tetra- to nonapeptides produced by partial hydrolysis of the hexadecapeptide antiamoebin I (Figure 6) were studied. The on-line FAB spectra provided molecular weight information and many sequence ions from six of the peptides, and, although the sample of antiamoebin I was thought to be pure, analysis of the spectra of the peptides indicated that antiamoebin III was present together with two new, related peptides. Larger peptides, antiamoebin I, and emericins IIA and IIB were also studied. Although the spectra obtained were of low intensity, molecular weight information was obtainable together with useful sequence ions. These studies show considerable potential for FAB LC/MS, but total ion current traces are poor as a result of what is thought to be the hydrophobic nature of the belt material. Some form of conditioning of the belt or use of an alternative belt material may assist in overcoming these problems.

Ac-Phe-Aib-Aib-Iva-Gly-Leu-Aib-Hyp-Gln-Iva-Hyp-Aib-Pro-Phol

FIGURE 6.   Structure of antiamoebin I.

Leu- and met-enkephalin have been studied by DLI LC/MS (107). Use of positive- and negative-ion mass spectra enabled the peptides to be readily identified and sequenced at the picomole level. Leu-enkephalin has also been studied by thermospray LC/MS (12,108); the spectra reported are much simpler than those obtained by DLI, consisting almost exclusively of $[M+Na]^+$ and $[M+1]^+$ ions with little sequence information. Larger peptides, including renin substrate (15), α-melanocyte stimulating hormone, and glucogen (109), have been studied by thermospray ionization. In the latter cases, the mass range of the quadrupole instrument used was insufficient to show singly charged ions in the molecular weight region, but multiply charged ions provided this information. The technique has been applied for peptide sequencing using columns containing immobilized enzymes, carboxypeptidase Y, and trypsin (110). The enzyme hydrolysates are carried into the thermospray ion source by a continuous flow of an aqueous buffer and are detected mainly as molecular ion species. These ions, which include C-terminal amino acids, residual N-terminal peptides, and tryptic fragments, provide useful sequence information about the parent peptides.

Although studies of underivatized peptides provide an attractive proposition, as yet none of the ionization techniques can be relied upon to provide total sequence information. Examination of N-acetyl-N,O-permethylated peptides using EI or CI still appears to be the method of choice for sequence studies. LC/MS is an attractive proposition for studies of this type because more complex mixtures can be examined than with the direct insertion probe approach. Studies of mixtures of N-acetyl peptide methyl esters with a DLI system showed the viability of LC/MS (111), although molecular weight information and sequence peaks were not always obtained. N-Acetyl-N,O-permethylated peptides have proved to be better derivatives: studies with moving-belt interfaces provide molecular weight information, and — if isobutane (112,113) or ammonia CI (114) is used — N- and C- terminal sequence ions are also present and permit unequivocal determination of the sequence of the peptides present in the mixture. One problem with the technique is the production of under- and overmethylation artifacts (114,115), but use of a combination of deuterated and nondeuterated reagents should enable ready identification of same. Currently, on the order of 40 nmol of peptide is required for derivatization and LC/MS study. Improvement in permethylation procedures and the use of spray deposition or microbore LC should enable sequencing to be performed with much smaller samples.

Carbohydrates, Glycosides, Nucleosides, and Nucleotides

Carbohydrates provide a convenient test for the range of compounds amenable to study with a particular LC/MS interface. Studies with moving-belt systems show that the older Finnigan MAT (67,91) and VG interfaces (67) can provide molecular weight information from mono- and disaccharides, whereas the latest Finnigan MAT interface provides such information from trisaccharides (67). Mono-

and disaccharides can also be handled by various types of DLI interfaces (116–120). Impressive data have been obtained by DLI LC/MS from reduced peralkylated saccharides obtained by partial hydrolysis of a variety of oligosaccharides (117,120). The technique has been used in combination with other techniques for sequence studies of oligosaccharides (120).

A variety of glycosides have been studied by both moving-belt (67,91) and DLI (118,119,121) systems. As yet there have been no examples of samples of this type being studied by LC/MS. Nucleosides are amenable to study with moving-belt (67,91) and DLI (33,122) techniques. Thermospray LC/MS (12,13,123), however, appears to be the method of choice, because much lower detection limits should be achieved and nucleotides can also be handled (12,13). The technique has been used to characterize nucleosides in the hydrolysate of rabbit liver tRNA[Val] (123).

Antibiotics

A mixture of spectinomycin and actinamine has been analyzed by positive-ion DLI LC/MS (117) and, in the negative mode, erythromycin A (124) and rifampicin (104) gave molecular ions. Negative-ion DLI LC/MS has been used to identify nodusmicin, nargenicin, and 18-deoxynargenicin in an extract from a fermentation broth (102). Aminoglycoside antibiotics are amenable to study with moving-belt systems (67), and the technique has been used to identify components of antimycin A (125). The author and colleagues have conducted extensive studies on extracts of *Penicillium patulum* and found moving-belt LC/MS an excellent method for the identification of griseofulvin and its metabolites (126,127). The technique has also been used to characterize impurities in pharmaceutical preparations of griseofulvin (127). The older versions of moving-belt systems fail to provide molecular weight information from pseudomonic acids and penicillins (128). The most recent Finnigan MAT system provides such information from pseudomonic acids, but not, however, from penicillins (129). Nonetheless, the author and colleagues have found both classes of compound are amenable to thermospray LC/MS, with low nanogram detection limits being obtained from penicillins (129). Although as yet only spectra of model compounds obtained by thermospray LC/MS have appeared in the literature (12,13,54), the technique has considerable potential for studies of such molecules.

Lipids

Moving-belt LC/MS has been used to characterize meibomian gland waxes (130). An impressive demonstration of the capabilities of this type of LC/MS interface is provided by analysis of derivatized sphingoid bases obtained by acid hydrolysis of brain sphingomyelin and gangliosides (131). A variety of phospholipids have also been analyzed with the same type of interface (132). The technique proved particularly useful for the analysis of phosphatidylcholines because these compounds provided both molecular weight and structurally useful fragment ions. Results obtained from the analyses of rat brain phospholipids agreed well with data obtained by GC/MS techniques. Phosphatidylcholines are also amenable to study by DLI LC/MS (103). The technique has also been used to analyze triacylglycerols from a variety of sources (133,134) and sn-1,2-diacylglycerol obtained by hydrolysis of phosphatidylcholines (135) derivatized as their t-butyldimethylsilyl ethers.

Other Classes of Compounds

An area of major utility for LC/MS is in the screening of natural extracts for new compounds with potential biological activity and in the identification and/or quantification of known natural products in such extracts. DLI LC/MS has been used for instance, for the identification of cannabinoids in an extract from Cannabis leaves (136) and of alkaloids from *Ochrosia balansae* (25). The same technique was used to quantify dethiobiotin and biotin in biological extracts (137) and to identify the mycotoxins nivalenol and deoxynivalenol in contaminated cereal samples (138). The potential of using this approach for the analysis of aflatoxins has also been demonstrated (139). DLI LC/MS using fused-silica columns has been used for the analysis of terpenes in pine resin (118), and conventional DLI LC/MS of a variety of liposoluble vitamins has been reported (140).

Moving-belt systems have the advantage of providing both EI and CI information, so they have been extensively used for studies of natural coumarins (60,141–143), Amaryllidaceae (141), ergot (144) and Chinchona (141) alkaloids, pepper and capsicum oleoresins (145), and gossypol (146). Tocopherols have also been identified in a maize germ oil by moving-belt LC/MS (82).

Four cytotoxic trichothecene mycotoxins and the fungal estrogen zearalenone have been studied by thermospray LC/MS. Detection limits using SIM varied between 0.5 ng and 20 ng, and the technique was used to analyze the compounds in porcine plasma and urine (147). The same technique has been used with stable isotope-labeled analogues to assay carnitine and short-chain acylcarnitines in biological samples (148).

CONCLUSIONS

Combined LC/MS is coming of age. With the availability of thermospray, DLI, and moving-belt interfaces, a wide range of problems can be tackled. As has been discussed, the various approaches have advantages and disadvantages; selection of interface depends on the problem to be addressed. There is still no universal system: thermospray LC/MS is the method of choice if low-level detection or quantification is to be undertaken and it handles components not amenable to the other techniques; DLI has advantages over moving-belt systems for studies of thermally labile compounds but because only CI information is provided it is not as good for structural studies as moving-belt systems that provide both EI and CI data. Surface ionization methods show considerable promise for extending the range of compounds amenable to study by moving-belt LC/MS, and electrospray ionization and monodipersive aerosol generation are exciting new possibilities for LC/MS. The recent record of development holds great promise for future achievements. Given the simplicity of spectra obtained from many types of compounds by thermospray ionization and atmospheric pressure ionization, there is no doubt that multisector mass spectrometers also will have considerable utility for LC/MS studies with these types of interfaces. In all, it seems that LC/MS will play a prominent and perhaps crucial role in the future of pharmaceutical development.

REFERENCES

(1) D.M. Desiderio and G.H. Fridland, *J. Liq. Chromatogr.* **7** (S-2), 317 (1984).

(2) D.M. Desiderio, *Analysis of Neuropeptides by Liquid Chromatography and Mass Spectrometry* (Elsevier, Amsterdam, 1984).

(3) H.-R. Schulten, *J. Chromatogr.* **251**, 105 (1982).

(4) R.J. Cotter, *Anal. Chem.* **52**, 1589A (1980).

(5) M.A. Baldwin and F.W. McLafferty, *Org. Mass Spectrom.* **7**, 1111 (1975).

(6) E.C. Horning, D.I. Carroll, I. Dzidic, and R.N. Stillwell, *Pure Appl. Chem.* **50**, 113 (1978).

(7) J.D. Henion, B.A. Thomson, and P.H. Dawson, *Anal. Chem.* **54**, 451 (1982).

(8) B.A. Thomson, J.V. Iribane, and P.J. Dziedzic, *Anal. Chem.* **54**, 2219 (1982).

(9) L.L. Mack, P. Krallik, A. Rheude, and M. Doyle, *J. Chem. Phys.* **52**, 4977 (1970).

(10) C.M. Whitehouse, R.N. Dreyer, M. Yamashita, and J.B. Fenn, *Anal. Chem.* **57**, 675 (1985).

(11) S.-T.F. Lai and C.A. Evans, Jr., *Biomed. Mass Spectrom.* **6**, 10 (1979).

(12) C.R. Blakley, J.J. Carmody, and M.L. Vestal, *J. Amer. Chem. Soc.* **102**, 5931 (1980).

(13) C.R. Blakley, J.J. Carmody, and M.L. Vestal, *Anal. Chem.* **52**, 1636 (1980).

(14) C.R. Blakley, J.J. Carmody, and M.L. Vestal, *Clin. Chem.* **26**, 1467 (1980).

(15) C.R. Blakley and M.L. Vestal, *Anal. Chem.* **55**, 750 (1983).

(16) R.D. Macfarlane, *Anal. Chem.* **55**, 1247A (1983).

(17) A. Benninghoven and W. Sichtermann, *Org. Mass Spectrom.* **12**, 595 (1977).

(18) M. Barber, R.S. Bordoli, G.J. Elliott, R.D. Sedgwick, and A.N. Tyler, *Anal. Chem.* **54**, 645A (1982).

(19) R.J. Cotter, *Anal. Chem.* **56**, 485A (1984).

(20) M.L. Vestal, *Mass Spectrometry Reviews* **2**, 447 (1983).

(21) G.D. Daves, Jr., *Acc. Chem. Res.* **12**, 359 (1979).

(22) N.M.M. Nibbering, *J. Chromatogr.* **251**, 93 (1982).

(23) P.J. Arpino and G. Guiochon, *J. Chromatogr.* **251**, 153 (1982).

(24) P.J. Arpino and G. Guiochon, *Anal. Chem.* **51**, 682A (1979).

(25) P.J. Arpino, in *Liquid Chromatography Detectors,* T.M. Vickrey, ed. (Marcel Dekker, New York, 1983), p. 243.

(26) Z.F. Curry, *J. Liq. Chromatogr.* **5**, 257 (1982).

(27) D.E. Games, in *Advances in Chromatography,* Volume 21, J.C. Giddings, E. Grushka, J. Cazes, and P.R. Brown, eds. (Marcel Dekker, New York, 1983), p. 1.

(28) W.H. McFadden, *J. Chromatogr. Sci.* **18**, 97 (1980).

(29) R.C. Willoughby and R.F. Browner, in *Trace Analysis,* Volume 2, J.F. Lawrence, ed. (Academic Press, New York, 1982), p. 69.

(30) P. Arpino, M.A. Baldwin, and F.W. McLafferty, *Biomed. Mass Spectrom.* **1**, 80 (1974).

(31) P.J. Arpino, B.G. Dawkins, and F.W. McLafferty, *J. Chromatogr. Sci.* **12**, 574 (1974).

(32) J.D. Henion, *Anal. Chem.* **50**, 1687 (1978).

(33) A. Melera, *Adv. Mass Spectrom.* **8B**, 1597 (1980).

(34) P.J. Arpino, P. Krien, S. Vajta, and G. Devant, *J. Chromatogr.* **203**, 117 (1981).

(35) P.J. Arpino, J.P. Bounine, M. Dedieu, and G. Guiochon, *J. Chromatogr.* **271**, 43 (1983).

(36) F.R. Sugnaux, D.S. Skrabalak, and J.D. Henion, *J. Chromatogr.* **264**, 357 (1983).

(37) J.D. Henion and G.A. Maylin, *Biomed. Mass Spectrom.* **7**, 115 (1980).

(38) N. Evans and J.E. Williamson, *Biomed. Mass Spectrom.* **8**, 316 (1981).

(39) A.P. Bruins and B.F.H. Drenth, *J. Chromatogr.* **271**, 71 (1983).

(40) K.H. Schafer and K. Levsen, *J. Chromatogr.* **206**, 245 (1981).

(41) J.D. Henion and T. Wachs, *Anal. Chem.* **53**, 1963 (1981).

(42) R.D. Voyksner, J.R. Hass, and M.M. Bursey, *Anal. Lett.* **15**, 1 (1982).

(43) R.D. Voyksner, C.E. Parker, J.R. Hass, and M.M. Bursey, *Anal. Chem.* **54**, 2583 (1982).

(44) T. Takeuchi, Y. Hirata, and Y. Okumuro, *Anal. Chem.* **50**, 659 (1978).

(45) Y. Yoshida, H. Yoshida, S. Tsuge, T. Takeuchi, and K. Mochizuki, *HRC&CC, J. High Res. Chromatogr. Chromatogr. Commun.* **3**, 16 (1980).

(46) H. Yoshida, K. Matsumoto, K. Itoh, S. Tsuge, Y. Hirata, K. Mochizuki, N. Kokubun, and Y. Yoshida, *Fresenius' Z. Anal. Chem.* **311**, 674 (1982).

(47) J.A. Apffel, U.A.Th. Brinkman, R.W. Frei, and E.A.I. Evers, *Anal. Chem.* **55**, 2280 (1983).

(48) R.G. Christensen, E. White, V.S. Meiselmann, and H.S. Hertz, *J. Chromatogr.* **271**, 61 (1983).

(49) D.I. Carroll, I. Dzidic, R.N. Stillwell, K.D. Haegele, and E.C. Horning, *Anal. Chem.* **47**, 2369 (1975).

(50) B. Shushan, J.E. Fulford, B.A. Thomson, W.R. Davidson, L.M. Danylewych, A. Ngo, S. Nacson, and S.D. Tanner, *Int. J. Mass Spectrom. Ion Phys.* **46**, 225 (1983).

(51) C.M. Whitehouse, R.N. Dreyer, M. Yamashita, and J.B. Fenn, *Anal. Chem.* **57**, 675 (1985).

(52) R.C. Willoughby and R.F. Browner, *Anal. Chem.* **56**, 2626 (1984).

(53) M.L. Vestal, *Anal. Chem.* **56**, 2590 (1984).

(54) J.R. Chapman, *J. Chromatogr.* **323**, 153 (1985).

(55) R.D. Voyksner and C.A. Haney, *Anal. Chem.* **57**, 992 (1985).

(56) D.E. Games and E.D. Ramsey, *J. Chromatogr.* **323**, 67 (1985).

(57) R.D. Voyksner, J.T. Bursey, and E.D. Pellizzari, *Anal. Chem.* **56**, 1507 (1984).

(58) W.H. McFadden, H.L. Schwartz, and S. Evans, *J. Chromatogr.* **122**, 389 (1976).

(59) D.E. Games, M.S. Lant, S.A. Westwood, M.J. Cocksedge, N. Evans, J. Williamson, and B.J. Woodhall, *Biomed. Mass Spectrom.* **9**, 215 (1982).

(60) N.J. Alcock, L. Corbelli, D.E. Games, M.S. Lant, and S.A. Westwood, *Biomed. Mass Spectrom.* **9**, 499 (1982).

(61) M.J. Hayes, H.E. Schwartz, P. Vouros, and B.L. Karger, *Anal. Chem.* **56**, 1229 (1984).

(62) R.D. Smith and A.L. Johnson, *Anal. Chem.* **53**, 739 (1981).

(63) M.J. Hayes, E.P. Lankmayer, P. Vouros, B.L. Karger, and J.M. McGuire, *Anal. Chem.* **55**, 1745 (1983).

(64) E.D. Hardin, T.P. Fan, C.R. Blakley, and M.L. Vestal, *Anal. Chem.* **56**, 2 (1984).

(65) P. Dobberstein, E. Korte, G. Meyerhoff, and R. Pesch, *Int. J. Mass Spectrom. Ion Phys.* **46**, 185 (1983).

(66) D.S. Millington, D.A. Yorke, and P. Burns, *Adv. Mass Spectrom.* **8B**, 1819 (1980).

(67) D.E. Games, M.A. McDowall, K. Levsen, K.H. Schafer, P. Dobberstein, and J.L. Gower, *Biomed. Mass Spectrom.* **11**, 87 (1984).

(68) E.D. Hardin, T.P. Fan, C.R. Blakley, and M.L. Vestal, *Anal. Chem.* **56**, 2 (1984).

(69) T.P. Fan, E.D. Hardin, and M.L. Vestal, *Anal. Chem.* **56**, 1870 (1984).

(70) J.G. Stroh, J. Carter Cook, R.M. Milberg, L. Brayton, T. Kihara, Z. Huang, and K.L. Rinehart, Jr., *Anal. Chem.* **57**, 985 (1985).

(71) R.D. Smith, J.C. Fjeldsted, and M.L. Lee, *J. Chromatogr.* **247,** 231 (1982).

(72) L.G. Randall and A.L. Wahrhaftig, *Rev. Sci. Instrum.* **52,** 1283 (1981).

(73) J.D. Henion, personal communication.

(74) C.G. Edmonds, J.A. McCloskey, and V.A. Edmonds, *Biomed. Mass Spectrom.* **10,** 237 (1983).

(75) L.E. Martin, J. Oxford, and R.J.N. Tanner, *Xenobiotica* **11,** 831 (1981).

(76) L.E. Martin, J. Oxford, and R.J.N. Tanner, *J. Chromatogr.* **251,** 215 (1982).

(77) M.S. Lant, L.E. Martin, and J. Oxford, *J. Chromatogr.* **323,** 143 (1985).

(78) J.K. Wellby, E.D. Ramsey, and D.E. Games, unpublished work.

(79) D.E. Games, E. Lewis, N.J. Haskins, and K.A. Waddell, *Adv. Mass Spectrom.* **8B,** 1233 (1980).

(80) D.E. Games, J.L. Gower, M.G. Lee, I.A.S. Lewis, M.L. Pugh, and M. Rossiter, in *Blood Drugs and Other Analytical Challenges,* E. Reid, ed. (Ellis Horwood, Chichester, 1978), p. 185.

(81) J.D. Baty and R.G. Willis, *Anal. Proc.* **19,** 251 (1982).

(82) J. Van der Greef, A.C. Tas, M.C. Ten Noever de Brauw, M. Höhn, G. Meijerhoff, and U. Rapp, *J. Chromatogr.* **323,** 81 (1985).

(83) D.P. Kirby, P. Vouros, B.L. Karger, B. Hidy, and B. Petersen, *J. Chromatogr.* **203,** 139 (1981).

(84) J.A. Apffel, U.A.Th. Brinkman, and R.W. Frei, *J. Chromatogr.* **312,** 153 (1984).

(85) F. Erni, *J. Chromatogr.* **251,** 141 (1982).

(86) S. Tsuge, Y. Hirata, and T. Takeuchi, *Anal. Chem.* **51,** 166 (1979).

(87) J.D. Henion, *J. Chromatogr. Sci.* **18,** 101 (1980).

(88) J.D. Henion and G.A. Maylin, *Biomed. Mass Spectrom.* **7,** 115 (1980).

(89) C. Eckers, D.S. Skrabalak, and J. Henion, *Clin. Chem.* **28,** 1882 (1982).

(90) J.B. Crowther, T.R. Covey, E.A. Dewey, and J.D. Henion, *Anal. Chem.* **56,** 2921 (1984).

(91) D.E. Games and E. Lewis, *Biomed. Mass Spectrom.* **7,** 433 (1980).

(92) T. Cairns and E.G. Siegmund, *Anal. Chem.* **54,** 2456 (1982).

(93) D.J. Dixon, *Analusis* **10,** 343 (1982).

(94) D.J. Liberato, C.C. Fenselau, M.L. Vestal, and A.L. Yergy, *Anal. Chem.* **55,** 1741 (1983).

(95) D. Watson, G.W. Taylor, and S. Murray, *Biomed. Mass Spectrom.,* in press.

(96) R.D. Voyksner, J.T. Bursey, J.W. Hines, and E.D. Pellizzari, *Biomed. Mass Spectrom.* **11,** 616 (1984).

(97) T.R. Covey, J.B. Crowther, E.A. Dewey, and J.D. Henion, *Anal. Chem.* **57,** 474 (1985).

(98) D.A. Catlow, *J. Chromatogr.* **323,** 163 (1985).

(99) J.D. Henion, B.A. Thomson, and P.H. Dawson, *Anal. Chem.* **54,** 451 (1982).

(100) E. Houghton, M.C. Dumasia, and J.K. Wellby, *Biomed. Mass Spectrom.* **8,** 558 (1981).

(101) T. Cairns, E.G. Siegmund, J.J. Stamp, and J.P. Skelly, *Biomed. Mass Spectrom.* **10,** 203 (1983).

(102) C. Eckers, J.D. Henion, G.A. Maylin, D.S. Skrabalak, J. Vessman, A.M. Tivert, and J.C. Greenfield, *Int. J. Mass Spectrom. Ion Phys.* **46,** 205 (1983).

(103) F.R. Sugnaux and C. Djerassi, *J. Chromatogr.* **251,** 189 (1982).

(104) D.J. Dixon, *Analusis* **10,** 343 (1982).

(105) K. Levsen, K.H. Schafer, and P. Dobberstein, *Biomed. Mass Spectrom.* **11,** 308 (1984).

(106) D. Pilosof, H.-Y. Kim, M.L. Vestal, and D.F. Dyckes, *Biomed. Mass Spectrom.* **11,** 403 (1984).

(107) C.N. Kenyon, *Biomed. Mass Spectrom.* **10,** 535 (1983).

(108) T. Covey and J. Henion, *Anal. Chem.* **55,** 2275 (1983).

(109) D. Pilosof, H.-Y. Kim, D.F. Dyckes, and M.L. Vestal, *Anal. Chem.* **56,** 1236 (1984).

(110) H.-Y. Kim, D. Pilosof, D.F. Dyckes, and M.L. Vestal, *J. Amer. Chem. Soc.* **106,** 7304 (1984).

(111) B.G. Dawkins, P.J. Arpino, and F.W. McLafferty, *Biomed. Mass Spectrom.* **5,** 1 (1978).

(112) T.J. Yu, H.E. Schwartz, R.W. Giese, B.L. Karger, and P. Vouros, *J. Chromatogr.* **218,** 519 (1981).

(113) T.J. Yu, H.E. Schwartz, S.A. Cohen, P. Vouros, and B.L. Karger, *J. Chromatogr.* **301,** 425 (1984).

(114) P. Roepstorff, M.A. McDowall, M.P.L. Games, and D.E. Games, *Int. J. Mass Spectrom. Ion Phys.* **48**, 197 (1983).

(115) T.J. Yu, H.E. Schwartz, R.W. Giese, B.L. Karger, and P. Vouros, *J. Chromatogr.* **218**, 519 (1981).

(116) P.J. Arpino, P. Krein, S. Vajta, and G. Devant, *J. Chromatogr.* **203**, 117 (1981).

(117) C.N. Kenyon, A. Melera, and F. Erni, *J. Anal. Toxicol.* **5**, 216 (1981).

(118) H. Alborn and G. Stenhagen, *J. Chromatogr.* **323**, 47 (1985).

(119) P. Hirter, H.J. Walther, and P. Dätwyler, *J. Chromatogr.* **323**, 89 (1985).

(120) M. McNeil, A.G. Darvill, P. Åman, L.-E. Franzén, and P. Albersheim, in *Complex Carbohydrates,* Methods in Enzymology, Volume 83, Part D, V. Ginsburg, ed. (Academic Press, New York, 1982), p. 3.

(121) R. Schuster, *Chromatographia* **13**, 1 (1980).

(122) E.L. Esmans, Y. Luyten, and F.C. Alderweireldt, *Biomed. Mass Spectrom.* **10**, 347 (1983).

(123) C.G. Edmonds and J.A. McCloskey, paper 789, presented at 31st Annual Meeting on Mass Spectrometry and Allied Topics, Boston, Massachusetts, 1983.

(124) M. Deieu, C. Juin, P.J. Arpino, and G. Guiochon, *Anal. Chem.* **54**, 2372 (1982).

(125) S.L. Abidi, *J. Chromatogr.* **234**, 187 (1982).

(126) D.E. Games, *Kemia-Kemi* **11**, 8 (1984).

(127) D.E. Games, M.S. Lant, and B.J. Woodhall, unpublished results.

(128) M.A. McDowall, D.E. Games, and J.L. Gower, *Int. J. Mass Spectrom. Ion Phys.* **48**, 157 (1983).

(129) D.E. Games, J.L. Gower, and M.A. McDowall, unpublished results.

(130) W.H. McFadden, D.C. Bradord, G. Eglinton, S.K. Hajlbrahim, and N. Nicolaides, *J. Chromatogr. Sci.* **17**, 518 (1979).

(131) F.B. Jungalwala, J.E. Evans, H. Kadowaki, and R.H. McCluer, *J. Lipid Res.* **25**, 209 (1984).

(132) F.B. Jungalwala, J.E. Evans, and R.H. McCluer, *J. Lipid Res.* **25**, 738 (1984).

(133) L. Marai, J.J. Myher, and A. Kukis, *Can. J. Biochem. Cell Biol.* **61**, 840 (1983).

(134) J.J. Myher, A. Kukis, L. Marai, and F. Manganaro, *J. Chromatogr.* **283**, 289 (1984).

(135) S. Pind, A. Kukis, J.J. Myher, and L. Marai, *Can. J. Biochem. Cell Biol.* **62**, 301 (1984).

(136) P.J. Arpino and P. Krien, *J. Chromatogr. Sci.* **18**, 104 (1980).

(137) M. Azoulay, P.-L. Desbene, F. Frappier, and Y. Georges, *J. Chromatogr.* **303**, 272 (1984).

(138) R. Tiebach, W. Blaas, M. Kellert, S. Steinmeyer, and R. Weber, *J. Chromatogr.* **318**, 103 (1985).

(139) R. Tiebach, W. Blaas, and M. Kellert, *J. Chromatogr.* **323**, 121 (1985).

(140) H. Milon and H. Bur, *LC, Liq. Chromatogr. HPLC Mag.* **2**, 455 (1984).

(141) C. Eckers, D.E. Games, E. Lewis, K.R.N. Rao, M. Rossiter, and N.C.A. Weerasinghe, *Adv. Mass Spectrom.* **8B**, 1396 (1980).

(142) C. Eckers, D.E. Games, M.L. Games, W. Kuhnz, E. Lewis, N.C.A. Weerasinghe, and S.A. Westwood, in *Recent Developments in Mass Spectrometry in Biochemistry, Medicine and Environmental Research,* Volume 7, A. Frigerio, ed. (Elsevier, Amsterdam, 1981), p. 189.

(143) N.J. Alcock, W. Kuhnz, and D.E. Games, *Int. J. Mass Spectrom. Ion Phys.* **48**, 153 (1983).

(144) C. Eckers, D.E. Games, D.N.B. Mallen, and B.P. Swann, *Biomed. Mass Spectrom.* **9**, 162 (1982).

(145) D.E. Games, N.J. Alcock, J. Van der Greef, L.M. Nyssen, H. Maarse, and M.C.T. Noever de Brauw, *J. Chromatogr.* **294**, 269 (1984).

(146) S.A. Matlin, R.H. Zhou, D.E. Games, A. Jones, and E.D. Ramsey, *HRC&CC, J. High Res. Chromatogr. Chromatogr. Commun.* **7**, 196 (1984).

(147) R.D. Voyskner, W.M. Hagler, Jr., K. Tyazkowska, and C.A. Haney, *HRC&CC, J. High Res. Chromatogr. Chromatogr. Commun.* **8**, 119 (1985).

(148) A.L. Yergey, D.J. Liberato, and D.S. Millington, *Anal. Biochem.* **139**, 278 (1984).

PART TWO

RECENT ADVANCES IN THE USE OF LIQUID CHROMATOGRAPHY IN THE PREPARATIVE SEPARATION OF DRUG SUBSTANCES

RECENT ADVANCES IN THE PREPARATIVE CHROMATOGRAPHY OF LOW MOLECULAR WEIGHT SUBSTANCES

Robert Sitrin, Peter DePhillips,
John Dingerdissen, Karl Erhard, and John Filan

Smith Kline & French Laboratories
Philadelphia, PA 19101

The pharmaceutical industry is a highly competitive business in which success often falls to the organization that can discover and develop new products in as short a time as possible. To facilitate the research and development process, the pharmaceutical industry has been prompt in taking advantage of the latest technical developments in biology and chemistry. Analytical high performance liquid chromatography (HPLC) is an example of a technology that has affected every area of drug research and development.

The area of preparative HPLC (PLC) is another example of an emerging technology undergoing rapid expansion. Although its use actually preceded the development of modern analytical HPLC, this technology has not received as much attention in the literature. In the few publications that exist, a variety of approaches to solving PLC problems have been described. Within the last five years, however, the availability of PLC instruments, columns, and bulk packings has resulted in increased publications on PLC applications. Although traditional preparative chromatography has been done in the normal-phase mode on silica or alumina supports, there is an increasing trend to use reversed-phase supports for preparative work. This parallels recent reports of extensive use of reversed-phase packings for analytical separations.

In the pharmaceutical industry, PLC has had an effect on all phases of drug discovery and development. Drug discovery has been facilitated by the extensive use of PLC in the preparation of synthetic intermediates and final products for biological testing. This has been especially true for modern compounds, such as leukotrienes and peptides, whose instability and complexity mandate the use of PLC in their preparation as opposed to other purification procedures. Because commer-

265

cial preparative LC units require a sizable capital investment, it is common practice to establish a PLC laboratory containing several chromatographs. Using such a laboratory, a few experienced individuals can support the purification needs of a number of synthetic chemists. In addition, in natural products research PLC has shortened the time needed to isolate and purify new bioactive molecules, especially polar, water-soluble compounds. In drug development, PLC has been applied to the preparation of primary standards and the isolation of contaminants and reaction by-products. Both of these have been required in order to meet increasingly rigid FDA specifications.

In preparing this chapter, the authors have drawn upon their own experience in medicinal chemistry and drug discovery, chemical process development, and natural products discovery and development. In most cases the applications of PLC — especially to compounds under commercial development — are proprietary and cannot be described in great detail in the literature. As a consequence, this chapter will use as examples applications from either the authors' own work, given without structural or chemical details if necessary, or from the industrial and nonindustrial literature. Descriptions will be limited to laboratory-scale work, because production-scale applications are discussed elsewhere in this volume. Discussion of protein purification by HPLC has also been omitted for the same reason.

Two distinct approaches to carrying out preparative HPLC have been described in the literature. One, labeled *semipreparative,* involves separations on 5–10 μm prepacked columns. It is essentially a direct scale-up of analytical HPLC and, as such, is run with high efficiency but limited loading. The other approach uses inexpensive, large-particle (40–60 μm), user-packed columns 2 in. in diameter or larger. Using these columns and appropriate pumping systems, multiple-gram quantities of products have been purified.

In this chapter, an overview of the authors' strategy for solving separation problems will be given because other authors seldom describe why they chose a particular column and instrument for an application. Next, recent advances in columns and instrumentation will be considered because the current growth in PLC could not have taken place without these developments. The applications sections will describe uses of PLC in the preparation and process development of synthetic compounds and the discovery of novel natural products. Finally, a discussion of future trends will be presented.

STRATEGY

Numerous articles have been written on preparative HPLC, describing various aspects of instrumentation, theory, column configuration, and loading capacity (1–23). A current review of PLC describing advances in theory, equipment, and column technology through 1984 also has recently appeared (24). Rather than giving a detailed discussion of theoretical aspects, this chapter instead will present a practical approach to PLC, one which has been helpful to the authors and others working in the area. When approaching a preparative separation, two questions can be asked: What column and instrument configurations should be used? How much can be loaded onto that column? Although some systematic approaches to answering these questions have appeared in the literature (6,9,12,17), investigators have seldom adequately described why a particular instrument or column was used for a particular separation problem. Many of the published separations are vaguely described in experimental sections and are of little help in developing a purification strategy. The authors have found a selectivity factor (α) and theoretical plate count (N) approach similar to that of Verzele and Geeraert (6) to be helpful in choosing appropriate columns and equipment.

HPLC has evolved mainly as an analytical technique with emphasis on improving column efficiency and supportive instrumentation. The development of 3-μm packings and high efficiency columns with 100,000 plates/m is an outstanding achievement in the analytical area. In the preparative mode, however, columns with such high efficiency are seldom necessary. Provided loading is kept at levels below 1 to 2 mg/g of packing, it is possible to calculate, using the resolution equation, how many plates are needed to effect a desired separation. Under these conditions of moderate loading, the well-known resolution and plate count expressions originally derived for analytical separations can still be used to approximate separation parameters. As loading is further increased, peak shapes usually become distorted and non-Gaussian, and, as a consequence, these expressions are no longer applicable. Therefore, the separation parameters observed at analytical loading levels are used to predict the initial choice of preparative column and instrument configuration. The performance at higher loading levels must be evaluated empirically on a case-by-case basis.

The relationships between the separation of two peaks and other chromatographic parameters are described by the following resolution equations:

$$R_s = 2(t_2 - t_1)/(w_1 + w_2) \qquad [1]$$

$$= \frac{(\alpha - 1)N^{1/2}}{4} \left(\frac{k'}{k' + 1} \right) \qquad [2]$$

where

R_s = resolution
t_1 and t_2 = the retention times
w_1 and w_2 = the base widths of two peaks
k' = their average capacity factor
α = the ratio of k' values for the two peaks being separated
N = their average plate count (32).

To achieve a preparative separation with little overlap between two peaks, a minimum resolution value of 1.0 is desirable. A resolution of less than 1.0 would result

TABLE I

Calculated N_{min} Values for
Various α Values

α	N_{min}
2	23
1.5	92
1.3	256
1.2	575
1.1	2300
1.05	9200
1.03	25,556
1.01	230,000

in losses of material in mixed cuts, requiring rechromatography or recycling of the fraction. Setting the value of R_s to 1.0 in Equation 2 and using a k' value of 5 — a reasonable value for preparative work — Equation 2 becomes:

$$N_{min} = 23/(\alpha - 1)^2 \qquad [3]$$

where

N_{min} = the minimal required plate count for a separation with a given α value. Some calculated values of N_{min} for various α values are shown in Table I. This listing is useful for classifying various types of separations based on their α values.

In practice, a separation is first carried out on an analytical column or a thin-layer plate (25) containing a support of identical chemistry to that contained in the corresponding preparative column. The separation conditions are varied to maximize the α value for the components of interest, and the derived α value is used to determine how efficient a column is needed for preparative work.

Gram-Scale Chromatography

If α is determined to be larger than 1.3, fewer than 250 plates are needed according to the above analysis (Table I). Such plate counts can be achieved using large-particle packings in the 40- to 60-μm range. These large-particle packings are now widely available at modest cost (less than \$30/kg for silica and less than \$1000/kg for a bonded phase) with a variety of surface chemistries and can be readily dry-packed by the user. In the authors' work, a glass column (50 cm × 2.4 cm) containing 160 g of reversed-phase packing (40–60 μm) and a steel column (50 cm × 4.8 cm) containing 600 g of the same packing yielded more than 300 plates for a test compound at light loading (1 mg/g) (26). Therefore, in these systems at 1 mg/g loading, purification of 160 or 600 mg of compound should be successful in each column. As loading for a compound is increased, values of α, N, and k' can be expected to change. For example, as loading increased to 10 mg/g in the above columns, plate counts were observed to fall to 50–100 plates (26). Such decreases in efficiency with increased loading have been well documented in the literature (1–4,6,14,18,22,27). This suggests that from 2 g to 6 g of material could be separated in the above columns if α values stayed greater than 1.5 at the higher loading. Because these large-particle, large-column systems are capable of purifying gram quantities of product in a single run, this type of chromatography is labeled *gram scale* in this chapter.

Generally, as loading increases, values of k' are also observed to fall (1,27), but the value of the ratio term in the resolution equation, $k'/(1 + k')$, does not fall as quickly and therefore has a relatively smaller effect on the resolution value. Furthermore, k' values can be readjusted by suitable cosolvent manipulation if necessary. Changes in the value of α as loading is increased cannot be predicted, but significant decreases in α will clearly limit loading. The term *overload* was originally defined in the literature as that level at which k' or N values drop by 10% (27). A recent suggestion for a new definition of *PLC overload* is perhaps more appropriate: that level at which "the separation can no longer be achieved . . . because of the increase in sample size" (24). In practice, loading is increased until the resolution is unacceptable, whether caused by increases in peak widths (losses in efficiency) or by a decrease in α (shifting of peaks together).

Semipreparative Chromatography

For more difficult separations — those with α values between 1.1 and 1.3 — higher efficiency of up to 2000 plates (Table I) is required. This restriction mandates the use of smaller particles (usually 10 μm) and is defined for this chapter as *semipreparative LC*. The columns are larger scale versions of analytical columns and are sold prepacked in sizes of 1–2.5 cm i.d. and 25–50 cm length. At analytical loading levels (5 μg/g or 0.6 mg) on a 50 cm × 2.2 cm column containing 120 g of packing, 4000 plates were obtained with ideal compounds (26). Unfortunately, columns with 5-μm and 10-μm packing are more sensitive to increases in loading than are columns packed with larger particles (14,16,28). Thus, at 250 mg loading (2 mg/g), the same column yielded less than 1000 plates for one compound studied (26). This efficiency loss at higher loading limits the amount that can be purified on this size column to several hundred milligrams of pure compound per injection. In fact, when loading reached 1 g for a glycopeptide antibiotic, column efficiency was no better than that achieved on the same size column packed with 40-μm packing (Figure 1) (26).

If larger amounts of product are required from a semipreparative separation, a larger column could be used while keeping loading low to maintain efficiency. Unfortunately, 10-μm columns larger than 2.5 cm in diameter are very expensive and not widely available. The usual way to get higher product yield on semipreparative systems is to carry out repeated separations at lower loading, often with automated equipment (see below). An alternative approach to producing larger

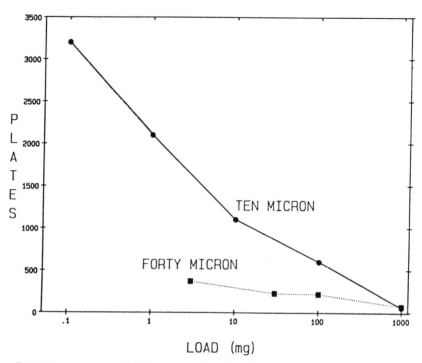

FIGURE 1.　Comparison of efficiency versus load for a glycopeptide antibiotic on 50 cm × 2.5 cm columns packed with 10 μm and 40 μm, Whatman ODS-3 packing. Mobile phase: 22% acetonitrile in 0.1 M phosphate, pH 6.0; detection: UV 210 nm (26).

amounts of materials is the use of less-expensive, 20-μm particles in 5–8 cm i.d. columns. Although such columns can in theory be user-packed by dry-tapping or slurry techniques, efficient packing of this particle size remains a challenge. Recently, reports have appeared describing the use of 15–20 μm spherical reversed-phase supports in radial compression cartridges to solve the packing problem (29–32).

When multiple-gram quantities of product are needed and α values are between 1.1 and 1.3, the use of recycling with large-particle systems has been also effective. By collecting the pure areas at the extremes of the peak and recycling the center mixed cuts back through the pump and column, larger quantities of pure product can be obtained.

When developing a preparative separation for which the α value is limiting and the quantity involved is several grams, it is essential that effort be applied to increasing the α value before commencing with the separation. This can often be done by changing the mobile phase or solid adsorbent in a manner analogous to procedures used in analytical work (32).

The above section on strategy should be considered as a guideline in choosing between 10-μm or larger particle columns for a given separation problem. Every compound and every separation is unique, and prediction of shifts in k' and α values at higher loading levels are difficult to make. In actual practice, scale-up to higher loading levels must be done empirically for each situation.

INSTRUMENTATION

This section will describe recent advances in equipment and instrumentation for carrying out PLC on the laboratory scale. The increasing use of PLC in industry has been possible in large part because of the increasing availability of equipment, supports, and instrumentation.

Semipreparative Chromatography

The area of semipreparative chromatography has evolved as an extension of analytical HPLC. Semipreparative columns are larger than standard analytical columns, with internal diameters of 8–25 mm and lengths varying from 15 cm to 50 cm. Like small-particle analytical columns (5–10 μm), they must be slurry packed, normally by the supplier, to provide comparably high column efficiencies.

These columns are designed to be compatible with most analytical HPLC systems and, therefore, offer the convenience of using existing instrumentation for preparative applications at relatively low additional cost. Analytical HPLC pumps can provide optimal flow rates for columns up to 10–15 mm i.d. with no modifications, as can be seen in Table II. Thus, a 10-mm i.d. column requires a flow of 9.4 mL/min to maintain equivalent linear velocity to that of an analytical column run at 2 mL/min.

As column diameter is increased beyond 10–15 mm, most analytical pumps require the addition of preparative heads to provide adequate flow rates (20–50 mL/min). Lower back pressure limitations (2000 psi) with some preparative heads may necessitate running the large-diameter (25 mm) columns at lower linear ve-

TABLE II

Flow Rate Required to Maintain Constant Linear Velocity as a Function of Column Diameter

Column diameter (mm)	Flow rate (mL/min)			
	Case 1	Case 2	Case 3	Case 4
4.6 (Analytical)	0.5	1.0	1.5	2.0
10.0	2.4	4.7	7.0	9.4
15.0	5.3	10.6	16.0	21.2
20.0	10.0	20.0	28.0	40.0
25.0	15.0	30.0	44.0	60.0

locities. For example, a flow rate of 15 mL/min on a 25-mm column is equivalent to a flow rate of only 0.5 mL/min on an analytical column. To achieve constant linear velocity at 2 mL/min (analytical) requires 60 mL/min on the preparative column. Some examples of HPLC pumps capable of providing flow rates in excess of 20 mL/min at high back pressures are listed in Table III.

The commercial availability of semipreparative columns is becoming more widespread. Most suppliers of analytical columns provide larger semipreparative versions containing packing of identical particle size and chemistry. This availability permits convenient scale-up of analytical separations to larger columns of identical selectivity. In addition, this increased availability has led to a general decrease in the cost of semipreparative columns.

Guard columns and solvent-saturation columns should be used at all times to protect these expensive semipreparative columns. Adequate contaminant protection can be difficult for some of the larger columns using low-capacity pellicular packings. More desirable, fully porous guard columns have recently become available for preparative work (Dynamax; Rainin Instruments, Woburn, Massachusetts).

There is an increasing trend to automate semipreparative HPLC instrumentation (33). Several instrument manufacturers have introduced HPLC systems designed specifically for automated semipreparative chromatography that are capable of production of gram quantities of products (Varex, Rockville, Maryland; Rainin-Gilson; Perkin-Elmer, Norwalk, Connecticut; Toyo Soda, Tokyo, Japan). These systems allow unattended, repetitive injection and collection of purified samples. Existing analytical HPLC instruments can also be modified by the user for automated work (5,34–36). A basic understanding of microprocessor operation, computer programming, and HPLC hardware is required to implement these changes.

TABLE III

High-Pressure Preparative-LC Pumps

Pump	Maximum flow rate (mL/min)	Maximum pressure (psi)
Rainin Rabbit HPX[*]	25	4000
(Gilson Model 303)[†]	50	2000
	100	1000
Perkin-Elmer Series 10[‡]	30	6000
Waters 590 EF[§]	45	3000
Eldex Model BBB-4[‖]	100	5000

[*]Rainin Instruments, Woburn, Massachusetts.
[†]Gilson, Middleton, Wisconsin.
[‡]Perkin-Elmer, Norwalk, Connecticut.
[§]Waters Associates, Milford, Massachusetts.
[‖]Eldex Laboratories, San Carlos, California.

Flash Chromatography/MPLC

Flash chromatography is an inexpensive, rapid purification technique requiring no instrumentation (37). Heavy-walled glass columns, available from suppliers such as Ace Glass (Vineland, New Jersey) and J.T. Baker Chemical Company (Phillipsburg, New Jersey) are dry-packed by the user with 40-μm supports. Mobile phase flow is controlled by the application of a small, positive nitrogen pressure to the head of the column, and fractions are usually collected manually. Safety shields should always be used during chromatography with glass columns. The use of bonded-phase packings in flash chromatography has been reported (38–40). Because the commonly used aqueous/organic solvent mixtures are more viscous than solvents used in normal-phase work, flow rates and column lengths must be reduced in order to ensure safe operation (pressures of less than 20 psi) (41).

Medium-pressure liquid chromatography (MPLC) is carried out with pumped flow on user-packed or prepacked glass columns (10,13,42). An MPLC system consists of a metering pump, pulse dampener, injector, pressure relief valve, glass column, detector, and fraction collector. Typically, 25-μm to 60-μm packings are used. In addition, MPLC operating pressures of 60–80 psi permit the practical use of reversed-phase packings at higher flow rates than are achievable in flash chromatography.

Gram-Scale Chromatography

As described in the strategy section, most gram-scale separations are run on 5-cm to 10-cm i.d. steel columns packed with large-particle (20–100 μm) normal- and reversed-phase supports. These large-particle packings are substantially less expensive when purchased in bulk quantities than are the corresponding 10-μm supports used in semipreparative work. Furthermore, because of their larger particle size they are amenable to on-site packing and repacking. In order to operate such column configurations, pumping and detection systems have been developed to obtain flow rates of 100–1000 mL/min at maximum back pressures of 500–2000 psi. Detailed descriptions of some of these instruments have been given in recent reviews (7,8).

Since its introduction in 1975, the Prep LC 500 (Waters Associates) has been a workhorse for the preparation of gram quantities of materials. This system uses radially compressed cartridges (30 cm × 5.7 cm i.d.) that eliminate wall effects. The cartridges are available prepacked from the manufacturer, with 37–75 μm silica (325 g) and 75–105 μm or 30-μm reversed-phase C$_{18}$ silica (350 g). Empty cartridges have been user-packed with XAD-4 (43) and various brands of 15–20 μm reversed-phase packings (29–31); if desired, two cartridges can be used in series to increase efficiency and loading. The chromatograph is equipped with an effluent flow splitter, recycling capability, and a refractive index detector. Flow rates from 50 to 500 mL/min at a maximum back pressure of 500 psi can normally be obtained. Optional features include a UV detector, a low-pressure gradient generator, and a computer-controlled automated system. Several reports have described the attachment of other columns such as the 50 cm × 4.8 cm i.d. Magnum 40 (Whatman, Clifton, New Jersey) to the Prep LC 500 pumping system (see below).

The Chromatospac Prep-100 (Jobin Yvon, available from Instruments SA, Metuchen, New Jersey), developed in the mid-1970s, uses nitrogen pressure to control solvent flow. Columns are user-packed by axial compression of a slurry using a pneumatic piston and are available in 4-cm and 8-cm i.d., the length of which is variable from approximately 20 to 80 cm depending upon the amount of packing used. Flow rates of 200–500 mL/min using 40-µm packings are possible with nonviscous solvents. This unit can be packed with a variety of supports such as silica, alumina, cellulose, ion-exchange, and bonded phases.

The PSLC 100 preparative LC unit (Varex Corporation) is a self-contained modular system that can be purchased in several configurations. The complete automated system consists of a solvent module, detector, low-pressure gradient module, autoinjector, and fraction collector. It is controlled by a Commodore 64 microcomputer for continuous unattended operation and is equipped with a color monitor, disk drive, and printer. The solvent module is available with a dual-head (0–190 mL/min, 5000 psi) or a choice of two triple-headed (0–275 mL/min, 300 psi; 0–580 mL/min, 1900 psi) diaphragm pumps. The gradient module is capable of generating continuous binary or ternary gradients. Injections can be made from a loop injector for sample volumes up to 50 mL; for larger volumes, injections are made through the pump. Detector options include fixed- and variable-wavelength UV and mass detection. A built-in five-port fraction collector selects cuts of unlimited volume.

Another recent entry into the preparative-LC market is the ST/Lab 800A system (Separations Technology, Wakefield, Rhode Island). This system is available with one or two dual-head Teflon diaphragm pumps capable of generating 800 mL/min at a back pressure of 1800 psi. This system also has gradient capability when purchased with two pumps, both ultraviolet and refractive index detectors, and plumbing for recycle and effluent splitting. Stainless-steel columns of 5 to 10 cm i.d. and 30 to 90 cm length are available prepacked or can be user-packed by slurry or dry-tapping techniques.

Preparative-LC Detection

As in analytical work, UV or RI detection is usually used in semipreparative HPLC. The variety of commercially available analytical detectors, however, offers the flexibility of alternate methods such as fluorescence, radioactivity, electrochemical, conductivity, or infrared detection. Some manufacturers offer preparative flow cells with short path lengths to decrease sensitivity.

UV and RI detectors are used most commonly in gram-scale LC. Of the two, RI detection is more popular, but its application is more difficult in gradient elution. UV detection should be preferable over RI because it is the most common detection system used in analytical work and would, therefore, give chromatograms comparable to those obtained in pilot analytical studies. Unfortunately, routine use of UV in preparative work has been limited because of its intrinsically high sensitivity. Analytical UV detectors are designed to detect microgram quantities of materials at 1–2 mL/min flow rates with minimal mixing (often 2–10 μL cell volumes). For preparative separations, modifications must be made to the cell to decrease sensitivity and to allow for increased flow rates. Stream splitters solve the flow rate limitations of analytical cells, but care must be taken to account for the time lag between collection and detector response when using a split flow. Splitting, however, will not solve the problem of high sample concentrations, which give absorptions outside the linear range of most UV detectors — typically 1–2 absorbance units (AU) for a 10-mm path length. This detector overload cannot be compensated for by electronic attenuation of the recorder or integrator signal. UV-detector overload is more common than is generally recognized in preparative LC and is often misinterpreted as column overload. Thus, a poorly defined preparative chromatogram with truncated or flattened peaks is not necessarily an indication of a poor separation.

Examination of Beer's law,

$$A = \varepsilon c d \qquad [3]$$

where

A = the UV absorption of a column effluent at a given wavelength

ε = the extinction coefficient

c = the concentration of product in the effluent

d = the cell path length

suggests two ways of avoiding detector overload. By changing the wavelength (*detuning*), ε can be substantially decreased. Contaminants with strong UV absorption at the new wavelength may, however, be disproportionally represented in the new chromatogram and give misleading results. An alternative approach is to shorten the path length as much as possible. Thus, for preparative work, flow cells with path lengths of 0.5 mm are available for some analytical UV detectors. Such a cell is 20 times less sensitive than a cell with a standard 10-mm path length. One variable-wavelength preparative detector, the Gow-Mac 80-850 unit (Bound Brook, New Jersey), has a cell with an 0.1-mm path length and is 100 times less sensitive than detectors with 10-mm cells. The cell has an open design and can handle preparative flows of up to 500 mL/min. It does not require a flow splitter but, by the nature of its design, is very sensitive to flow pulsations.

When a preparative detector is not available, the column effluent can be monitored at frequent intervals by analytical HPLC or thin-layer chromatography (TLC). The analysis time must be shorter than the peak width if optimal cuts are to be taken during the separation.

Sample Introduction

Injections in semipreparative work can be made by a sample loop, if solubility permits. If solubility mandates large injection volumes, the sample can be introduced through the pump providing it is dissolved in a weaker solvent than the mobile phase. Under such conditions, solutes will band at the inlet of the column (29,44). For gram-scale work, the large injection volumes frequently encountered — as much as several liters (45) — can be routinely injected through the pump.

APPLICATIONS TO RESEARCH AND DEVELOPMENT OF SYNTHETIC PRODUCTS

Preparative liquid chromatography is used routinely in the pharmaceutical industry to solve a variety of commonly encountered separation problems such as the purification of synthetic intermediates, final products, reference standards, trace and major impurities, radiolabeled compounds, and metabolites. Many biological tests require highly purified compounds for meaningful results. Another application of PLC is in the isolation and identification of minor components or impurities in pharmaceutical preparations that are required to conform to the more stringent FDA specifications resulting from the use of increasingly sophisticated analytical techniques. In addition, the isolation and characterization of by-products can aid in the mechanistic understanding of important chemical reaction steps in a synthetic process — information that can then be used to optimize the yield of a chemical sequence by minimizing reaction by-products. The use of PLC accelerates the development of pharmaceutical processes and allows for the purification of compounds not amenable to traditional separation techniques.

Following its original introduction, chromatography was occasionally used by chemists when required for the purification of synthetic products. Recent improvements in PLC have made it indispensable for the purification of molecules of importance to the pharmaceutical industry, including *cis/trans* and positional isomers, diastereomers, peptides, steroids, and unstable molecules such as leukotriene analogues. Many of these compounds have quite similar physical properties — such as boiling point and solubility — making separation by classical techniques such as crystallization, distillation, and extraction impossible.

Drug Discovery

An important application of preparative LC is the purification of synthetic analogs for biological evaluation. The leukotrienes are one class of compounds for which the use of PLC is indispensable. Naturally occurring leukotrienes C_4 (LTC$_4$) and D_4 (LTD$_4$) are thought to be the main mediators in human asthma (46). These molecules are unstable in the presence of heat, light, and air and cannot be purified by traditional methods. Highly purified natural and synthetic leukotrienes have

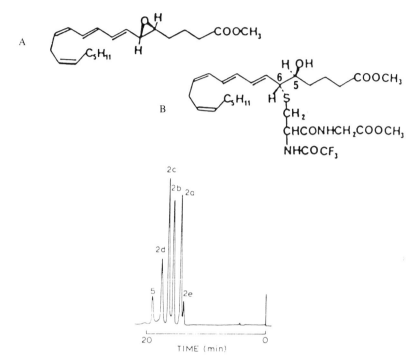

FIGURE 2. Structures (Figure 2a) of leukotriene A_4 methyl ester (A) and leukotriene D_4 synthetic intermediate (B); separation (Figure 2b) of isomers of leukotriene A methyl ester (2a–2e) and tetraene by-product (5). Column: Spherisorb S5W, 50 cm × 8 mm; mobile phase: hexane/diethyl ether/triethylamine (95:5:0.5); detection: UV 276 nm. (Reproduced, with permission, from Reference 49.)

been obtained by preparative HPLC (47). Because of the potency of these compounds in vitro (effective concentrations are 10^{-9} M), semipreparative LC can supply adequate quantities for biological testing.

Leukotriene A_4 (LTA$_4$) methylester is a key intermediate in the syntheses of LTC$_4$ and LTD$_4$. The double-bond stereochemistry of natural LTA$_4$ ester is shown in Figure 2a. One of the original synthetic routes to racemic LTA$_4$ ester produced an isomeric mixture (48) that could not be resolved on silica-gel TLC and proved to be a difficult separation on analytical HPLC ($\alpha = 1.06$–1.10) — as shown in Figure 2b (49). Racemic LTA$_4$ ester with the natural double-bond stereochemistry is designated as isomer 2b in Figure 2b. By using semipreparative HPLC, 3 mg of crude mixture was resolved on a Spherisorb S5W silica column (50 cm × 8 mm i.d.) (Phase Separations, Queensferry, Clwyd, United Kingdom). Four major LTA$_4$ double-bond isomers were purified (>99%) and characterized by UV, mass, and NMR spectrometry. Using repetitive injections, the process could be scaled up to yield 5–10 mg of each isomer for further synthetic modification.

Reaction of racemic LTA$_4$ ester (isomer 2b) with N-trifluoroacetylcysteinyl-glycine methyl ester produced a mixture of the natural 5S, 6R LTD$_4$ ester (Figure 2a) and its 5R, 6S diastereomer. In this case, the selection of the bonded-phase adsorbent was important for the successful purification of the diastereomeric mixture (50). No separation of isomers could be achieved on the Spherisorb S5W silica column using a dichloromethane/methanol mobile phase. By selecting a Spherisorb S5NH amino column, an α value of 1.2 was obtained. Using a semi-preparative column, pure LTD$_4$ diastereomers were obtained in milligram quantities for hydrolysis and biological testing.

When larger quantities of synthetic products are needed and low α values mandate the use of semipreparative systems, multiple injections are required. The practicality of automation to solve this problem was demonstrated for a four-component mixture of benzyl-D-glucosides, which was resolved in two steps on a Partisil column (Whatman) (25 cm $\times$ 9.4 mm i.d.) (34). The first step used 50-mg injections to resolve the mixture into two pairs of isomers. Each binary mixture was then resolved on the same column using a different mobile phase and 40 mg/injection. With this methodology, 3 g of purified isomers could be processed during a 24-h period.

Other analogous applications of semipreparative HPLC to drug discovery include the purification of anthracycline derivatives (51), steroids (52,53), aromatic retinoids (54), phosphonium salts (55), phospholipids (56), and bilirubin isomers (57).

Preparation of Synthetic Intermediates

If large enough α values can be achieved, MPLC or gram-scale preparative LC can be used for the production of large amounts of synthetic products or intermediates of sufficient purity for further transformations. The separation of an isomeric benzazepine mixture in the authors' laboratories illustrates a large-scale separation and demonstrates the importance of polar modifier concentration in maintaining adequate resolution on the preparative scale (58). Analytical HPLC using 0.1% diethylamine to minimize tailing of these basic compounds gave clean separation of the isomers with an α value of 1.5 (Figure 3). Preparative HPLC with 0.1% diethylamine in the mobile phase, however, produced severe tailing of compounds I and II (Figure 4), even though the preparative column had been deactivated by a 5% diethylamine wash before use. This tailing could be completely eliminated by increasing the diethylamine concentration in the mobile phase to 1%. In this manner, 10 g of isomeric mixture ($\alpha = 1.9$) was separated on two Prep PAK silica cartridges (Waters Associates), as is seen in Figure 5. Recovery of each isomer was in excess of 80% with greater than 98% purity. Because of the quantities available, the center mixed cuts were discarded rather than recycled. Failure to maintain high amine modifier concentration led to serious tailing and mixed fractions in an analogous literature separation (59).

In another example of the use of PLC in synthetic chemistry, the successful

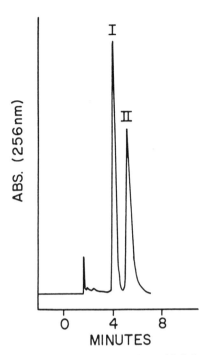

FIGURE 3. Analytical HPLC of isomeric benzazepine mixture labeled as compounds I and II. Column: Ultrasphere 5-μm silica, 25 cm $\times$ 4.6 mm; mobile phase: isopropanol/hexane/diethylamine (25:75:0.1); flow rate: 2 mL/min; detection: UV 265 nm.

scale-up of a reversed-phase, paired-ion HPLC method was used for the multigram separation of synthetic epimeric alkaloids (60). This work was made practical by the use of inexpensive camphor sulfonic acid as the ion-pairing agent. Injection of 10 g of mixture onto a μBondapak C$_{18}$ preparative column (Waters Associates) gave a baseline separation of epimers. The pure alkaloids were recovered by adjusting the pH of the eluent and subsequent extraction.

Preparative MPLC and HPLC have also been used extensively in the carbohydrate research for the purification of synthetic intermediates (10).

PLC can also be used to facilitate the preparation of compounds whose regio- or stereospecific synthesis is either impractical or impossible (61–64). In this approach, PLC is used to isolate a desired product from a complex mixture prepared by a nonstereospecific route. An example from the authors' laboratories is the isolation of benzazepine sulfate analogues (Figure 6a) from a crude synthetic reaction mixture (65). These compounds were originally isolated in small amounts as bioactive metabolites and were desired in larger amounts for further testing. The separate synthesis of each of the desired isomers would have been difficult and

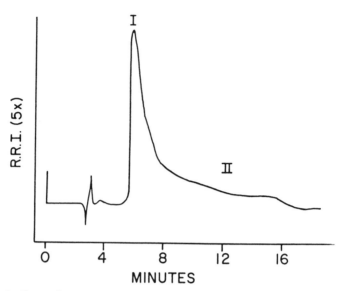

FIGURE 4.   Preparative separation of isomeric benzazepine mixture (compounds I and II, 5 g). Column: Waters PrepPAK silica; mobile phase: isopropanol/hexane/diethylamine (25:75:0.1); flow rate: 500 mL/min; detection: refractive index.

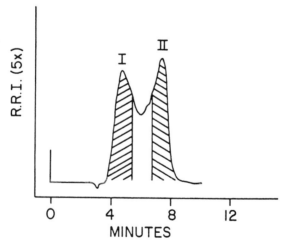

FIGURE 5.   Preparative separation of isomeric benzazepine mixture (compounds I and II, 10 g). Chromatographic conditions, same as in Figure 4 except mobile phase: isopropanol/hexane/diethylamine (25:75:1).

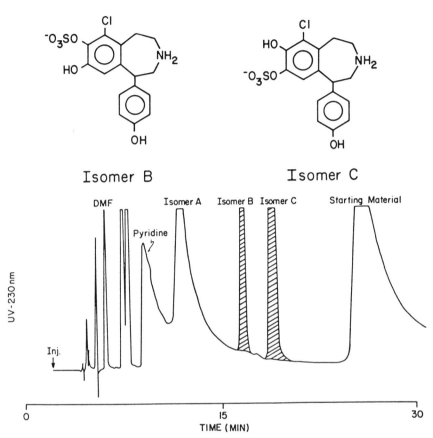

FIGURE 6.   Structures of benzazepine sulfate esters (Figure 6a) and analytical chromatogram of crude synthetic reaction mixture (Figure 6b). Column: Ultrasphere ODS, 5-μm, 25 cm × 4.6 mm; mobile phase: 15% methanol in pH 4, 0.05 M NH_4OAc; flow rate: 2 mL/min; detection: UV 230 nm.

time-consuming. Instead, the authors were able to carry out a nonstereoselective reaction and then separate the desired products in gram quantities.

Figure 6b shows the analytical chromatogram of the crude reaction mixture that contains the starting material, isomeric products, reagents, and solvents. This proved to be a difficult separation because the two products of interest were positional isomers, present at only 4% and 8% by weight and extremely insoluble. A two-step preparative system was developed to solve this problem. The first step, a reversed-phase system on a Whatman Partisil 40 ODS-3 column (Figure 7), isolated the desired isomers as a mixture that is shown in the analytical chromatogram of the first product (Figure 8). The second step, a normal-phase separation (Figure 9), separated the two isomers in greater than 99% purity using a PrepPAK silica column. The solvent system was chosen for optimum resolution after extensive

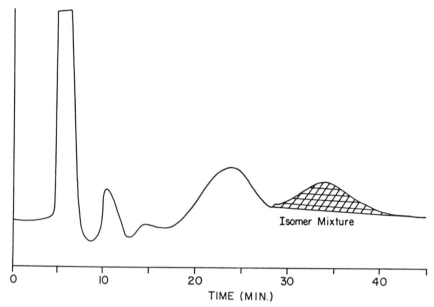

FIGURE 7.   Preparative separation of 1 g of crude synthetic reaction mixture. Column: Partisil 40 ODS-3 in a Whatman Magnum 40, 50 cm × 4.8 cm; mobile phase: methanol/water (20:80); flow rate: 200 mL/min; detection: refractive index. Shaded area is mixture of desired isomers.

TLC screening and was transferred directly to a preparative scale system. As a result of this effort, the products were available for testing in a significantly shorter time than would have been required by developing a complex synthesis.

Purification of Synthetic Peptides

Synthetic peptides are an emerging class of important therapeutic compounds. In the course of their synthesis, peptide products are contaminated with closely related by-products resulting from deletion and epimerization of amino acids and incomplete deblocking. Because of these similarities, α values between peptide products and contaminants are typically small. Preparative reversed-phase LC has become particularly applicable to the purification of underivatized synthetic peptides and has generally replaced ion-exchange and gel-filtration techniques because of its high resolution capabilities. Semipreparative LC has found extensive use on the laboratory scale for the purification of small amounts of synthetic peptides (66,72), and two monographs on HPLC of peptides have been published (73,74).

For larger-scale work, high resolution 15–20 μm packing is necessary because the 40–60 μm particle packings do not provide sufficient plate counts. For example, in the authors' work, a 60 cm × 5 cm i.d. steel column, slurry-packed with 15–20 μm reversed-phase packing, afforded more than 1000 plates with a low mo-

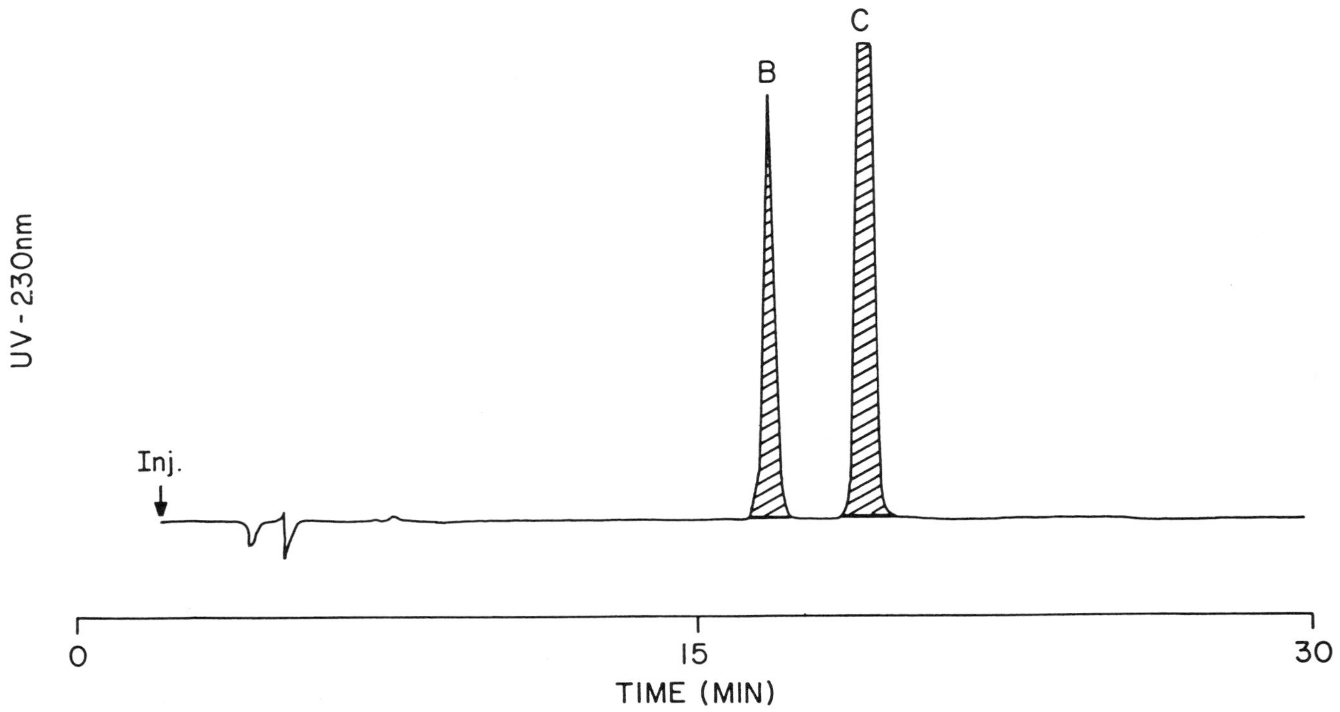

FIGURE 8. Analytical chromatogram of shaded pool from Figure 7. Chromatographic conditions same as in Figure 6.

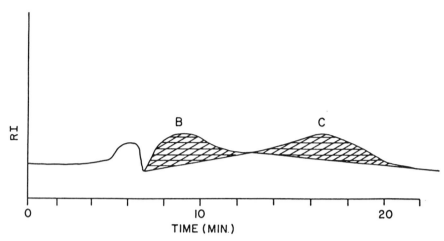

FIGURE 9.   Preparative separation of shaded pool from Figure 7. Column: Waters PrepPAK silica, 30 cm × 5 cm; mobile phase: acetonitrile/methanol/water/ammonia (1030:30:20:5); flow rate: 200 mL/min; detection: refractive index.

lecular weight test compound. Synthetic, biologically active peptides of 15–40 amino acid residues have been purified routinely in gram quantities by researchers at the Salk Institute (29). These researchers have developed a detailed strategy for the optimization of preparative HPLC conditions based on gradient-elution analytical HPLC. Wide-pore bonded-phase silica gels (Vydac 300 Å, 15–20 μm; The Separations Group, Hesperia, California) were found to be excellent chromatographic supports and could be successfully dry-packed into empty polyethylene cartridges for radial compression on the Waters Prep LC 500. Peak shapes and recoveries of peptides were found to be very good. Furthermore, supports were shown to be highly efficient and durable for multiple separations. Moreover, for difficult separations selectivity could be increased by using supports with other surface chemistries as well as by changing mobile phase composition. The wide-pore packing was necessary to obtain adequate permeability for the relatively high molecular weight peptides.

As an example of this separation strategy, a crude gonadotropin-releasing hormone analogue was purified on a Vydac C$_{18}$ column using a gradient of acetonitrile in aqueous triethylammonium phosphate. Figure 10 shows an analytical-gradient chromatogram of the initial deblocked synthetic peptide. This was scaled up to a 1.2-g run, as shown in Figure 11. Appropriate fractions (at a retention time of 50 min) were then collected and desalted in two separate steps on the same column using 0.1% TFA/CH$_3$CN. After lyophilization, 128 mg of 99.7% pure peptide was obtained with 70% recovery, as is seen in the analytical chromatogram in Figure 12. Gradient elution is particularly helpful in the purification of peptides because of the very steep response of k' to changing organic modifier composition indicative of this

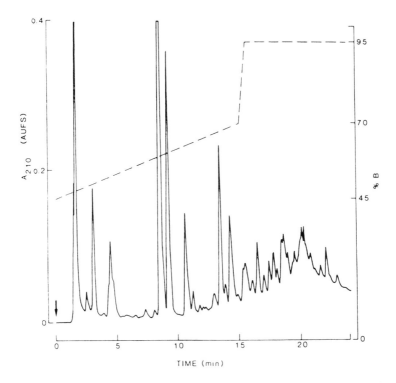

FIGURE 10.   Analytical separation of crude synthetic peptide mixture (gonadotropin-releasing hormone analogue). Column: Vydac C_{18} 5-μm, 25 cm × 4.6 mm; mobile phase: triethylammonium phosphate/acetonitrile gradient; flow rate: 2 mL/min; detection: UV 210 nm. (Reproduced, with permission, from Reference 29.)

class of large molecules. Other investigators have also described the use of radially compressed 15–20 μm columns for similar separations (29–31). In the authors' laboratories, synthetic peptides are routinely purified on a gram scale using analogous conditions (that is, wide-pore packing and similar solvent systems).

Process Development of Pharmaceuticals

Preparative LC is of great use in facilitating process development, as is exemplified in the authors' own work on the semisynthetic cephalosporin antibiotic cefonicid (75). Analytical HPLC analysis of a preparation of cefonicid run at high sensitivity indicated the presence of several minor impurities that needed to be isolated and characterized in order to meet FDA specifications and to support process improvement (Figure 13). Because these contaminants were present only in small amounts (less than 5% each), a large-scale purification was required in order to obtain sufficient quantities for characterization. The impurities were isolated in 200-mg quantities by peak shaving on a Waters Prep 500 C_{18} cartridge using water

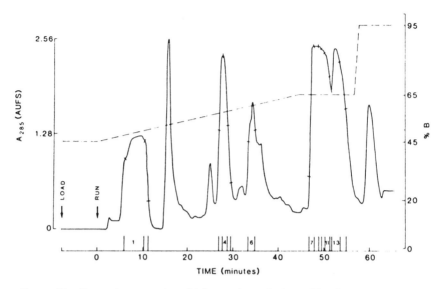

FIGURE 11. Preparative separation of 1.2 g crude synthetic peptide mixture (gonadotropin-releasing hormone analogue). Column: Vydac 17-μm C$_{18}$ packing in a Waters Prep-500 system. Mobile phase: triethylammonium phosphate/acetonitrile (%B in figure); flow rate: 75 mL/min; detection: UV 285 nm. (Reproduced, with permission, from Reference 29.)

as the mobile phase. Two of the isolated compounds were identified as contaminants in the starting materials and several others as reaction by-products. Levels of these impurities were then minimized by appropriate quality control of starting materials and by proper optimization of reaction conditions.

Preparative LC is also routinely used to prepare primary reference standards. These materials are required to establish and perform analytical procedures necessary for monitoring purity of synthetic intermediates and final products. An example is the case of the synthetic antimicrobial agent propylene phenoxetol (76). Standard manufacturing batches of this material were found to contain approximately 10% of an unknown compound. A series of peak shaving/recycling steps on the Waters Prep 500 was used to obtain, from 40 g of impure material, a quantity of 15 g of pure propylene phenoxetol of 99.9% purity for use as a primary standard. In the course of this work, the isomeric contaminant was also recovered and identified.

APPLICATIONS TO NATURAL PRODUCTS CHEMISTRY

In the pharmaceutical industry a variety of natural sources, usually plants or fermentation broths, are screened in the search for novel drug substances such as antibacterials, antifungals, immunoregulators, antineoplastics, and enzyme inhibitors. In many cases, screening has been done for decades, and thousands of such

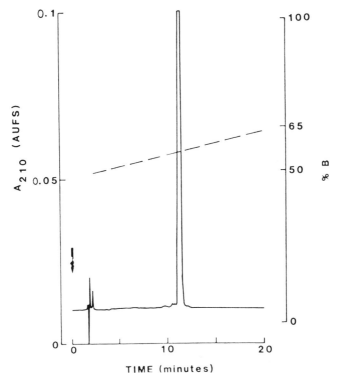

FIGURE 12.   Analytical chromatogram of final purified peptide. Column: Vydac 5-μm C$_{18}$, 25 cm × 4.6 mm; mobile phase: 0.1% TFA/acetonitrile; flow rate: 1.8 mL/min; detection: UV 210 nm. (Reproduced, with permission, from Reference 29.)

natural products have been reported in the literature. For a screening program to be effective, these known compounds must be efficiently isolated, identified, and eliminated at an early stage. When a new entity appears to be present, its facile purification for structure elucidation and biological evaluation is crucial to the success of the program.

Natural-product chemistry has always depended on chromatographic purification techniques to obtain pure compounds for structure elucidation and biological evaluation. Traditionally, natural products were isolated from plants or fermentation broths by extraction and purified by open-column chromatography on silica or alumina. As more-polar products were encountered, long and complex purification schemes using ion-exchange or gel-permeation techniques or columns packed with adsorbents such as charcoal, cellulose, and polyamide were frequently required. These complex, time-consuming separation schemes can now be streamlined and done more efficiently using PLC, particularly with reversed-phase supports. This is exemplified by increasing reports on the use of PLC in new

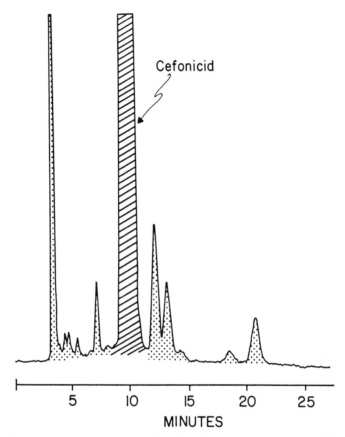

FIGURE 13.   Analytical chromatogram of cefonicid antibiotic preparation. Column: Waters C_{18}, 30 cm × 4 mm; mobile phase: 12% methanol in 0.05 M $NH_4H_2PO_4$; flow rate: 1.5 mL/min; detection: UV 254 nm. Large peak is antibiotic; small shaded peaks are contaminants.

product isolations published in recent years in the *Journal of Antibiotics* and the *Journal of Chromatography.* Because of this need for chromatography in the natural products area, there are more numerous examples of published detailed preparative chromatograms than exist in the synthetic area.

Semipreparative Isolations for Structure Identification

Thousands of natural products have been identified, so a key objective in a natural products screen is the early detection of known substances in a new crude extract. Often a natural product can be identified as a known substance on the basis of its UV, IR, and mass spectra determined on less than 1 mg of compound. Although large quantities are not required for these studies, the isolate must be pure enough to obtain reliable data. Frequently, a separation on a 250 mm × 4.6 mm, 10-μm analytical column can provide enough material for such a structural assignment (5,9). When novelty is suspected for a substance, larger quantities are required for structure elucidation (usually several milligrams) and biological evaluation (possibly several hundred milligrams). Semipreparative chromatography offers an efficient and convenient approach to separations on this scale.

An example of such a separation in the authors' laboratories is the purification of a complex mixture of closely related glycopeptide antibiotics shown in Figure 14 (26). Samples of the individual components, especially the small peaks, were needed for structural studies as well as biological evaluation in vitro and in vivo. These mixtures had previously been partially purified by chromatography on low-efficiency MPLC systems but could not be well resolved from each other because of their small α values (1.1–1.2). A high-efficiency 50 cm × 2.2 cm, 10-μm semi-preparative column (ODS-3; Whatman) was used to separate the components. Attempts to purify large amounts of mixtures (600 mg) in a single run were not successful, so repeated injections of smaller amounts (approximately 200 mg of mixture) were used. Because the minor components were only present at levels of

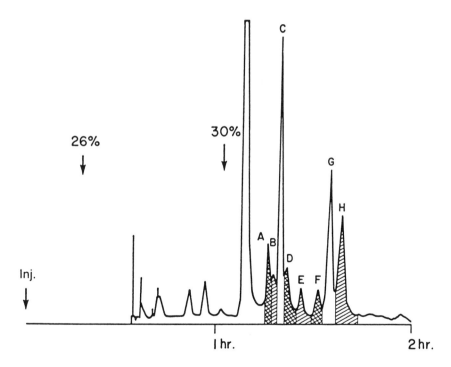

FIGURE 14.   Preparative chromatogram of 200 mg of glycopeptide antibiotic mixture. Desired peaks are labeled A to H. Column: Whatman Magnum 20, 10-μm, 50 cm × 2.3 cm; mobile phase: 28% to 32% acetonitrile in 0.1 M phosphate, pH 6.0; flow rate: 12 mL/min; detection: UV 290 nm.

10–50 mg/injection, their loading was low enough to maintain the high efficiency necessary for their successful fractionation. After many runs, 50–100 mg of each antibiotic was eventually prepared in >95% purity. These amounts were adequate for complete structural characterization by NMR techniques and for biological evaluation.

In other work in the natural products area, semipreparative chromatography was used to separate milligram quantities of four naturally occurring potato gly-coalkaloids (77). Previously reported methods using TLC and open-column chromatography were unable to produce products of sufficient purity for biochemical and toxicological studies. Two crude glycoalkaloid mixtures, obtained by extraction of potato blossoms, were resolved on a 25 cm × 9.4 mm Zorbax NH₂ column (Du Pont Company, Wilmington, Delaware). Figure 15 shows the resulting semipreparative chromatogram and conditions for the separation of α-chaconine and α-solanine and Figure 16 for that of another pair, demissine and commersonine. In both examples the pairs of compounds had identical aglycones but differed in their carbohydrate composition. Resolution was sufficient to obtain all four glycoalka-loids at high purity (99%). Solubility considerations limited loading to 4 mg (total glycoalkaloids)/injection for the α-chaconine and α-solanine mixture and to 0.75 mg for demissine and commersonine. By repeated injections, isolation of 30–60

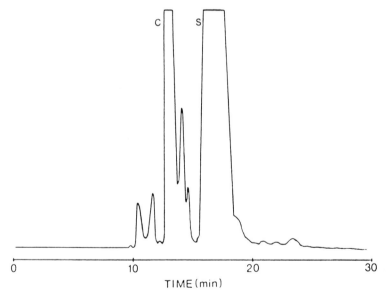

FIGURE 15.   Semipreparative separation of 4 mg of potato glycoalkaloid [α-chaconine (C) and α-solanine (S)] mixture. Column: Zorbax NH₂ 25 cm × 9.4 mm; mobile phase: tetrahydrofuran/water/acetonitrile (55:20:25); flow rate: 1 mL/min; detection: UV 215 nm. (Reproduced, with permission, from Reference 77.)

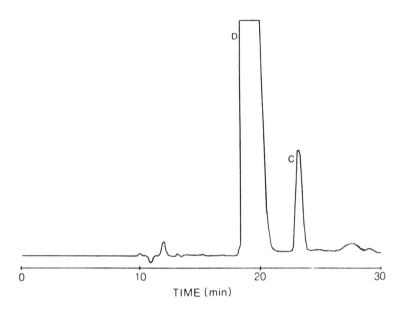

FIGURE 16. Semipreparative separation of 0.75 mg of potato glycoalkaloid [demissine (D) and commersonine (C)] mixture; chromatographic conditions same as in Figure 15. (Reproduced, with permission, from Reference 77.)

mg each of α-chaconine and α-solanine, 13 mg of demissine, and 3 mg of commersonine were possible in an 8-h period.

In other examples, four isomers of lysergic acid α-hydroxyethylamide and their decomposition products, obtained by microbial fermentation and extraction, were purified on a 50 cm × 8 mm, 10-μm MicroPak amino column (Varian Instruments, Palo Alto, California), as is shown in Figure 17 (78). Although details of loading were not given, sufficient quantities were obtained for spectroscopic characterization by mass spectrometry and NMR spectroscopy. In another case, a mixture of heptaene antibiotics was fractionated on 5–20 μm reversed-phase packing using a solvent containing DMSO to enhance solubility (79).

Other natural products purified by semipreparative LC include macrolide antibiotics (80), soybean isoflavones (81), the anthracycline antibiotic decilorubicin (82), (S)-α-benzylmalic acid (a carboxypeptidase inhibitor) (83), a new carbapenem antibiotic 6643-X (84), ferricrocin (85), plant growth regulators (86), the glycopeptide antibiotics teicoplanin (87) and OA-7653 (88), sesquiterpene lactones (89,90), N-acetylsporavidins (91), new antiplatelet agents (92), polypeptides (93), and steroidal plant growth regulators purified on a 3-μm column (94). In some of these cases, detailed chromatograms showing separations were not published, but the use of semipreparative conditions was indicated in the experimental sections. In all of these semipreparative examples, only milligram quantities of pure compounds were prepared.

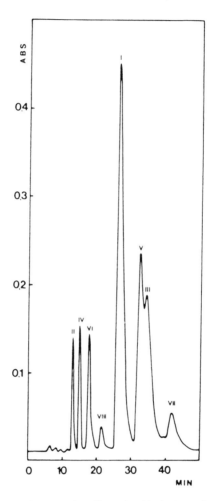

FIGURE 17.   Semipreparative separation of lysergic acid α-hydroxyethylamide isomers and decomposition products. Column: MicroPack NH$_2$, 10-μm, 50 cm × 8 mm; mobile phase: diethyl ether/ethanol (9:1); flow rate: 4 mL/min; detection: UV 310 nm. (Reproduced, with permission, from Reference 78.)

Intermediate-Scale MPLC Separations

When greater quantities of natural products are required for biological evaluation, MPLC systems are frequently used. MPLC is also helpful as an intermediate step to clean-up products for semi-preparative purification (95). An example of this use is the successful fractionation of 11 novel granulation-inhibiting principles from *Eucalyptus globulus* (96). A two-step process was used in which a rough separation was obtained on a Lobar RP-8 Column (E. Merck, Darmstadt, West Germany), yielding

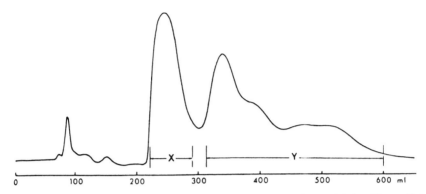

FIGURE 18.   Preparative separation of euglobals isolated from *Eucalyptus globulus* (50 mg). Column: Merck LoBar LiChroprep RP-8, 31 cm × 25 mm i.d.; mobile phase: acetonitrile; flow rate: 2.4 mL/min; detection: UV 254 nm. (Reproduced, with permission, from Reference 96.)

two major cuts labeled *X* and *Y* (Figure 18). This was followed by repeated runs at low loading (20–50 mg) on an automated, high-resolution, semipreparative system (Figure 19). By this procedure, a total of several hundred milligrams of each of the components was prepared for spectroscopic and biological evaluation. MPLC was also used to purify new β-lactam antibiotics with a 120 cm × 8 mm μBondapack C_{18}/Porasil B column (Waters) (97), a mixture of streptogramin antibiotics (silica column, 23 mm i.d. × 30 cm) (98), and a mixture of alkaloids (normal-phase, ion-pair chromatography (50 cm × 25 mm i.d.) (99).

A non–silica-based reversed-phase support, MCI gel CHP20P (HP-20; Mitsubishi Chemical Co., Tokyo, Japan) was used in glass MPLC columns to purify new monobactams SQ 28,332 (100), SQ 28,502, and SQ 28,503 (101) as well as the broad spectrum antibiotic EM5519 (102).

Gram-Scale Separations

In the natural products area, gram-scale chromatography is used for purifying sufficient quantities of materials for extensive biological testing or for preparation of semisynthetic analogues. A few papers have appeared describing details of the application of gram-scale preparative chromatography to the purification of natural products (103,114).

In the authors' work in the glycopeptide research, multiple-gram quantities of pure components were needed for extensive degradation, spectroscopic, and biological studies of these polar, nonextractable antibiotics (45). Initial separations by reversed-phase HPLC and semipreparative LC yielded pure components but only in limited quantities. Because the α values were moderate in this case — as shown by the analytical chromatogram in Figure 20 (Peaks I, II, and III with α values of 1.6 and 1.9) — the use of a large-particle system was warranted. In order to use the

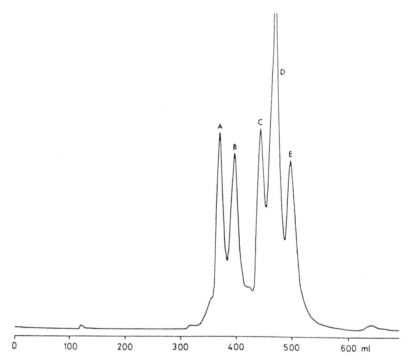

FIGURE 19.   Rechromatography of fraction X from Figure 18 (30 mg). Peaks A to E are desired products. Column: TSK-LS401KG, 50 cm × 21 mm; mobile phase: acetonitrile; flow rate: 15 mL/ min; detection: UV 254 nm. (Reproduced, with permission, from Reference 96.)

same packing as had been used in early semipreparative work, a Waters Prep-500 pumping system was modified for use with a Magnum 40 column (50 cm × 4.8 cm i.d.) dry-packed with Partisil ODS-3 (37–63 μm). In preliminary studies at low loading (2-g complex), effective baseline separations of the three major components could be obtained, as is illustrated in Figure 21. Using the Gow-Mac 80-850 UV detector, appropriate fractions were collected, as shown by the shaded areas in Figure 21. After desalting on XAD-7, the three components were obtained at 70% to 90% recovery and were >95% pure.

   With suitable parameters established, the separation of crude extract (25 g) was further scaled up to produce multiple-gram quantities of each component in >95% purity (Figure 22) for extensive in vivo, in vitro, and semisynthetic studies. The isolation and purification of these complex antibiotics was easily accomplished in two steps from fermentation broth, obviating an alternative multistep fractionation scheme that was impractical on this scale. When scaled up to 50 g, the extensive amount of front-eluting colored contaminants interfered with the preparative separation and resulted in overlap of peaks I and II (Figure 23). Rechromatography of the mixed cuts from this run on a fresh column gave efficient separation

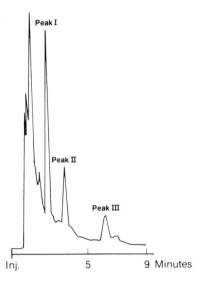

FIGURE 20.   Analytical separation of a crude glycopeptide complex. Column: Beckman Ultra-sphere ODS, 5-μm, 150 mm × 4.6 mm; mobile phase: 35% acetonitrile in 0.1 M KH$_2$PO$_4$, pH 3.2; flow rate: 1.5 mL/min; detection: UV 220 nm. (Reproduced, with permission, from Reference 45.)

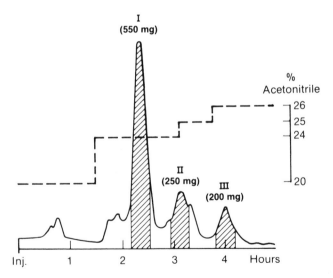

FIGURE 21.   Pilot-scale preparative separation of 2 g of glycopeptide complex. Column: Magnum 40 (50 cm × 4.5 cm), Whatman Partisil Prep 40 ODS-3 (37–60 μm); mobile phase: 20% to 26% aceto-nitrile in 0.1 M KH$_2$PO$_4$, pH 6.0; flow rate; 100 mL/min; detection: UV 210 nm. (Reproduced, with permission, from Reference 45.)

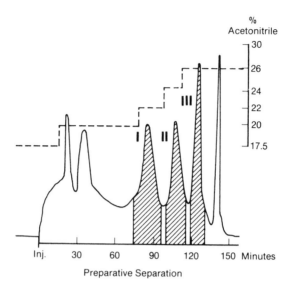

FIGURE 22.   Scale-up preparative separation of 25 g of glycopeptide complex. Column: same as in Figure 21; mobile phase: 17.5% to 26% acetonitrile in 0.1 M KH$_2$PO$_4$, pH 6.0; flow rate: 250 mL/min; detection: UV 210 nm. (Reproduced, with permission, from Reference 45.)

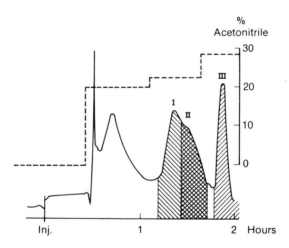

FIGURE 23.   Scale-up preparative separation of glycopeptide complex (50 g in 6 L buffer). Column: same as in Figure 21; mobile phase: 0% to 28% acetonitrile in 0.1 M KH$_2$PO$_4$, pH 6.0; flow rate: 250 mL/min; detection: UV 210 nm. (Reproduced, with permission, from Reference 45.)

(Figure 24), allowing for the preparation of 3–6 g of each component. It is notable that this column is being used for two purposes: The first step effects a crude clean-up to provide a window on fractions enriched in the desired products; the second step, run on a fresh column, gives the actual separation of the antibiotic factors. In separate loading studies with the pure antibiotic, approximately 50 plates could be achieved for an 8-g load (13 mg/g), which is probably the maximum amount that can be purified on a single run on this column. As is often a problem in the scale-up preparative area, insufficient quantities were available to test higher loadings than 8 g.

In other work, gram quantities of monensins A and B polyether antibiotics were obtained for extensive physicochemical and biological studies by chromatography on a Chromatospac Prep 100 system using a 34 cm × 8 cm i.d. column packed with 15-μm Separon C_{18} packing (111). This reversed-phase separation was superior to previously reported normal-phase column and TLC procedures and reduced purification times from hours to minutes as described in Figure 25. Loading was studied from 50 mg to 2 g and evaluated in terms of P values, an alternative measure of preparative resolution (111). Using this ratio as a guide during system development, the ideal loading level was determined to be 1 g of complex/run.

A Waters Prep 500 LC system in the recycle mode was used to purify mixtures of ansamitocin antibiotics containing analogues differing in their fatty acid content (112). Preparative separation of up to 5 g of mixture was carried out in good yield (90%) using a radially compressed cartridge dry-packed with Kieselgel 60 (40–60 μm) and ethyl acetate/hexane as the mobile phase. Because of the closeness of the peaks in the preparative run, extensive recycling and peak shaving were required to obtain products of >95% purity. Figure 26 shows a typical preparative run on 5

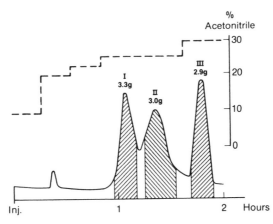

FIGURE 24. Rechromatography of pooled fractions from Figure 23. Column: same as in Figure 21; mobile phase: 10% to 28% acetonitrile in 0.1 M KH_2PO_4, pH 6.0; flow rate: 250 mL/min; detection: UV 210 nm. (Reproduced, with permission, from Reference 45.)

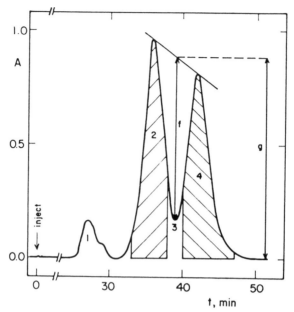

FIGURE 25.　Preparative separation of 1 g of monensins A and B. Column: 33.6 cm × 8 cm, 15-μm Separon C_{18}; mobile phase: methanol/water (88:12); flow rate: 80 mL/min; detection: UV 215 nm. Peaks 2 and 4 are monensins A and B, respectively. (Reproduced, with permission, from Reference 111.)

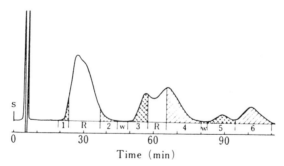

FIGURE 26.　Preparative separation of 5 g of ansamitocins P-3 and P-2 with two recycles. Column: Waters Prep-500 Cartridge packed with Merck Kieselgel 60 (40–63 μm); mobile phase: hexane/ethyl acetate (1:7); flow rate: 100 mL/min; detection: refractive index. Peaks 1, 3, and 5 are ansamitocin P-3, and peaks 2, 4, and 6 are ansamitocin P-2; R is recycled material; W is waste. (Reproduced, with permission, from Reference 112.)

g of a mixture carried through two recycles. Peaks 1, 3, and 5 are one component; 2, 4, and 6 are the other component. This separation was again developed to improve on previously reported procedures involving TLC and conventional open-column chromatography.

In a detailed study of gram-scale chromatography loading and operation, analytical-scale columns packed with preparative packing were used to optimize the purification of a crude mixture of polymyxin antibiotics before scale-up on a Waters Prep LC 500A system (114). Six separations of 1 g of crude mixture on a single PrepPAK C_{18} cartridge yielded 1.2 g of biologically active peptide having greater than 98% purity. This paper also describes a comparable purification of the peptide ACTH, which contains 39 amino acids, and compares the effects of loading on the α value for isocratic and gradient separations.

Other applications of preparative-scale LC were published without preparative chromatograms. The compound classes purified include naphthoquinones (103,110), anthraquinones (104), peptides (105), macrolides (106,107), carbapenems (108,109), and antitumor antibiotics (115,116).

FUTURE TRENDS

New Applications

As the purification of pharmaceutical products becomes more difficult (for example, with synthetic peptides and recombinant DNA products), the use of preparative HPLC will, no doubt, increase. Indeed, the sophisticated analytical methodology for monitoring the purification of these products, such as HPLC and gel electrophoresis, will mandate higher standards of purity for final products. Furthermore, because these newer biological products are usually very potent, dosage levels are often in the submilligram scale. This implies that production levels will be in the gram to kilogram scale as opposed to the multiple-ton production levels common for antibiotics and other high-dose therapeutics (117). Thus, it is possible to foresee the use of preparative LC techniques on the production scale for such high value/low dose products. Instruments for production-scale LC are currently available from Waters, Elf-Aquitaine, and Separations Technology, and their use in process scale can be expected to increase. One example from the authors' laboratories has been recently described (118).

Displacement Chromatography

Another future trend for the use of preparative LC is the area of displacement chromatography as developed by Horváth and co-workers (119–122). In this technique, the sample mixture is applied to the column at high loading in a noneluting solvent. The various components of the mixture displace each other down the column in order of their k' values (their affinity for the packing surface). After loading is complete, a new eluent, the *displacer,* is pumped on in order to elute all of the preadsorbed compounds sequentially. The resulting chromatogram appears as a bar graph (see Figure 27 for an example using polymixin antibiotics), with each major peak (B_1 and B_2) being a highly concentrated band of pure component (120).

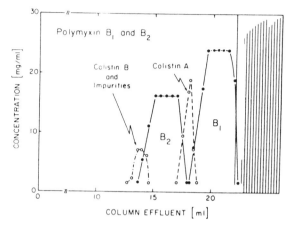

FIGURE 27.   Preparative displacement chromatography of 150 mg of polymyxin mixture. Column: LiChrosorb RP-8, 5-μm, 25 cm × 4.6 mm; displacer: 0.05 M octyldodecyldimethylammonium chloride/10% acetonitrile; flow rate: 0.1 mL/min; detection: HPLC analysis of 0.5 mL cuts. (Reproduced, with permission, from Reference 120.)

Using this procedure on an analytical 25 cm × 4.6 mm, 5-μm column, loading exceeded 100 mg. Clearly this procedure has the greatest potential for scale-up on process level LC where development time is available.

Other Developing Areas

At the present time, almost every column manufacturer sells a 5- or 10-μm prepacked preparative column for semipreparative work. Because loading of such columns is limited, multiple injections are needed to produce reasonable amounts of products for difficult separations. A key breakthrough will be in the wide application of inexpensive user-packed 15–20 μm supports in large-bore (2- to 4-in.) columns. With these columns, multigram quantities of mixtures with low α values could readily be fractionated at high efficiency by keeping loading low. Of course, the prices of these packings would also need to decrease to acceptable levels (less than $1000/kg). This price reduction will presumably occur in a competitive marketplace.

Other fields of potential growth of preparative high performance liquid chromatography, especially in the area of biologicals or peptides, are the affinity (123), sizing (124), chiral (125), and ion-exchange techniques that are finding increasing application in the high-performance mode. Increasing references in the biological literature are appearing on the application of wide-pore reversed-phase and ion-exchange packings to the purification of proteins. Future growth will no doubt be in these areas, along with the continued evolution of preparative LC as an important tool in pharmaceutical research.

Acknowledgment

The authors wish to acknowledge the excellent assistance of Betty Bullock in the preparation of this manuscript.

REFERENCES

(1) L. Snyder and J. Kirkland, *Introduction to Modern Liquid Chromatography*, Second Edition (John Wiley & Sons, New York, 1979), p. 615.

(2) P. Haywood and G. Munro, *Developments in Chromatography: 2*, C.F.H. Knapman, ed. (Applied Science Publishers, London, 1981), p. 33.

(3) J. DeStefano and J. Kirkland, *Anal. Chem.* **47**, 1103A (1975).

(4) J. DeStefano and J. Kirkland, *Anal. Chem.* **47**, 1193A (1975).

(5) K. Hupe, H. Lauer, and K. Zech, *Chromatographia* **13**, 413 (1980).

(6) M. Verzele and E. Geeraert, *J. Chrom. Sci.* **18**, 559 (1980).

(7) D. Nettleton, *J. Liq. Chromatogr.* **4** (Supp. 1), 141 (1981).

(8) D. Nettelton, *J. Liq. Chromatogr.* **4**, 359 (1981).

(9) M. Verzele, C. Dewaele, J. Van Dijck, and D. Van Haver, *J. Chromatogr.* **249**, 231 (1982).

(10) D. Bundle, T. Iverson, and S. Josephson, *Am. Lab.* **12**, 93 (1980).

(11) P. Gareil and R. Rosset, *Analysis* **10**, 397 (1982).

(12) P. Gareil and R. Rosset, *Analysis* **10**, 445 (1982).

(13) F. Eisenbeiss and H. Henke, *HRC&CC, J. High Resolut. Chromatogr. Chromatogr. Commun.* **2**, 733 (1979).

(14) A. De Jong, H. Poppe, and J. Kraak, *J. Chromatogr.* **209**, 432 (1981).

(15) A. De Jong, H. Poppe, and J. Kraak, *J. Chromatogr.* **148**, 127 (1978).

(16) A. De Jong, J. Kraak, H. Poppe, and R. Nooitgedacht, *J. Chromatogr.* **193**, 181 (1980).

(17) K. Hupe and H. Lauer, *J. Chromatogr.* **203**, 41 (1981).

(18) B. Coq, G. Cretier, and J. Rocca, *Anal. Chem.* **54**, 227 (1982).

(19) A. De Jong, J. Smit, H. Poppe, and J. Kraak, *Anal. Proc.* **17**, 508 (1980).

(20) R. Scott and P. Kucera, *J. Chromatogr.* **119**, 467 (1976).

(21) H. Poppe and J. Kraak, *J. Chromatogr.* **225**, 395 (1984).

(22) G. Cretier and J. Rocca, *Chromatographia* **18**, 623 (1984).

(23) R. Majors, J. Barth, and C. Lochmüller, *Anal. Chem.* **56**, 300R (1984).

(24) M. Verzele and C. Dewaele, *LC, Liq. Chromatogr. HPLC Mag.* **3**, 22 (1985).

(25) P. Rahn, M. Woodman, W. Beverung, and A. Heckendorf, "Preparative Liquid Chromatography and Its Relationship to Thin Layer Chromatography" (Waters Associates, Milford, Massachusetts, 1974).

(26) R. Sitrin, P. DePhillips, G. Chan, J. Dingerdissen, and K. Snader, paper presented at Eighth International Symposium on Column Liquid Chromatography, May 20, 1984, New York, New York.

(27) L. Snyder, *Anal. Chem.* **39**, 698 (1967).

(28) A. Wehrli, *Fresenius' Z. Anal. Chem.* **277**, 289 (1975).

(29) J. Rivier, R. McClintock, R. Galyean, and H. Anderson, *J. Chromatogr.* **288**, 303 (1984).

(30) J. Rivier, J. Spiess, M. Perrin, and W. Vale, *J. Chromatogr. Sci.* **10**, 223 (1979).

(31) C. Bishop, D. Harding, L. Meyer, W. Hancock, and M. Hearn, *J. Liq. Chromatogr.* **4**, 661 (1981).

(32) L. Snyder and J. Kirkland, *Introduction to Modern Liquid Chromatography*, Second Edition (John Wiley & Sons, New York, 1979).

(33) J. DiCesare and F. Vandemark, *Ind. Res. and Dev.* **24**, 138 (1982).

(34) A. Hadfield, R. Dreyer, and A. Sartorelli, *J. Chromatogr.* **257**, 1 (1983).

(35) F. Sugnaux and C. Djerassi, *J. Chromatogr.* **248**, 37 (1982).

(36) D. Berger and B. Gilliard, *J. Chromatogr.* **210**, 33 (1981).

(37) W. Still, M. Kahn, and A. Mitza, *J. Org. Chem.* **43**, 2923 (1978).

(38) T. Kuhler and G. Lindsten, *J. Org. Chem.* **48**, 3589 (1983).

(39) M. Zief, L. Crane, and J. Howath, *Am. Lab.* **14**, 144 (1982).

(40) M. Zief, L. Crane, and J. Howath, *J. Liq. Chromatogr.* **5**, 2271 (1982).

(41) L. Crane, M. Zief, and J. Howath, *Am. Lab.* **13**, 128 (1981).

(42) A. Meyers, R. Smith, and C. Whitten, *J. Org. Chem.* **44**, 2247 (1979).

(43) D. Pietrzyk and W. Cahill, *J. Liq. Chromatogr.* **5**, 781 (1982).

(44) K. Gazda and J. Kowalczyk, *HRC & CC, J. High Resolut. Chromatogr. Chromatogr. Commun.* **6**, 264 (1983).

(45) R. Sitrin, G. Chan, P. DePhillips, J. Dingerdissen, J. Valenta, and K. Snader, in D. LeRoith, J. Shiloach, and T. Leahy, eds., *Purification of Fermentation Products* (ACS Symposium Series No. 271, Washington, DC, 1985), p. 71.

(46) B. Samuelsson, *Trends Pharmacol. Sci.* **1**, 227 (1980).

(47) R. Lewis, K. Austen, J. Drazen, D. Clark, A. Marfat, and E. Corey, *Proc. Nat. Acad. Sci. U.S.A.* **77**, 3710 (1980).

(48) J. Gleason, D. Bryan and C. Kinzig, *Tetrahedron Lett.* 1129 (1980).

(49) S. McKay, D. Mallen, P. Shrubsall, J. Smith, S. Baker, W. Jamieson, W. Ross, S. Morgan, and D. Rackham, *J. Chromatogr.* **214**, 249 (1981).

(50) S. McKay, D. Mallen, P. Shrubsall, J. Smith, S. Baker, and R. Koenigsberger, *J. Chromatogr.* **219**, 325 (1981).

(51) P. Scourides, R. Brownlee, D. Phillips, and J. Reiss, *J. Chromatogr.* **288**, 127 (1984).

(52) R. Megges and H. Grobe, *J. Chromatogr.* **241**, 193 (1982).

(53) F. Gasparrini, S. Cacchi, L. Caglioti, D. Misiti, and M. Giovannoli, *J. Chromatogr.* **194**, 239 (1980).

(54) S. Mohanraj, *J. Liq. Chromatogr.* **7**, 1455 (1984).

(55) P. Jandik, U. Deschler, and H. Schmidbaur, *Fresenius' Z. Anal. Chem.* **305**, 347 (1981).

(56) W. van Kessel, M. Tieman, and R. Demel, *Lipids* **16**, 58 (1981).

(57) J. DiCesare and F. Vandemark, *Chromatogr. News.* **9**, 7 (1981).

(58) K. Erhard, unpublished results.

(59) D. Musso and N. Mehta, *J. Liq. Chromatogr.* **4**, 1417 (1981).

(60) P. Salva, G. Hite, and J. Henkel, *J. Liq. Chromatogr.* **5**, 305 (1982).

(61) L. Weidolf, K. Hoffman, S. Carleson, and K. Borg, *Acta Pharm. Suec.* **21**, 209 (1984).

(62) S. Matin and R. Zhou, *J. High Resolut. Chromatogr.* **7**, 629 (1984).

(63) W. Goodman and B. Adams, *J. Chromatogr.* **294**, 447 (1984).

(64) C. Bishop, T. Kitson, D. Harding, and W. Hancock, *J. Chromatogr.* **208**, 141 (1981).

(65) J. Filan, unpublished results.

(66) G. Tyler and M. Rosenblatt, *J. Chromatogr.* **266**, 313 (1983).

(67) D. Knighton, D. Harding, J. Napier, and W. Hancock, *J. Chromatogr.* **249**, 193 (1982).

(68) D. Pietrzyk, W. Cahill, and J. Stodola, *J. Liq. Chromatogr.* **5**, 443 (1982).

(69) M. Knight, Y. Ito, and T. Chase, *J. Chromatogr.* **212**, 356 (1981).

(70) M. Hearn, B. Grego, and C. Bishop, *J. Liq. Chromatogr.* **4**, 1725 (1981).

(71) C. Bishop, L. Meyer, D. Harding, and W. Hancock, *J. Liq. Chromatogr.* **4**, 661 (1981).

(72) L. Meyer, D. Harding, and W. Hancock, *J. Liq. Chromatogr.* **4**, 80 (1981).

(73) M. Hearn, "High Performance Liquid Chromatography of Peptides," in *High-Performance Liquid Chromatography: Advances and Perspectives,* Volume 3, Cs. Horváth, ed. (Academic Press, New York, 1983), p. 88.

(74) M. Hearn, F. Regnier, and T. Wehr, eds., *High Performance Liquid Chromatography of Proteins and Peptides* (Academic Press, New York, 1983).

(75) J. Filan et al., manuscript in preparation.

(76) P. Taylor and P. Braddock, *J. Liq. Chromatogr.* **5**, 2155 (1982).

(77) R. Bushway and R. Storch, *J. Liq. Chromatogr.* **5**, 731 (1982).

(78) M. Flieger, P. Sedmera, J. Vokoun, A. Ricicova, and Z. Rehacek, *J. Chromatogr.* **236**, 453 (1982).

(79) P. Gareil, G. Salinier, M. Caude, and R. Rosset, *J. Chromatogr.* **208**, 365 (1981).

(80) G. Werner and H. Hagenmaier, *J. Antibiot.* **37**, 110 (1984).

(81) E. Farmakalidis and P. Murphy, *J. Chromatogr.* **295**, 510 (1984).

(82) K. Ishii, S. Kondo, Y. Nishimura, M. Hamada, T. Takeuchi, and H. Umezawa, *J. Antibiot.* **36**, 451 (1983).

(83) T. Tanaka, H. Suda, H. Naganawa, M. Hamada, T. Takeuchi, R. Aoyage and H. Umezawa, *J. Antibiot.* **37**, 682 (1984).

(84) S. Tanabe, M. Okuchi, M. Nakayama, S. Kimura, A. Iwasaki, T. Mizoguchi, A. Murakami, H. Itoh, and T. Mori, *J. Antibiot.* **35**, 1237 (1982).

(85) H. Fiedler, *J. Chromatogr.* **209**, 103 (1981).

(86) R. Mitchell, T. Mawhinney, G. Cox, H. Garrett, and J. Hopfinger, *J. Chromatogr.* **284**, 494 (1984).

(87) A. Borghi, C. Coronelli, L. Faniuolo, G. Allieve, R. Pallanza, and G. Gallo, *J. Antibiot.* **37**, 615 (1984).

(88) T. Kamogashira, T. Nishida, and M. Sugawara, *Agric. Bio. Chem.* **47**, 499 (1983).

(89) F. Belliardo and G. Appendino, *J. Liq. Chromatogr.* **4**, 1601 (1981).

(90) F. Belliardo, T. Sacco, and G. Appendino, *J. Liq. Chromatogr.* **6**, 543 (1983).

(91) K. Harada, I. Kimura, E. Takami, and M. Suzuki, *J. Antibiot.* **37**, 976 (1984).

(92) K. Umehara, K. Yoshida, M. Okamoto, M. Iwami, H. Tanaka, M. Kohsaka, and H. Imanaka, *J. Antibiot.* **37**, 1153 (1984).

(93) B. Cavalleri, H. Pagani, G. Volpe, E. Selva, and F. Parenti, *J. Antibiot.* **37**, 309 (1984).

(94) T. Yokata, J. Baba, S. Koba, and N. Takahashi, *Agric. Biol. Chem.* **48**, 2529 (1984).

(95) R. Martin and H. Becker, *Fresenius' Z. Anal. Chem.* **318**, 247 (1984).

(96) T. Amano, R. Komiya, M. Hori, M. Goto, M. Kozuka, and T. Sawada, *J. Chromatogr.* **208**, 347 (1981).

(97) M. Nakayama, S. Kimura, T. Mizoguchi, S. Tanabe, A. Iwasaki, A. Murakami, M. Okuchi, H. Itoh, and T. Mori, *J. Antibiot.* **36**, 943 (1983).

(98) E. Grell, E. Lenitzki, S. Dehal, I. Oberbaumer, F. Raschdorf, and W. Richter, *J. Chromatogr.* **290**, 57 (1984).

(99) H. Huizing, F. DeBoer, and T. Malingre, *J. Chromatogr.* **214**, 257 (1981).

(100) P. Singh, J. Johnson, P. Ward, J. Wells, W. Trejo, and R. Sykes, *J. Antibiot.* **36**, 1245 (1983).

(101) R. Cooper, K. Bush, P. Principe, W. Trejo, J. Wells, and R. Sykes, *J. Antibiot.* **36**, 1252 (1983).

(102) E. Meyers, R. Cooper, W. Trejo, N. Georgopapadakou, and R. Sykes, *J. Antibiot.* **36**, 190 (1983).

(103) R. Peterson and M. Grove, *Appl. Environ. Microbiol.* **45**, 1937 (1983).

(104) M. Patel, A. Horan, V. Guilo, D. Loebenberg, J. Marquez, G. Miller, and J. Waitz, *J. Antibiot.* **37**, 413 (1984).

(105) M. Nakajima, A. Torikata, Y. Ichidawa, T. Katayama, A. Shiraishi, T. Haneishi, and M. Arai, *J. Antibiot.* **36**, 961 (1983).

(106) R.G. Takesako and T. Beppu, *J. Antibiot.* **37**, 1161 (1984).

(107) T. Smitka, R. Bunge, R. Bloem and J. French, *J. Antibiot.* **37**, 823 (1983).

(108) R. Evans, H. Ax, A. Jacoby, T. Williams, E. Jenkins, and J. Scannell, *J. Antibiot.* **36**, 213 (1983).

(109) J. Shoju, H. Hinoo, R. Sakazaki, N. Tsuji, K. Nagashima, K. Matsumoto, Y. Takahashi, S. Kozuki, T. Hattori, E. Konda, and K. Tanaka, *J. Antibiot.* **35**, 15 (1982).

(110) I. Vinezawa, H. Oka, K. Komiyama, K. Hagiwara, S. Tomisaka, and T. Miyano, *J. Antibiot.* **36**, 1144 (1983).

(111) M. Beran, J. Tax, V. Schon, Z. Vanek, and M. Podojil, *J. Chromatogr.* **268**, 315 (1983).

(112) M. Izawa, K. Haibara, and M. Asai, *Chem. Pharm. Bull.* **28**, 789 (1980).

(113) R. Dixon, R. Evans, and T. Crews, *J. Liq. Chromatogr.* **7**, 177 (1984).

(114) R. Burgoyne, D. Bowles, and A. Heckendorf, *Biotechnol. Lab.* **1** (3), 38 (1984).

(115) S. Stampwala, R. Bunge, T. Hurley, N. Willmer, A. Brankiewicz, C. Steinman, T. Smitka, and J. French, *J. Antibiot.* **36**, 1601 (1983).

(116) R. Bunge, T. Hurley, T. Smitka, N. Willmer, A. Bronkiewicz, C. Steinman, and J. French, *J. Antibiot.* **37**, 1566 (1984).

(117) J. Dwyer, *Biotechnology* **2**, 957 (1984).

(118) A. Cantwell, R. Calderone, and M. Sienko, *J. Chromatogr.* **316**, 133 (1984).

(119) Cs. Horváth, A. Nahum, and J. Frenz, *J. Chromatogr.* **218**, 365 (1981).

(120) H. Kalasz and Cs. Horváth, *J. Chromatogr.* **215**, 295 (1981).

(121) H. Kalasz and Cs. Horváth, *J. Chromatogr.* **239**, 423 (1982).

(122) Cs. Horváth, J. Frenz, and Z. Rassi, *J. Chromatogr.* **255**, 273 (1983).

(123) P. Larsson, M. Glad, L. Hansson, M. Mansson, S. Ohlson, and K. Mosbach, in *Advances in Chromatogr.* Volume 21 (Marcel Dekker, New York, 1983), p. 41.

(124) W. Yan, J. Kirkland, and D. Bly, *Modern Size Exclusion Liquid Chromatography* (John Wiley & Sons, New York, 1979).

(125) W. Pirkle, M. Hyun, and B. Bank, *J. Chromatogr.* **316**, 585 (1984).

A SURVEY OF THE USES OF LIQUID CHROMATOGRAPHY IN THE ISOLATION OF NATURAL PROTEINS

Stanley Stein[*]

Hoffmann–La Roche, Inc.
Nutley, NJ 07003

With the advent of recombinant DNA technology, proteins now can be produced on a large scale at a reasonable cost. Biologically active proteins, which previously could be isolated from natural sources only in minute quantities, are now reproducible enough to be used as pharmaceuticals. In order to take advantage of this technology, however, it is first necessary to have structural information about the protein. In many instances it has been possible to deduce the structure of the protein by recombinant DNA procedures rather than determine it through traditional protein purification and analysis procedures. It is absolutely essential, however, to confirm the structure of the physiologically active form of the protein, because posttranslational processing of the primary transcript is unpredictable.

There have been major advances in the methodology for analyzing proteins. High-sensitivity procedures now available require only 1 nmol (that is, 20 µg of a protein with a 20,000-dalton molecular weight) or less in order to obtain amino acid sequence data. From these data, it is possible to synthesize peptides and raise polyclonal or monoclonal antibodies as well as to synthesize oligodeoxynucleotide probes. These reagents may then be used to obtain the complete structure of the protein and to prepare the recombinant version of that protein, with considerable savings in time and cost over traditional approaches. The requirements for smaller amounts of the purified protein are well-suited to the use of high performance liquid chromatography (HPLC). The application of HPLC and other liquid chromatographic techniques to specific examples of proteins of biomedical interest will be presented in this chapter. Other considerations relevant to the successful isolation of biologically active proteins will also be discussed.

[*] Current affiliation: Schering Corp., Bloomfield, NJ 07110

PRELIMINARY CONSIDERATIONS

In order to follow the isolation of a biologically active protein, one must have a satisfactory assay for that protein's activity. The assay must be specific, should not give false positive or negative results with respect to extraneous factors, and should be reasonably quantitative so that percent recovery can be calculated from each purification step and the stability of the bioactive form can be followed. The resolving power achievable with modern HPLC columns requires that numerous samples (about 40) be assayed at one time. The sensitivity should be such that little of the sample is consumed for assay; sufficient material then will be available at the end of the purification procedure for chemical analysis. High sensitivity is also advantageous in that samples may simply be diluted into the assay buffer; possible interference from solvent or acid in the column eluent is eliminated by the large dilution factor. The final criterion is rapidity of the assay. A turnaround time of more than two to three days can lead to a purification project lasting several months or longer.

Another consideration is the stability of the biological activity with respect to factors such as pH, temperature, solvents, and salt concentration. If organic solvents destroy the biological activity, for example, then the use of reversed-phase chromatography may be precluded. The proper conditions of storage must be determined, and these conditions may vary depending on the stage of purification of the protein.

The source of the crude material from which the active protein is to be isolated may also be a factor in the progress of the project. In the isolation of opioid peptides from the adrenal gland, for example, the chromaffin granules were first prepared from the homogenized tissue by differential centrifugation (1). In this way, a high purification factor was obtained even before any protein separations were made. Many proteins are isolated from conditioned-medium; in the case of human β-interferon, fibroblasts grown in culture are induced by challenge with a virus. During this overnight induction period the cells are maintained under serum-free conditions (2). Thus, the minute quantities of β-interferon released into the medium are not contaminated by massive amounts of extraneous proteins found in bovine serum, which is typically used to support growth of cells in culture.

It should be kept in mind that each protein and each project have unique characteristics. Unlike human β-interferon, which exists as a single form, human leukocyte interferon comprises a family of structurally related proteins that can be resolved using HPLC into multiple protein peaks that have biological activity (3). By contrast, the multitude of atrial natriuretic peptides represent different intermediate stages in processing of a single precursor protein (4), so chromatographic and bioassay data should be interpreted carefully.

CHROMATOGRAPHIC SUPPORTS FOR PROTEIN SEPARATIONS

All of the modes of chromatographic separation are applicable to proteins. These include gel-permeation (also called gel-filtration), ion-exchange, adsorption, reversed-phase, and a variety of special affinity procedures. The advantages that HPLC has brought to the technique of liquid chromatography are also achievable for protein separations. The original HPLC supports were designed for low molecular weight substances. The development of wide-pore supports into which large proteins could freely diffuse has improved chromatographic properties (5,6). Silica, which is the matrix for most HPLC supports, must be properly end-capped when prepared for the reversed-phase mode and must be properly coated when prepared for use in the gel-permeation or ion-exchange mode (7). Polymer-based supports, such as the Mono-S and Mono-Q ion-exchangers from Pharmacia Fine Chemicals (Piscataway, New Jersey), are becoming more popular for protein chromatography.

Affinity chromatography can be especially useful in protein purifications. Typically, the ligand is covalently coupled to a polymer-based support such as agarose, but silica-based supports have also been used. In this mode, one takes advantage of a special affinity that the protein has for a ligand. Enzymes have specific high-affinity binding sites for substrates and cofactors. This approach was successfully applied to the resolution of isoenzymes on a silica-based support to which adenosine monophosphate had been coupled (8). Certain dyes, such as Cibacron blue, will bind to certain proteins with high affinity by unknown mechanisms. Kits composed of different dyes immobilized on chromatographic supports, such as the one from Amicon Corporation (Danvers, Massachusetts), are convenient for determining the usefulness of this approach in a particular protein isolation project. Immobilized antibodies can be used to obtain nearly homogeneous protein from a crude mixture in a single step. The application of some of these chromatographic techniques will be discussed later in this chapter.

DETECTION TECHNIQUES

Proteins in the column effluent are generally monitored on-line by ultraviolet (UV) absorption. The characteristic absorption maximum at 280 nm is attributable to the presence of aromatic amino acids, namely, tyrosine, phenylalanine, and tryptophan. Greater detection sensitivity may be achieved at shorter wavelengths (206–215 nm) in cases in which absorption is based on the peptide bond. Furthermore, the infrequent polypeptide that does not contain any aromatic amino acid residues could only be detected at the shorter wavelengths.

A different approach to on-line monitoring of proteins in the column effluent — based on a postcolumn reaction with the fluorometric reagent, fluorescamine — has been used in several laboratories (9). The advantages of fluorescamine are its high sensitivity, which can surpass that of short wavelength UV detection, and its specificity for primary amino groups (that is, the lysine side chains and the amino terminal), which allows the discrimination of polypeptides from extraneous UV-absorbing compounds. Detection with fluorescamine allows the use of eluents containing solvents and salts that are not UV-transparent. The instrumentation used for this approach is described in Reference 10. In the author's laboratory, columns are often monitored in series by both UV absorption and fluorescamine reaction.

EXTRACTION OF PROTEINS FROM CRUDE MIXTURES

In the preliminary stage of a protein isolation, one deals with large volumes of conditioned-medium or large amounts of homogenized tissue. Cellular debris is removed by centrifugation and/or filtration. A protein cut is typically made by precipitation with ammonium sulfate, ethanol, or trichloroacetic acid, then liquid chromatographic procedures are applied to achieve the purification. An alternative approach for the isolation of proteins from crude extracts and conditioned-medium was developed in the author's laboratory; it uses low-cost (200–400 mesh) silica and silica-based adsorbents for batch extraction, followed by either batch or column elution (Figure 1). This method is more rapid, more efficient, and simpler than the traditional procedures and has been applied to the various proteins noted in Reference 11; examples from this work are described below.

Human interleukin-2 (IL-2) was stored as a trichloroacetic acid precipitate prepared from serum-free conditioned-medium. The protein was dissolved in sodium bicarbonate buffer, pH 8.0, to give a 10-fold concentrate compared with the original medium. About 1–2 L of this concentrate was mixed with 25 mL of Lichroprep RP-8 (reversed-phase; E. Merck and Associates, Darmstadt, West Germany) support and stirred in a microcarrier vessel with a Teflon paddle for 2 h at room temperature. When the stirring was stopped, the beads settled rapidly and the supernatant was decanted. The protein-laden support was suspended in 100 mL of pyridine/acetate buffer, pH 4.0, and transferred into a 25-mL stainless steel column. Gradient elution with increasing n-propanol was begun (Figure 2), and the IL-2 eluted after most of the other proteins at about 30% propanol. Even though the adsorbed proteins were distributed throughout the 25 mL of column bed, it was possible to effect good chromatographic separation and obtain the IL-2 in a volume of 8 mL. Typically 10 μg of IL-2 was obtained per preparation, which represented a yield of close to 100% of the biological activity and a purification factor of about 100.

Adsorption chromatography using acid-activated silica was applied to the isolation of human immune interferon. Cells were removed from the conditioned-medium by centrifugation; removal of cell debris by filtration was not necessary and, in retrospect, removal of the cells may not even have been necessary. A 50-mL volume of acid-activated silica was mixed with 12.5 L of conditioned-

FIGURE 1. Batch adsorption of proteins from crude mixtures. Silica or a modified silica support (reversed-phase or ion-exchange) having an average pore size of 30 nm and particle diameter of about 40 μm is suspended in conditioned-medium obtained from cell culture. The support is maintained in suspension for several hours by gentle stirring with a Teflon paddle in a specially constructed glass microcarrier vessel (Bellco Glass, Vineland, New Jersey) on a magnetic mixer. When the stirring is halted, the support, now laden with proteins, sediments within 1 min. The solution of nonbound proteins, as well as cellular debris, is easily decanted. The support may be washed in this batch fashion with a weak eluent and is then eluted in a batch-wise manner with a small volume of a strong eluent or is packed into an open tube or closed column for gradient elution. The support is chosen to suit the chemical properties of the protein to be isolated. Available supports include acid-activated silica, reversed-phase, cation-exchange, and anion-exchange silicas. It is possible to couple virtually any type of ligand, including those with special affinities such as antibodies, lectins, or dyes, to the silica matrix, which provides the desirable mechanical properties for this procedure. Four 16-L bottles, each containing 8 L of solution, may be simultaneously handled on the magnetic stirrer depicted.

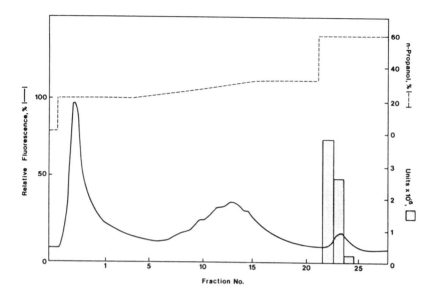

FIGURE 2.   Reversed-phase chromatography of human IL-2. Twenty-five milliliters of 40-μm Li-chroprep RP-8 was suspended in 1.5 L of 10-times concentrated, serum-free conditioned-medium from the JURKAT cell line, pH 8.0, at room temperature for 2 h. The support was collected by sedi-mentation, suspended in 100 mL of 0.9 M pyridine/acetate buffer, pH 4.0, and transferred to polypro-pylene tubes with constant swirling. After decanting, the support was then transferred to a 25-mL stainless steel column (6.6 cm × 2.2 cm) in several portions. After each addition, the support was packed down by applying suction to the outlet of the column with a syringe. The column was sealed and then eluted at 1.0 mL/min with an increasing gradient of n-propanol in the pyridine/acetate buffer. The volume of each fraction was 4.0 mL. (Reprinted, with permission, from Reference 11.)

medium overnight in the cold. The protein-laden adsorbent was collected and washed with phosphate-buffered saline. The rapid settling of the adsorbent al-lowed the cellular debris and nonbound proteins to be poured off easily. The inter-feron was batch-eluted with 100 mL of 1 M tetramethylammonium chloride. A re-covery of 86% of the original biological activity was achieved with a 100-fold reduction in volume and a 100-fold increase in specific activity. In an alternate pu-rification scheme, partially purified immune interferon was adsorbed to sulfopropyl-silica (cation-exchange; Diagnostic Specialties, Metuchen, New Jer-sey). The support was packed into an open column and eluted with a gradient of in-creasing concentration of tetramethylammonium chloride.

In the above applications the adsorbent was chosen to suit the properties of the protein; that is, a reversed-phase support was chosen for the hydrophobic protein IL-2. Immune interferon, however, is denatured by organic solvent, which pre-cluded the use of reversed-phase chromatography; still, it is a highly basic protein and binds strongly to silanol groups on silica and to the cation-exchange support. It is possible to couple various affinity ligands, ranging from dyes to antibodies, onto the silica matrix. These procedures are analogous to those using agarose-based ion-exchange or affinity supports, except that the silica has advantages in ease of handling of the support and has better chromatographic properties. Many variations, such as the Sep-Pak cartridges from Waters Associates (Milford, Massachusetts) or the Bond Elut columns from Analytichem International (Harbor City, California), are available with this approach.

The use of an agarose-based affinity support to purify human fibroblast interferon is depicted in Figure 3. Protein was secreted into the medium under serum-free conditions (2). After removal of the fibroblast cells by centrifugation, the conditioned-medium was passed through a column of Cibacron blue–agarose. The fibroblast interferon was eluted with 50% ethylene glycol in 1 M sodium chloride at pH 7. This highly enriched solution was ready for final purification of the interferon by HPLC, as described in the following section.

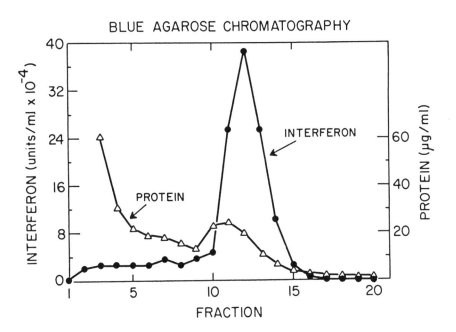

FIGURE 3.   Cibacron-blue chromatography of crude human fibroblast interferon. Fibroblasts were induced to produce interferon by overnight incubation in serum-free medium in the presence of an infectious virus. Sodium chloride was added to the conditioned-medium to a final concentration of 1 M and the solution was pumped through a 25-mL column of Blue Sepharose CL 6B (Pharmacia Fine Chemicals) at 2.5 mL/min. The column was washed with 250 mL of dilute phosphate buffer, pH 7.2, containing 1 M sodium chloride and 30% ethylene glycol (fractions 1–9). The interferon was eluted with the same solution containing 50% ethylene glycol. The volume of each fraction was 25 mL. (Reprinted, with permission, from Reference 2.)

FINAL PURIFICATION OF PROTEINS BY HPLC

After an initial purification step, it is next possible to use HPLC for the final purification. If the sample is still particularly impure, it is often desirable to do a size cut using gel-filtration chromatography; HPLC columns are most useful with less-than-milligram quantities of protein. It is usually possible to prepare a homogeneous protein in just a few HPLC runs. Affinity-purified fibroblast interferon was in fact purified to homogeneity by a single step of reversed-phase HPLC (Figure 4). The solution was pumped onto the column, which was then eluted with a gradient of increasing n-propanol concentration.

The procedure of Rinderknecht et al. for the purification of human immune interferon used controlled-pore glass (silica adsorption), concanavalin A-Sepharose (affinity for carbohydrate groups of glycoproteins), Sephacryl S-200 (gel-filtration), and reversed-phase HPLC (12). As discussed above, immune interferon is denatured in the eluents commonly used for reversed-phase HPLC. In this case, the rate of loss of biological activity was decreased by carrying out the chromatography at pH 7 with a dioxane gradient. Aliquots of the fractions were assayed immediately; the denaturation was halted by dilution into the assay buffer. To keep these preparations active, the fractions were freed of the dioxane by dialysis or gel filtration immediately after chromatography.

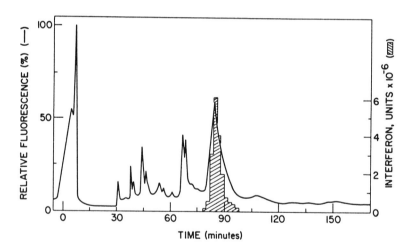

FIGURE 4.   HPLC of human fibroblast interferon. The fractions from the Blue Sepharose column containing interferon (see Figure 3) were pumped through a 25 cm × 0.46 cm Lichrosorb RP-8 column. The column was eluted at a flow rate of 22 mL/h with a gradient of increasing n-propanol in pyridine/acetate buffer, pH 4.2. A portion (3%) of the column effluent was stream split for automated monitoring of protein with fluorescamine. The interferon peak represents about 50 μg of protein. (Reprinted, with permission, from Reference 2.)

Controlled-pore glass, desalting and concentration by ultrafiltration, and cation-exchange HPLC on a Mono-S column were used in the purification procedure for human immune interferon reported by Friedlander et al. (13). The ultrafiltration step was required to make the sample suitable for application to the ion-exchange column. Chromatographic separation was achieved with a gradient of sodium chloride in 10 mM sodium phosphate buffer and 20% (v/v) ethylene glycol (Figure 5). The overall recoveries of biological activity reported for these last two

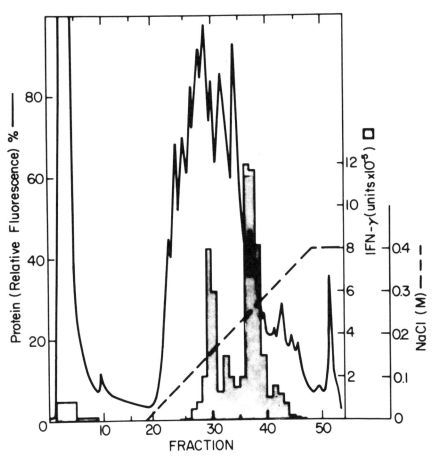

FIGURE 5.   HPLC of human immune interferon. A Mono-S cation-exchange column (50 mm × 5 mm) was equilibrated with 10 mM sodium phosphate buffer, pH 7.0, containing 20% ethylene glycol. Partially purified interferon was loaded onto the column in this buffer and eluted with a gradient of increasing concentration of sodium chloride at a flow rate of 0.5 mL/min. The volume of each fraction was 1.0 mL. Analysis of the major band of biological activity by gel electrophoresis indicated that fraction 38 was the 26,000-dalton subtype and fraction 39 was the 21,000-dalton subtype of immune interferon. The larger subtype has an additional carbohydrate moiety (12). (Reprinted, with permission, from Reference 13.)

procedures were 5% and 7%, respectively. The lymphocytes were induced in both cases under serum-free conditions.

The research of Sporn and colleagues about β-type transforming growth factor (TGF-β) illustrates many of the concepts described above. The ability to induce cells, which are normally anchorage-dependent, to grow in soft agar is the basis for the TGF-β assay. This ability, along with other morphological and biological responses that are characteristic of the transformed (cancer-like) phenotype, require the presence of both TGF-β and an α-type TGF, such as epidermal growth factor (14). Transforming growth factors have been indicated to play a physiological role in wound healing (15). TGF-β has been isolated from human placenta (16), bovine kidney (17), and human blood platelets (18) and is now known to be a 25,000-dalton, acid-stable protein that is composed of two identical 12,500-dalton subunits covalently attached by disulfide bonds. Similar purification schemes were used to isolate TGF-β from each of the three types of tissues.

Human placentas were frozen soon after delivery and extracted by homogenization in acid/ethanol solution that contained protease inhibitors. After centrifugation, the supernatant was adjusted with alkali to pH 3 and the protein was precipitated with ether and ethanol; the precipitate was collected and redissolved in 1 M acetic acid. Separation according to size was achieved on a Bio-Gel P-30 column (Figure 6), and individual fractions were assayed for biological activity. According to the elution positions of the marker proteins, the activity plot indicated a molecular weight of only a few thousand daltons (Figure 6). In retrospect, this assumption was incorrect and the spurious retardation in elution position can be explained by a retentive interaction of the TGF-β with the polyacrylamide-based support. The pattern of the activity plot in Figure 6 also had an unusual valley. It was determined that pool B actually contained the bulk of the transforming activity but gave a low response because of the presence of an inhibitory substance, which could be removed on a Bio-Gel P-6 column (Table I). It should be noted that the bed of the Bio-Gel P-30 column was on the order of 80 L. This traditional (and in this case laborious) approach of using gel-permeation chromatography in the first step gave a 20-fold increase in purity. More important, the high molecular weight proteins, which can often be troublesome in later chromatographic steps, were removed in this step.

Pools B and C were separately lyophilized and redissolved in low-salt buffers at pH 4.5 for chromatography on a CM-Trisacryl M cation-exchange column (Figure 7). Elution was accomplished with a gradient of increasing salt concentration. The fractions containing the biological activity were pooled, adjusted to pH 2.0 and to 10% acetonitrile, and pumped onto a C_{18} reversed-phase column (data not shown) and eluted with a gradient of increasing acetonitrile in aqueous 0.1% TFA. Fractions that corresponded to the single peak of biological activity obtained on this column were pooled and lyophilized. The sample was dissolved in 30% n-propanol/0.1% TFA and chromatographed on a cyanopropyl reversed-phase column using a gradient of increasing n-propanol (Figure 8). The sequential use of

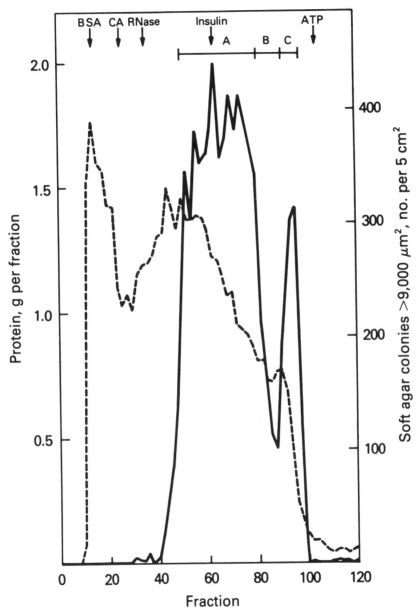

FIGURE 6.   Gel-permeation chromatography of the crude acid/ethanol extract of placenta. The residue (132 g) from an extract of 4.4 kg of placenta was applied in 2.6 L of 1 *M* acetic acid to a 90 cm × 36
cm column of Bio-Gel P-30. The column was eluted at a flow rate of 1.6 L/h with 1 *M* acetic acid; fractions of 800 mL were collected. The markers were: BSA (68,000 daltons), CA (25,000 daltons),
RNAse (13,800 daltons), insulin (6000 daltons), and ATP (551 daltons). The fractions containing biological activity (A, B, and C) were pooled as indicated and carried through the purification separately.
(Reprinted, with permission, from Reference 16.)

TABLE I

Purification of TGF-β from Human Placenta*

Step	Procedure	Protein recovered (mg)	ED$_{50}$ (ng/mL)[†]	Specific activity (units/μg)[‡]	Total activity (units × 10³)	Degree of purification (fold)	Recovery of activity (%)
1	Crude extract	239,000	7600	0.09	21,510	1.0	100
2	Bio-Gel P-30						
	Pool A	73,900	15,000	0.05	3695	0.6	17
	Pool B	27,720	—	—	—	—	—
	Pool C	1900	360	2.0	3800	22	18 (100)[∥]
3	Bio-Gel P-6						
	Pool B	8700	410	1.7	14,790	19	69
4	Ion-exchange						
	Pool B[§]	140	62	11.5	1610	128	31
	Pool C	46.3	85	8.4	390	93	1.8 (10)
5	HPLC-C$_{18}$						
	Pool B	0.27	0.10	7142	1928	79,000	37
	Pool C	0.26	1.2	595	155	6610	0.7 (4.1)
6	HPLC-CN						
	Pool B	0.025	0.072	9920	248	110,000	4.8
	Pool C	0.022	0.064	11,160	245	124,000	1.1 (6.4)

* Data taken from Reference 16.

† Defined as the concentration (ng/mL) of TGF-β required to give a response of 1 unit in the presence of EGF (1 unit of activity gives 50% maximal response = 1000 colonies >3000 μm² per plate).

‡ Defined as [1 unit/ED$_{50}$ (total mL per Petri dish)] × 1000.

§ Twenty-four percent of pool B from step 3 was used for further purification.

∥ Numbers in parentheses are recoveries for pool C.

reversed-phase chromatography with different columns and/or eluents often leads to the successful purification of a protein because this technique takes advantage of minor changes in selectivity in the different reversed-phase systems.

The peak of protein that corresponded to the biological activity from the last reversed-phase step was homogeneous, as is indicated by a single band on polyacrylamide gel electrophoresis (Figure 8). When an unstained track of the gel was cut into slices that were then extracted, biological activity was found to correspond to the position of the stained band. The TGF-β was purified more than 100,000-fold with a recovery of about 6% of the activity present in the crude extract (Table I).

The purification of TGF-β from bovine kidney followed the same scheme and resulted in about the same yield with a similar degree of purification. In the case of platelets, however, TGF-β was found to be enriched about 100-fold over the concentration found in placenta and kidney. Consequently, a less rigorous purification procedure was required: the same extraction and protein precipitation procedures were applied to the platelets, but on a smaller scale. Gel-permeation chromatog-

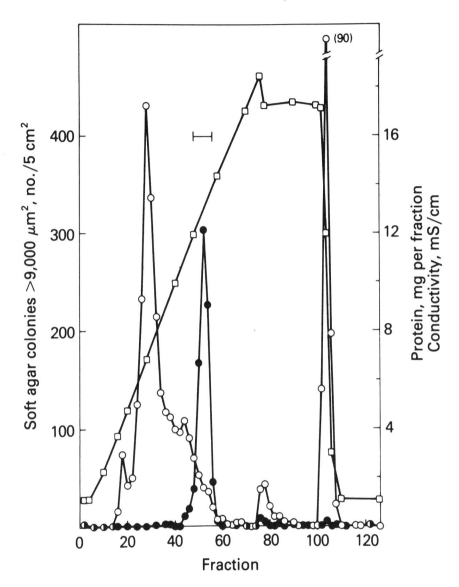

FIGURE 7.   Cation-exchange chromatography of placental TGF-β. Pool C (see Figure 6) was lyoph-ilized and dissolved in 60 mL of 0.01 M acetic acid and adjusted to pH 4.5. The solution was applied to a 10 cm $\times$ 5 cm column of CM-Trisacryl M, preequilibrated with 0.05 M sodium acetate, pH 4.5. The column was eluted at a flow rate of 145 mL/h with a gradient of increasing concentration of sodium chloride at pH 4.5. The volume of each fraction was 29 mL. (Reprinted, with permission, from Reference 16.)

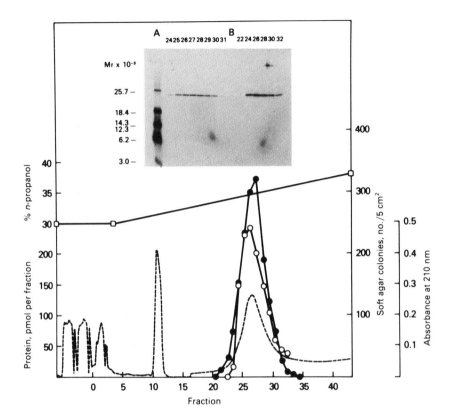

FIGURE 8.   Reversed-phase HPLC of placental TGF-β. The fractions from the cation-exchange column (see Figure 7) were further purified on a reversed-phase (C₁₈) HPLC column (not shown). The
fractions containing the biological activity were pooled, lyophilized, and redissolved in 30%
n-propanol/0.1% TFA and applied to a 30 cm × 0.38 cm μBondapak CN column (Waters Associates)
equilibrated with the same solvent mixture. The column was eluted at a flow rate of 1.1 mL/min with a
gradient of increasing n-propanol in 0.1% TFA. The volume of each fraction was 2.2 mL. Polyacrylamide gel electrophoresis indicated the homogeneity of the protein in the fractions corresponding to the
biological activity. (Reprinted, with permission, from Reference 16.)

raphy was done with a column of Bio-Gel P-60 about one-fiftieth the size used in
the previous procedure. As expected, the TGF-β eluted spuriously with polypeptides at around 8000–10,000 daltons. Upon rechromatography of the active fractions on another Bio-Gel P-60 column — this time in the presence of 8 M urea —
the TGF-β eluted at a position that corresponded to its molecular weight of 25,000
daltons and was well separated from the lower molecular weight polypeptides.
Pure TGF-β was thus obtained by using the Bio-Gel column in two modes, the first
based on an unusual affinity with the support and the second being the usual
permeation mode.

CONCLUSIONS

There are many possible approaches to take when designing a purification scheme for a biologically active protein. Naturally, the chromatographic procedures chosen are often based on the previous experience of the investigator and on the available equipment. By describing a variety of options for protein purification, this chapter is intended to be thought-provoking. The eventual success of a project will depend on careful consideration of the properties of the protein and on the establishment of a sound bioassay for activity in the purified protein.

REFERENCES

(1) R.V. Lewis, A.S. Stern, J. Rossier, S. Stein, and S. Udenfriend, *Biochem. Biophys. Res. Commun.* **89**, 822–829 (1979).

(2) S. Stein et al., *Proc. Natl. Acad. Sci. USA* **77**, 5716–5719 (1980).

(3) M. Rubinstein et al., *Proc. Natl. Acad. Sci. USA* **76**, 640–644 (1979).

(4) M.G. Currie et al., *Science* **223**, 67–69 (1984).

(5) R.V. Lewis, A. Fallon, S. Stein, K.D. Gibson, and S. Udenfriend, *Anal. Biochem.* **104**, 153–159 (1980).

(6) S.H. Chang, K.M. Gooding, and F.E. Regnier, *J. Chromatogr.* **125**, 103–114 (1976).

(7) S.H. Chang, R. Noel, and F.E. Regnier, *Anal. Chem.* **48**, 1839–1845 (1976).

(8) S. Ohlson, L. Hansson, P. Larsson, and K. Mosbach, *FEBS Lett.* **93**, 5–10 (1978).

(9) P. Bohlen, S. Stein, J. Stone, and S. Udenfriend, *Anal. Biochem.* **67**, 438–445 (1975).

(10) S. Stein and J. Moschera, in *Methods in Enzymology,* S. Pestka, ed. (Academic Press, New York, 1981), Volume 79, pp. 7–15.

(11) R.A. Wolfe, J. Casey, P.C. Familletti, and S. Stein, *J. Chromatogr.* **296**, 277–284 (1984).

(12) E. Rinderknecht, B.H. O'Connor, and H. Rodriguez, *J. Biol. Chem.* **259**, 6790–6797 (1984).

(13) J. Friedlander, D.G. Fischer, and M. Rubinstein, *Anal. Biochem.* **136**, 115–119 (1984).

(14) A.B. Roberts, C.A. Frolik, M.A. Anzano, and M.B. Sporn, *Fed. Proc.* **42**, 2621–2626 (1983).

(15) M.B. Sporn et al., *Science* **219**, 1329–1331 (1983).

(16) C.A. Frolik et al., *Proc. Natl. Acad. Sci. USA* **80**, 3676–3680 (1983).

(17) A.B. Roberts et al., *Biochem.* **22**, 5692–5698 (1983).

(18) R.K. Assoian et al., *J. Biol. Chem.* **258**, 7155–7160 (1983).

PART THREE

APPLICATION OF AUTOMATED COUPLED COLUMN TECHNIQUES TO THE PHARMACOLOGICAL TESTING OF DRUG SUBSTANCES

AN AUTOMATED LC APPROACH TO THE DETERMINATION OF THE PHARMACOLOGICAL FATE OF DRUG SUBSTANCES

Willy Roth

Pharmacokinetics and Metabolism Group
Dr. Karl Thomae GmbH
Biberach/Riss, FRG

It is a common principle in research to relate any type of effect to variables that are readily accessible experimentally and that also provide an easy basis for the interpretation of scientific data. This principle can be directly applied to the field of drug monitoring in both therapy and drug development.

In 1953, Dost initiated the field of *pharmacokinetics,* a discipline of biopharmaceutic drug research that exists today at the interface of biochemistry, toxicology, pharmacology, and medicine (1). Pharmacokinetics describes the time course of drugs in the body. It is generally accepted that systemic effects of drug molecules can only be observed if the active drug is distributed within the body. This, in turn, implies transportation and distribution via the blood stream (2). Thus, pharmacological activity is closely related to the presence of the active form of the drug within the body; the definite time course of a drug begins with administration and distribution and terminates with biochemical deactivation and/or physiological elimination.

Determination of the pharmacokinetic parameters of a drug substance is of interest not only to pharmacologists but also to clinicians, toxicologists, and others involved in the formulation and development of pharmaceuticals. They are key parameters for the distinct characterization and comparison of drug formulations as well as for the interpretation of the biological action of molecules. Moreover, kinetic parameters of a drug substance and its metabolites, in addition to pharmacodynamic data, provide an important basis for the description and understanding of species differences in drug response.

323

BACKGROUND

The measurement of biologically active compounds in body fluids is an important prerequisite in the field of therapeutic drug monitoring (3) and also comes into play during industrial drug development. A new drug entity is no longer solely characterized by its effect, molecular structure, and physicochemical properties: an increasing part of a substance's pharmacological profile is its time course in the body, which is characterized by its adsorption, distribution, metabolism, and excretion. Included in this profile are the time courses of the substance's metabolites, which often have different pharmacological effects. The growing importance of a drug substance's physiological time course has led to an increased pressure and need for analytical methods with the capability to quantify drug concentrations to nanogram per milliliter levels in the major biological fluids (4) — plasma, urine, saliva, bile, blister fluids, cerebrospinal fluids, and milk.

The main methodological requirements for clinicians as well as for others working in drug development laboratories, clinical pharmacological institutes, and drug metabolism units include:
- uncomplicated handling
- selective separation and sensitive quantitation
- simultaneous detection of drugs with different polarities
- high reproducibility
- uniqueness of detection to cover most types of pharmaceutical compounds
- high sample throughput
- labor saving
- integrated data handling
- consistency and full automation.

It is in this light that the problems involved in, and a methodology for, setting up an automated liquid chromatography system for the determination of drug substances are discussed in this chapter. This chapter will also review the literature generated between 1970 and 1984.

THE NATURE OF BIOLOGICAL SAMPLES

Plasma and urine are the biological fluids most often assayed in quantitative drug analysis. Depending on study design and the characteristic properties of a drug, however, assays from bile, saliva, cerebrospinal fluid, blister fluid, or even tissue homogenates are required. All these biological samples have a common characteristic: a very low amount of drug is carried among abundant endogenous constituents, and trace amounts of one or more components are embedded in this complicated biological matrix. In urine the matrix is composed mainly of salts, aromatic acids, and acidic components from endogenous catecholamine metabolism as well as acidic conjugates (5,6,7). In plasma it consists of a bulk of basic, large molecular weight proteins such as albumins, globulins, and minor amounts of acidic α-glycoproteins that have extended properties for the reversible binding of drugs. In addition, plasma contains lipids, fats, hormones, and a whole endogenous pool of small molecular weight compounds (8).

SAMPLE PRETREATMENT AND CLEANUP TECHNIQUES

The traditional approaches to the separation of lipophilic drugs from their aqueous biological surroundings are wet chemical methods. These methods have been applied in automated continuous flow techniques or autoanalyzer methods that use the liquid-liquid extraction of compounds that are then directly assayed or assayed after admixing of appropriate derivatization reagents. Liquid-liquid extraction was the most widespread methodology used for determination of organic compounds, particularly drugs and metabolites in plasma (10,11,12). With the invention of XAD absorber resins, however, solid-phase extractions began to become important; XAD ion-exchange resins have a macroreticular structure composed of styrene-divenylbenzene copolymers with high porosity and surface area. Fujimoto and Wang (13) were among the first to use glass columns packed with Amberlite XAD as a solid preconcentration tool for the enrichment of barbiturate and synthetic morphine derivatives from plasma and urine before gas chromatography (GC). Use of XAD resins for drug isolation is now a common and powerful off-line technique.

With the rapid improvement of HPLC hardware — powerful, constant-flow pumps; a wide variety of ion-exchange and reversed-phase column packing materials; various detection systems; and highly sophisticated low-cost microprocessors — new and flexible methods for drug analysis have been created during the last decade. The aim of all of these improvements was a higher degree of automation in order to be able to manage an increasing number of samples within reasonable time and at reasonable cost.

The classical steps in sample pretreatment (Figure 1) were carried out manually and more or less all in off-line steps. They were susceptible to the introduction of a large amount of error because of the many step-wise processes, including pipetting, adjusting the pH with buffer, and so forth. In addition, there was an increased risk of loss of compound, especially when working in the nanogram range, by adherence of the compound to the glass walls of the various containers and the filter paper and during evaporation. Sensitivity could also be reduced by inefficient extraction efficiency. In addition, these procedures were time-consuming and often needed the full and concentrated attention of a well-trained analyst during a whole day.

Within the last few years, the trend to use of solid-phase extraction became stronger because of its greater convenience and also because of the possibility of using solid columns packed with phases of different selectivity in automated chromatographic systems. Thus, on-line extraction in connection with column-

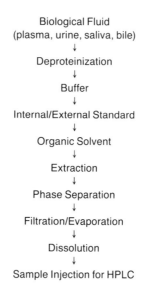

Biological Fluid
(plasma, urine, saliva, bile)
↓
Deproteinization
↓
Buffer
↓
Internal/External Standard
↓
Organic Solvent
↓
Extraction
↓
Phase Separation
↓
Filtration/Evaporation
↓
Dissolution
↓
Sample Injection for HPLC

FIGURE 1.   Classical steps in the pretreatment of body fluids prior to HPLC analysis.

switching techniques became possible (17). The need for such systems stemmed above all from the fact that the precision and accuracy of the analytical assay is, in essence, as dependent on sample pretreatment as on chromatographic conditions.

One of the first automated systems was described by Williams and colleagues, who developed an automated extractor/concentrator for the concentration of anticonvulsant drugs (18). The system, called PREP 1 (Du Pont, Wilmington, Delaware), is a microprocessor-controlled, centrifugally based extractor/ concentrator using extraction cartridges consisting of a resin column, an effluent cup, and a sample-recovery unit. The resin may be any sorbent suitable for pre-concentrating the target drug. The system showed a high extraction efficiency (90%–100% for phenobarbital, phenytoin, carbamazepine, primidone, clonaze-pam, and others) from serum. Applications are also available for the preconcen-tration of tricyclic antidepressants such as amitriptyline, doxepin, imipramine, nortriptyline, and desimipramine (19) as well as for analgesics and barbiturates (20). The system facilitates the enrichment of drugs from body fluids with high ex-traction efficiencies, although in its present stage of development this sample pre-treatment method can only be run off-line from the HPLC. The appropriate use of robotics, however, may facilitate the on-line automation of the extractor.

CONTINUOUS FLOW ANALYSIS/LIQUID CHROMATOGRAPHY

An elegant approach that combines the advantage of automated solvent mixing and separation in extraction coils was the connection of continuous flow analysis

(CFA) with the selectivity and specificity of liquid chromatography (CFA/LC). This approach was first described for the anticoccidiosis agent roberidine (21). A modified version, which became available as the Fast-LC system (Fully Automated Sample Treatment with Liquid Chromatography; Technicon Industrial Systems, Tarrytown, New York) (22,23) was initially described by Dolan and colleagues (24) for the separation and quantification of theophylline and anticonvulsants such as primidone, phenobarbital, phenytoin, carbamazepine, phenyl ethyl malonamide, and metabolites from serum. Figure 2 is a flow diagram for the system.

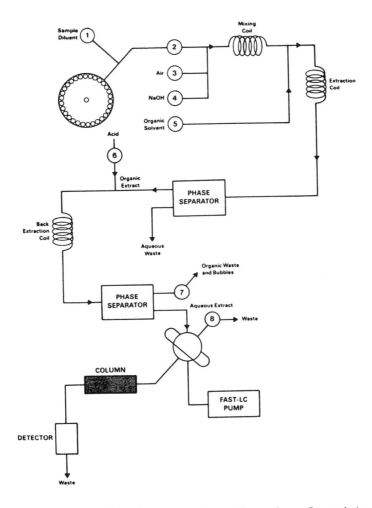

FIGURE 2. Flow diagram of sample pretreatment by applying continuous flow analysis prior to LC separation and detection using Technicon Fast-LC system. (Reprinted, with permission, from Reference 25.)

With a comparable system, Bannister and co-workers separated and quantitated the tricyclic antidepressants amitriptyline, nortriptyline, imipramine, desimipramine, doxepin, and protriptyline. After an automated alkaline extraction and subsequent acidic back-extraction using a continuous segmented flow procedure, samples were submitted to HPLC separation and quantitation (25). The determination of the diuretic and antihypertensive agent hydrochlorothiazide in blood, plasma, and serum with a limit of detection down to 10 ng/mL was described by Weinberger et al. (26) using the Technicon Fast-LC system and a moving-wire concentrator described elsewhere (24). Similar systems were applied for fat-soluble vitamins (27), acetaminophen, and caffeine (28). Automated postcolumn cleanup steps after liquid chromatographic separation using continuous extraction of the analyte with an organic solvent in an autoanalyzer are described for the antitumorogenic agents etoposide and teniposide (29) and for secoverine (30). Figure 3 illustrates the system used in the secoverine analysis (30).

AUTOMATED LIQUID CHROMATOGRAPHY FOR DRUG MONITORING

In the current literature, different meanings are attached to the term *automated liquid chromatography*. Various authors describe automated LC as the addition of an autosampler to a normal HPLC arrangement (31–35) so that mechanized sample injection is guaranteed. For biological samples that need rather time-consuming manual sample cleanup and pretreatment steps before HPLC analysis, however, the author would like to restate the term as *fully* automated LC with respect to the quantitation of samples from biological matrices. *Fully automated LC* is defined as an LC analytical technique in which the equipment itself man-

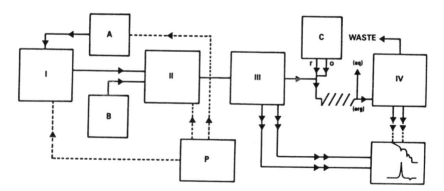

FIGURE 3.   Apparatus for automated preconcentration and HPLC analysis. A, B = pumps; P = programmer; C = peristaltic pump; I = autosampler; II = column switching apparatus; III = UV detector; IV = fluorescence detector; r = reagent solution; o = organic phase. (Reprinted, with permission, from Reference 30.)

ages, without any manual help, an analytical assay from genuine body fluids. The system performs sample pretreatment from any given matrix, introduces the sample by an automated sampling device, and handles extraction and enrichment, chromatographic separation, detection, quantitation, and the data.

AUTOMATED SAMPLE PRETREATMENT ON SOLID MATRICES AND THE COLUMN SWITCHING TECHNIQUE

Nanogram amounts of lipophilic drugs in biological surroundings can only be quantified with sufficient sensitivity if a reasonable preconcentration can be achieved. The preconcentration method of choice is reversible adsorption or distribution onto hydrophobically modified surfaces. This approach can only be adapted to automated HPLC analysis, however, if these preconcentration matrices are packed into small columns that may easily be connected to LC systems. Therefore, a second prerequisite for system development was the design of high pressure, multiport valves with low dead volumes and high performance characteristics for switching from preconcentration columns to an analytical column.

A theoretical paper covering the practical aspects of column switching was published by Huber and colleagues (36). Using a two column/one switching valve arrangement, they demonstrated the possible modes of operation of that system, including such operations as bypassing one column, stripping a part of the injected sample, and separating complex mixtures using columns with different phase ratios. Column switching or multicolumn chromatography together with reversed-phase materials of different pore sizes, surface areas, shapes, and modifications and various solvent systems opened up new techniques for sample cleanup.

In general, a sample probe consisting of lipophilic and hydrophilic material is flushed by means of an eluent with low elution capacity onto a small preconcentration column filled with a lipophilic matrix that extracts the lipophilic drug. A subsequent change of the eluent to one with higher elution strength manages the transfer from the enrichment column as a backflush to an analytical column for quantitation.

Cantwell, Dolphin, and others were among the first to use short precolumn packed with ground XAD-2 and Partisil for the preconcentration of p-hydroxybenzoate in pharmaceuticals and of organochlorine pesticides in milk (37,38). Gfeller and associates applied this precolumn concentration technique to the determination of fluorproquazone in medicated feed used for long-term toxicological studies (39), a very interesting application that may be generalized to the quantitation of drugs administered in food mixtures.

A successful application of this method to the determination of the metabolic profiles of a drug from body fluids, specifically plasma and urine, has been reported for aminopyrene (40). This work demonstrated another advantage of precolumn enrichment: a higher extraction efficiency (100%) compared to an efficiency of 30% when liquid-liquid extraction was applied (40).

A critical point in this approach, as alluded to by de Jong, may be the choice of an appropriate packing material for the precolumn in conjunction with the analytical column. This was pointed out in an assay developed for the determination of the antidepressants clovoxamine and fluvoxamine (41).

A diagram of the system used for the determination of p-hydroxybenzoates in complex pharmaceuticals is presented in Figure 4. In this approach, the sample is injected via Valve 2 (V2) onto the precolumn (C1) and is flushed with solvent 1 for

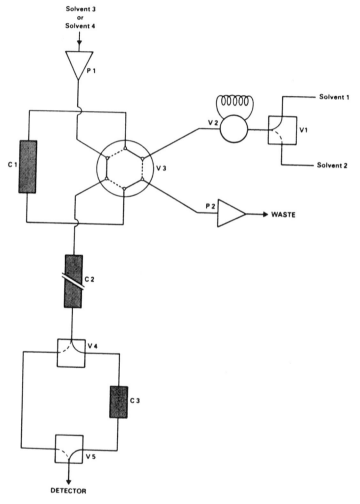

FIGURE 4.   Diagram of the liquid chromatographic system used to determine p-hydrobenzoates. P1 = main pump; P2 = auxiliary peristaltic pump; C1 = precolumn; C2 = analytical column (6.3 mm i.d.); C3 = trap column; V2 = sample injection valve; other components described in text. (Reprinted, with permission, from Reference 37.)

4 min. Valve 3 (V3) is then switched to place the precolumn in series with the analytical column (C2), and the retained sample is eluted with a solvent of stronger elution capacity onto C2 for chromatography. If the samples contain interfering aldehydes, V1 is switched instead of V3 and solvent 2 (bisulfite reagent) is allowed to react with and to destroy the aldehyde. Then, after switching V1 back to the initial position, solvent 1 is delivered; V3 then switches the precolumn back into series with C2. The advantages of this arrangement included the fact that a sample could be loaded and preconcentrated on the precolumn while the previous sample was still being separated on the analytical column and that sample pretreatment could be performed on-column without manual separation steps.

This arrangement was used in a similar but improved way by applying the backflush technique for the determination of PCBs in milk (38), methotrexate in plasma (in connection with an ion-exchanger) (42), the antiinflammatory linazelac (43), and the nonsteroidal antiphlogistic agent suprofen in human plasma and urine (44). In these papers, column switching was still performed manually. Then, Willmott and co-workers described a microprocessor-controlled column switching device for the on-column preconcentration of PCBs, DDE, and DDT (45). In this approach, switching events were controlled by a timing program in order to improve accuracy and to permit exact switching events for repeated, serial analyses.

The first implementation of precolumn cleanup together with electrochemical detection and HPLC was described by Koch and Kissinger. They concentrated the biogenic amine serotonin and its major metabolite, 5-hydroxyindolacetate, from brain tissue on a small guard column filled with reversed-phase C_{18} material. The column was placed in the injection loop of a six-port valve (46). The electrochemical detection of the antihypertensive agent indoramin was achieved after extraction of the drug on directly injected plasma. A CN-bonded silica column was used in connection with a valve switching system and cut plasma impurities that had interfered with the detector by passivating the electrode (47).

Erni and colleagues presented an interesting and flexible automated column switching setup that may prove advantageous for the separation of complex mixtures of samples that contain components of different polarity or lipophilicity (48). Such applications would include generation of metabolic profiles for drugs in plasma, urine, or bile. The two columns are both put into the loop of a six-port valve and are mounted in series (Figure 5). By using a gradient elution profile, it is possible to separate selectively and preconcentrate compounds on the first reversed-phase column and switch the fraction of interest onto the second column or into waste. These cutting techniques can be tailored to the separation by using appropriate filling materials for both columns and by choosing the optimal gradient elution profile.

Depending on sample composition and the peaks of interest, the system may cut specific fractions from the first column for chromatography on the second column (front cut, heart cut, or end cut). For example, this system is especially useful for metabolic investigations in which large amounts of polar components may interfere by tailing with the peaks of interest. In this case, an end-cut technique, passing the polar components parallel to the second column, will eliminate such separation problems (39).

In routine application, the columns in this setup can be used for hundreds of injections if milliliter amounts of urine are injected. This system was presented above in Figure 5. A modified version of that system involves the additional mounting of a preenrichment column in the loop position of the valve for on-column concentration (Figure 6) and backflushing and use of two analytical col-

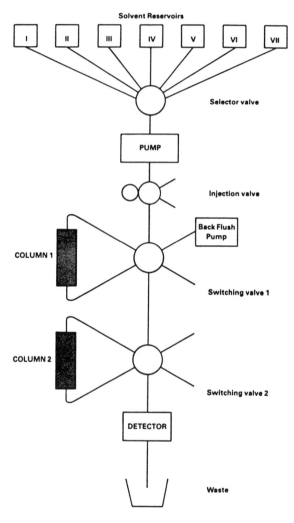

FIGURE 5.   Diagram of the column switching system with a step gradient elution solvent delivery system. (Reprinted, with permission, from Reference 48.)

umns to increase the separation power of the system by applying an additional cutting step from one analytical column to the second analytical column (49,50). This arrangement was used for the automated determination of the immunosuppressive cyclosporin A in blood and plasma (49). Another interesting application of an automated, on-line, multidimensional chromatographic technique involved use of microparticulate, aqueous-compatible, steric exclusion columns as the primary separation step coupled to either RP, normal, or ion-exchange columns as the sec-

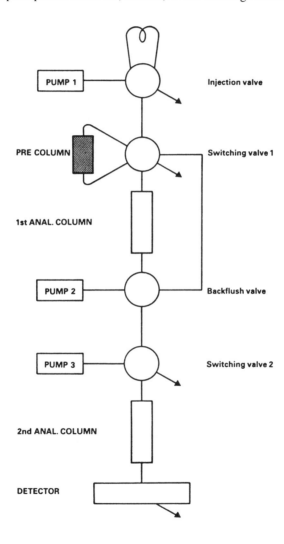

FIGURE 6.   Schematic diagram of automated precolumn concentration with a coupled column system. (Reprinted, with permission, from Reference 50.)

ondary step (51). The technique, which was automatically controlled for routine purposes by a microprocessor-based liquid chromatograph with time-programmable events, was applied to the analysis of theophylline, caffeine in biological fluids, catecholamines in urine, vitamins in food supplements, and sugars in molasses and candy bars.

A fully automated analytical LC system for routine drug monitoring should work not only with directly injected body fluids using as little technician effort as possible, but it should also analyze as many samples per unit time as is possible. All of the systems discussed above and described in the literature solved the general problem of preconcentrating lipophilic components on reversed-phase material and also possessed the degree of automation needed for switching valves and controlling such an HPLC system. Unsolved problems persisted, however, especially for routine drug monitoring. Those challenges included finding a way to handle hundreds of directly injected plasma samples in order to preconcentrate selectively both free and protein-bound drug in plasma without interference from plasma proteins themselves as well as the challenge of optimizing the time of analysis.

The author and co-workers (52,53) created a novel HPLC system for the determination of drugs and metabolites for routine drug analysis with a setup that has met all requirements for a modern, fully automated LC treatment of directly injected body fluids. Operating with two preenrichment columns used in an alternated mode for preconcentration, the system is the first that can be used for the direct analysis of injected plasma without any classical sample pretreatment step. This system is presented in Figure 7.

In this system, precolumn 2 of Figure 7A is equilibrated with solvent (pump A) after the injected sample is enriched, but plasma components such as salts, sugars, proteins, and endogenous polar material are flushed through by the solvent of pump A. Once that process is completed, precolumn 2 is switched into the eluent stream of pump B (Figure 7B) and is placed in series with the analytical column. The enriched compound is backflushed on top of the analytical column. Precolumn 1, which was backflushed with eluent of high elution strength, is then equilibrated with the eluent of pump A before the next sample is enriched (52).

The valve switching unit for valves 2 and 3 may be a time event function of an integrator and is responsible for equilibration of the precolumns, on-column enrichment, and sample cleanup as well as backflushing from the precolumns onto the analytical column. These working cycles, valid for both precolumns, are arranged so that a minimum of time is needed for loading one precolumn while backflushing the other (Figure 8).

With injection volumes of 10–100 μL of plasma, up to 1000 samples can be run without changing the guard column, making the system ideally suited to automated pharmacokinetic and clinical drug monitoring in animals and humans. Especially for samples of biological origin — and above all when genuine biological samples are injected — the reversed-phase material for the precolumns should be of a large pore size (25–70 μm) and packed between glass fiber filters. Another important feature in the alternated precolumn enrichment system is backflushing from the precolumns, which retains column performance because the enriched drug is eluted in a minimal volume and thus reapplied as a small-volume bolus on top of the analytical column. In order to modify normal isocratic HPLC equipment for alternated precolumn switching, commercially available valve switching units, either pneumatically or electrically activated, may be used to operate the system automatically. (Table I).

If gradient elution must be applied in order to optimize the separation with respect to time, a modified version of the alternated precolumn enrichment system (52) in which a second analytical column is implemented may be an appropriate solution (Figure 9). The second analytical column is mounted in such a way that, while one analytical column is reconditioned with the initial composition of the starting eluent, the other column is in the chromatographic mode — and vice versa (53).

Jürgens (55) applied alternated precolumn switching techniques (52) to the determination of 5-(p-hydroxyphenyl)-5-phenylhydantoin (p-HPPH), which is the main urinary metabolite of the anticonvulsive agent phenytoin. A commercially available column switching unit was used (56), and the precision of the technique was demonstrated with alternated precolumn enrichment. Studies generated mean deviations of 1.4% within a day and 2.75% between days for the quantitation of p-HPPH in human urine. The precolumns were changed after 400–500 30-μL injections of urine, and the analytical columns could be used for up to 800 injections.

Schöneshöfer and co-workers (57) designed an automated LC system for the routine quantitation of low concentrations of triamcinolone from urine (Figure 10). The system is designed for injection of higher volumes of genuine body fluids and accomplishes this by applying selective washing steps with a step gradient (solvents I–VI). The solvents move via a pump (P_1) onto a precolumn (PC 1) and then, applying the principle described by Lindner and colleagues (58), the elution strength is changed after the precolumn after the enriched compounds have been desorbed. A second pump, P_2, delivers water for lowering the elution capacity via a mixing

TABLE I

Commercially Available Column Switching Devices

Switching unit	Manufacturer
Switching module SM2	Gyncotek, Munich
7067 Tandem	Rheodyne, Cotati, California
HMV Switchingvent	Latek, Heidelberg
MCS 670 (Tracer)	Kontron, Zurich

chamber (MC) in order to enrich the sample selectively on a second precolumn (PC 2) before backflushing and gradient elution using two other pumps, P_3 and P_4. Chromatographic separation on the analytical column (AC) is then performed.

Essers (59) has described an automated HPLC method for quantitation of the aminoglycosides amikacin, dibekacin, glutamycin, netilemicin, sisomycin, and tobramycin. Here, an automated column switching unit is used for two sequential purposes: preconcentration and cleanup followed by automated on-column derivatization with o-phthalaldehyde with Valve 1 in the load position (L) and Valve 2 in the inject position (I) (Figure 11). The 30-μL serum sample is transferred via the

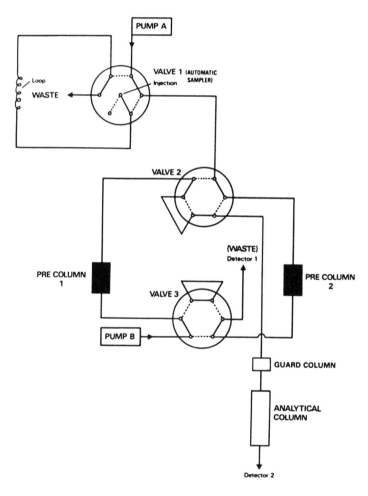

FIGURE 7.   Alternating precolumn switching technique for sample enrichment using an autosampler (valve 1) and two six-port valves that connect two short precolumns (for example, 2 cm × 0.46 cm

eluent of Pump A onto the precolumn for enrichment and cleanup. During that time, the derivatization reagent is recirculating. After rinsing for 3 min, Valve 1 is switched to I and Pump B delivers the derivatizing agent to the precolumn on which the enriched aminoglycosides are derivatized. Switching both Valve 1 and Valve 2 to L, the OPA-glycoside is transferred by the eluent stream of Pump C to the analytical column, which consists of a short anion exchanger (Nucleosil SA, 5 μm) and a reversed-phase system (Nucleosil RP-18, 5 μm). Switching of Valve 2 to I terminates the cycle.

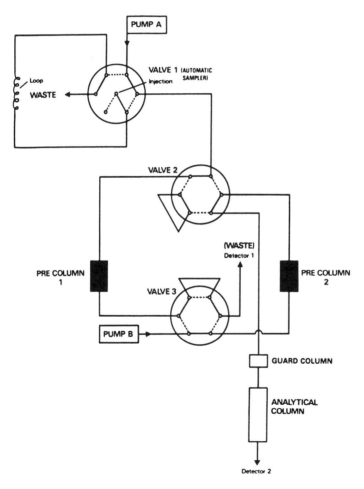

i.d.) with an analytical column (12 cm × 0.46 cm i.d.) and an additional guard column. (Reprinted, with permission, from Reference 52.)

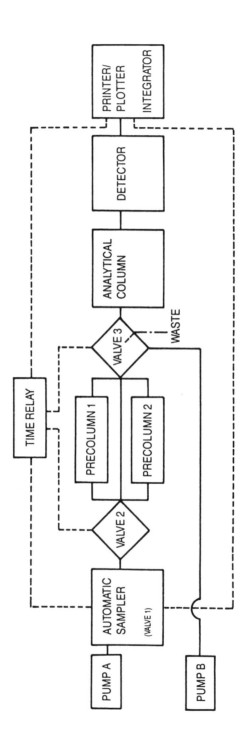

FIGURE 8.   Flow chart of the automated HPLC system with alternating precolumn sample enrichment (------- = electric connections). (Reprinted, with permission, from Reference 52.)

CONCLUSION AND PROSPECTS

The rapid development of HPLC within the last decade and its predominant role as the unique chromatographic methodology in drug research is mainly based on two factors: the rationale for monitoring free drug levels in any accessible body compartment in clinical research and preclinical drug development; and the uniqueness of chromatographic equipment in addition to its ability to discriminate and quantitate drugs and metabolites simultaneously and within minutes. Another

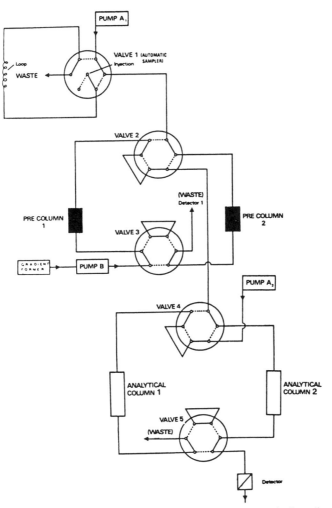

FIGURE 9.   HPLC switching technique for gradient elution chromatography from directly injected body fluids by using alternated precolumn enrichment and tandem switching for the analytical columns. (Reprinted, with permission, from Reference 54.)

very important aspect is ease of automation: not only the LC methods but also sample cleanup and enrichment procedures can be automated. These latter procedures historically have been the most time-consuming and error-creating steps leading up to bioanalytical assays (60,61).

Current LC instrumentation and column-packing materials and the accessibility of low-cost microprocessors allow nearly unlimited possibilities for LC configurations tailored to each analytical problem. It is possible to work out strategies for selected assays (4) and to run computer-aided method development in order to assist physicians, toxicologists, and pharmacologists in the management and inter-

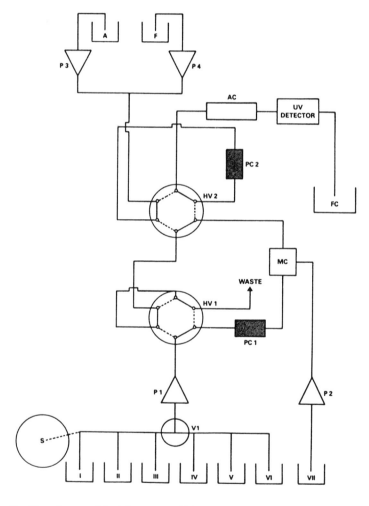

FIGURE 10.   Flow diagram of the fully automated LC assay of triamcinolone in human urine. (Reprinted, with permission, from Reference 57.)

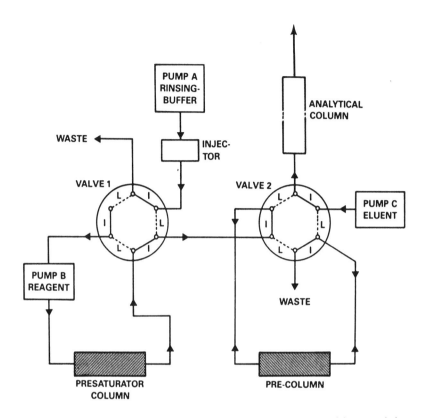

FIGURE 11.   Diagram of the precolumn switching system for precolumn enrichment and cleanup and concomitant on-precolumn derivatization with *o*-phthalaldehyde. (Reprinted, with permission, from Reference 59.)

pretation of data. Commercially available HPLC kits, which might include the column, the ready-to-use eluent, and a description of the assay, could help to increase applications of LC to drug monitoring in the clinical laboratory.

With the invention of robotics — the first on the European market was recently introduced by Zymark of Hopkinton, Massachusetts — new possibilities will be developed, including the automated integration of off-line steps such as centrifugation, changing of vials, filtration processes, and various other cleanup procedures for HPLC and GC.

The development of more potent and selectively active drugs that consequently will be administered in lower doses represents a challenge for future LC technologies. Increased sensitivity of detection will be needed and may be represented by laser spectrophoto- and fluorometers and by the improvement of micro-LC in connection with mass spectrometry and electrochemical detection. Such progress will adapt the outstanding features of automated LC, its uniqueness, and above all its precision to the needs of pharmaceutical development in the future.

REFERENCES

(1) F.H. Dost, *Der Blutspiegel* (Georg Thieme, Leipzig, 1953).

(2) D.S. Davies and B.N.C. Prichard, eds., *Biological Effects of Drugs in Relation to Their Plasma Concentrations* (Macmillan Press Ltd., London, 1973).

(3) J.A.F. de Silva, *J. Chromatogr.* **273**, 19–42 (1983).

(4) A.P. de Leenheer and H.J.C.F. Nelis, *The Analyst* **106**(1267), 1025–1035 (1981).

(5) C.D. Scott, *Clin. Chem.* **14**, 521–528 (1968).

(6) I. Molnar, Cs. Horváth, and P. Jatlow, *Chromatographia* **11**, 260–265 (1978).

(7) M. Spiteller and G. Spiteller, *J. Chromatogr.* **164**, 253–317 (1979).

(8) K. Diem and C. Lentner, eds., "Wissenschaftliche Tabellen" (Ciba-Geigy AG, Basel, 8 August 1979).

(9) L.J. Skeggs, *Am. J. Clin. Pathol.* **28**, 311–322 (1957).

(10) M.J. Stewart, in *Drug Metabolite Isolation and Determination*, E. Reid and J.P. Leppard, eds. (Plenum Press, New York, 1983), p. 41.

(11) E. Reid, *The Analyst* **101**(1198), 1–18 (1976).

(12) H. Ko and E.N. Petzold, in *GLC and HPLC Determination of Therapeutic Agents*, Part I, K. Tsuij and W. Morozowitch, eds. (Marcel Dekker, New York, 1978).

(13) J.M. Fujimoto and R.I.H. Wang, *Toxicol. Appl. Pharmacol.* **16**, 186–193 (1970).

(14) G. Machata and W. Vycudilik, *Arch. Toxicol.* **33**, 115–122 (1975).

(15) H.P. Gelbke, T.H. Grell, and G. Schmidt, *Arch. Toxicol.* **39**, 211–217 (1978).

(16) H.L. Bradlow, *Steroids* **11**, 265–272 (1968).

(17) B.L. Karger, R.W. Giese, and L.R. Snyder, *Trends in Anal. Chem.* **2**(5), 106–109 (1983).

(18) R.C. Williams and J.L. Viola, *J. Chromatogr.* **185**, 505–513 (1979).

(19) P. Koteel, R.E. Mullins, and R.H. Gadsden, *Clin. Chem.* **28**(3), 462–466 (1982).

(20) E.V. Repique, H.J. Sacks, and S.J. Farber, *Clin. Biochem.* **14**(4), 196–200 (1981).

(21) J.B. Zagar, P.P. Ascione, and G.P. Chrekian, *J. Assoc. Offic. Anal. Chem.* **58**, 822–827 (1975).

(22) Sj. van der Wal and L.R. Snyder, *Clin. Chem.* **27**(7), 1233–1240 (1981).

(23) L.R. Snyder and H.J. Adler, *Anal. Chem.* **48**(7), 1017–1022 (1976).

(24) J.W. Dolan, Sj. van der Wal, S.J. Bannister, and L.R. Snyder, *Clin. Chem.* **26**(7), 871–880 (1980).

(25) S.J. Bannister, Sj. van der Wal, J.W. Dolan, and L.R. Snyder, *Clin. Chem.* **27**(6), 849–855 (1981).

(26) R. Weinberger and T. Pietrantonio, *Ann. Chim. Acta* **146**, 219–226 (1983).

(27) J.W. Dolan, J.R. Gant, N. Tanaka, R.W. Giese, and B.L. Karger, *J. Chromatogr. Sci.* **16**, 616–622 (1978).

(28) Sj. van der Wal., S.J. Bannister, and L.R. Snyder, *J. Chromatogr. Sci.* **20**, 260–265 (1982).

(29) C.E. Werkhoven-Goewie, U.A.Th. Brinkman, and R.W. Frei, *J. Chromatogr.* **276**, 349–357 (1983).

(30) C.E. Werkhoven-Goewie, C. de Ruiter, U.A.Th. Brinkman, R.W. Frei, G.J. de Jong, C.J. Little, and O. Stahel, *J. Chromatogr.* **255**, 79–90 (1983).

(31) P.P. Ascione and G.P. Chrekian, *J. Pharm. Sci.* **64**(6), 1029–1033 (1975).

(32) M. Uihlein, *Chromatographia* **12**(6), 408–411 (1979).

(33) V. Rovei, G. Remones, J.P. Thenot, and P.L. Morselli, *J. Chromatogr.* **231**, 210–215 (1982).

(34) L. Duranti, M. Caracciolo, and G. Oriani, *J. Chromatogr.* **277**, 401–407 (1983).

(35) G.L. Diggory and W.R. Buckett, *J. Pharmacol. Meth.* **11**, 207–217 (1984).

(36) J.F.K. Huber, R. van der Linden, E. Ecker, and M. Oreans, *J. Chromatogr.* **83**, 267–277 (1973).

(37) F.F. Cantwell, *Anal. Chem.* **48**, 1854–1859 (1976).

(38) R.J. Dolphin, F.W. Willmott, A.D. Mills, and L.P.J. Hoogeveen, *J. Chromatogr.* **122**, 259–268 (1976).

(39) J.G. Gfeller and M. Stockmeyer, *J. Chromatogr.* **198**, 162–168 (1980).

(40) W. Voelter, T. Kronbach, K. Zech, and R. Huber, *J. Chromatogr.* **239**, 475–482 (1982).

(41) G.J. de Jong, *J. Chromatogr.* **183**, 203–211 (1980).

(42) J. Lankelma and H. Poppe, *J. Chromatogr.* **149**, 587–598 (1978).

(43) R. Huber, K. Zech, M. Wörz, Th. Kronbach, and W. Voelter, *Chromatographia* **16**, 233–235 (1982).

(44) H. Müller, H.W. Zullinger, B.A. Dörig, and R.A. Egli, *Arzneim.-Forsch./Drug Res.* **32** (3), 257–260 (1982).

(45) F.W. Willmott, F. Mackenzie, and R.J. Dolphin, *J. Chromatogr.* **167**, 31–39 (1978).

(46) D.D. Koch and P.T. Kissinger, *Life Sciences* **26**, 1099–1107 (1980).

(47) D.E. Leelavathin, E.F. Soffer, D.E. Dressler, and J. Knowles, *Pharmacologist* **25**, 174 (1983).

(48) F. Erni, H.P. Keller, C. Morin, and M. Schmitt, *J. Chromatogr.* **204**, 65–76 (1981).

(49) K. Nussbaumer, W. Niederberger, and H.P. Keller, *HRC&CC, J. High Resolut. Chromatogr. Chromatogr. Comm.* **5**, 424–427 (1982).

(50) K.A. Ramsteiner and K.H. Böhm, *J. Chromatogr.* **260**, 33–40 (1983).

(51) J.A. Apffel, T.V. Alfredson, and R.E. Majors, *J. Chromatogr.* **206**, 43–57 (1981).

(52) W. Roth, K. Beschke, R. Jauch, A. Zimmer, and F.W. Koss, *J. Chromatogr.* **222**, 13–22 (1981).

(53) W. Roth, *J. Chromatogr.* **278**, 347–357 (1983).

(54) W. Roth and K. Beschke, *J. Pharmac. Biomed. Anal.* **2**(2), 289–296 (1984).

(55) U. Jürgens, *J. Chromatogr.* **275**, 335–343 (1983).

(56) H. Riggenmann, *Labor Praxis,* July/August 190 (1981).

(57) M. Schöneshöfer, A. Kage, and B. Weber, *Clin. Chem.* **29**(7), 1367–1371 (1983).

(58) W. Lindner and H. Ruckendorfer, paper presented at the Seventh International Symposium on Column Liquid Chromatography, Baden-Baden, FRG, 1983 (submitted for publication to *J. Chromatogr.*, 1985).

(59) L. Essers, *J. Chromatogr.* **305**, 345–352 (1984).

(60) H.M. Abdou, *Pharm. Technol.* **4**(5), 71–79 (1980).

(61) H. Bethke, *Chromatographia* **12**(6), 335–337 (1979).

COUPLED COLUMNS IN BIOANALYTICAL WORK

Lars-Erik Edholm and Lars Ögren

Bioanalytical Section
AB Draco
Lund, Sweden

The bioanalytical laboratory involved as a service function in the development of new drugs is faced with a number of analytical problems (Figure 1). Methods must be developed for initial explorative work on research animals, and the process must be pursued all the way to the final registration of the drug. All chosen or developed methods must be selective and sensitive because the drug concentration is often very low (pmol/mL) and the sample matrix — plasma, urine, or tissue — is usually complex.

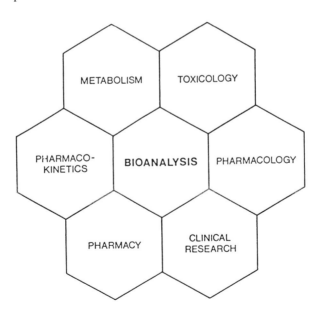

FIGURE 1. The bioanalytical laboratory working with drug development provides analytical services to several research disciplines.

Initial studies in drug development often require the development of several methods because many different substances must be analyzed in varying concentrations and in different kinds of biological material. Often very short series are analyzed, and most of the time is spent on development. A rationalization of this kind of work is needed and would mean the saving of both time and effort. There is also a need for automation and high sample throughput, especially in the later stages of drug development in which large-scale clinical studies generate large numbers of samples.

This chapter will show that column liquid chromatography (LC), which is probably the most versatile analytical technique available for use in bioanalytical work, can be applied to the situation described above. For example, LC can be applied to all types of biological matrices, and, in addition, small and large molecules, such as proteins, can be separated and quantitated with high accuracy and precision. The problems of time and effort encountered when large numbers of samples are involved can also be addressed by LC because of two characteristics. First, LC methods can easily be *automated* and, second, *sample work-up,* which is often a rate-limiting step in bioanalytical work, can be simplified and easily automated in combination with LC (1). Often a simple extraction is sufficient, and in some cases even direct injection is possible.

BACKGROUND

The most common approach in LC is to use a single column and to run it in an isocratic mode (2). This mode of operation is extremely useful because there are many possibilities to alter the selectivity of the system. For example, the type of packing material and composition of the mobile phase can easily be changed to obtain the desired separation (2,3). The usefulness of this LC approach can be further increased with the use of selective detectors and postcolumn and precolumn derivatization techniques (4).

LC can be even more powerful if coupled columns (CC) are used (2). In this chapter, some examples will be given to show how LC with CC can be most efficiently used for bioanalytical work in conjunction with drug development. Coupled column chromatography (CCC) can be performed in many different ways, both on- and off-line; this discussion will be restricted to on-line techniques. Many configurations are possible (2), and the choice of which to use depends upon the problem.

Although CCC has been used for many years (2), the potential of this technique has only begun to be recognized in practical work. Several recent papers have pointed out the benefit of using CC for the analysis of complex samples (5,6,7). One important factor that facilitates work with CC is that the *whole system* is easily automated with the use of pneumatically or motor-driven valves, microprocessors, automated injectors, and on-line data handling systems.

Because several analytical problems can be handled by CCC, it is impossible to cover all aspects of the approach in this discussion. Instead, the discussion will be focused on how CCC can be used to solve some important analytical problems facing bioanalysts working on drug development, including sample work-up, screening of drug candidates, metabolic studies, validation, quantitative analysis, separation of enantiomers, and increasing sample throughput.

Sample Work-up

In bioanalytical work, sample work-up is needed for several reasons — for example, to eliminate substances that might clog the analytical system, to release tissue-bound substances, or to enhance the selectivity of the whole analytical procedure. Sample work-up, however, is often a tedious and rate-limiting factor; therefore, automation of sample preparation is necessary.

The most common approach in sample work-up is liquid-liquid extraction (3), which can be done manually or by automated procedures off- or on-line with a liquid chromatograph (1). Liquid-solid chromatography is also used for sample work-up (3). A recent approach uses disposable prepacked columns. In this way, several samples can be worked up in parallel. CCC offers another possibility for sample work-up (1,8,9): because of the inherent selectivity of CCC, sample preparation can be simplified or the samples can be directly injected and worked up on-line in the CC system itself (1,8,9). In addition, all of the procedures can be easily automated. Simplified work-up procedures also mean the minimization or elimination of problems that can occur during sample work-up — such as losses attributable to adsorption or chemical degradation.

Screening

In early stages of drug development, it is often necessary to screen a number of substances to determine their concentration in biological matrices in order to establish their pharmacokinetic profile. In such cases it could be necessary to develop one new method for each compound. CCC can be used in such situations to save a great deal of work because samples can often be directly injected and worked up on-line. The necessary selectivity can be obtained within the same system merely by changing one or more of the chromatographic parameters.

Metabolic Studies

There are several problems associated with metabolic studies that can be more efficiently solved using CCC. To date little has been done in this area, but it is believed that CCC has great potential in these cases. Partially or fully automated CCC could be used for direct injection of biological samples and subsequent analysis of the parent drug and all of its metabolites in a single run; as an alternative to gradient elution to cope with the general elution problem (2); for faster analysis; for optimization of resolution; and for identification purposes.

Validation of Analytical Methods

Quantitative analysis of drug substances and/or their metabolites — often at low concentrations and in a complex matrix — requires selective analytical methods with low detection limits. An important part in the validation of any analytical method is to prove its accuracy. This step can be done by using another method based upon completely different principles for separation and detection — for example, gas chromatography/mass spectrometry. CCC offers another alternative for testing the accuracy of a method because of the extreme selectivity that can be obtained with such systems.

NOMENCLATURE

Coupled Column Chromatography

The designation *coupled column chromatography* has been chosen to cover all analytical applications in which a fraction (or fractions) from one chromatographic column is transferred on-line to one or more chromatographic columns. There are several other designations found in the literature for this technique — column switching, mode switching, and valve switching are a few. Figure 2 examines some important operations and effects related to CCC.

Peak Compression

When a fraction (or fractions) is transferred from one chromatographic column to another, it is necessary to avoid band broadening as much as possible. This can be accomplished by compression of the band (10), which in turn can be accomplished in one of two ways: *differential retentions* or *precolumn concentration*. In the former case, the fraction to be analyzed is transferred via a switching valve to the analytical column. If the compound (or compounds) of interest is more retarded on the analytical column than on column 1, the peak (or peaks) will be compressed on the top of the analytical column. The volume of mobile phase containing the compound i, V_i, has been contracted according to Equation 1 (11):

$$V_i = V_0 \left(\frac{1}{1 + k_i^{ci}} \right) \qquad [1]$$

where
   V_0 = sample volume (trapped volume from column 1)
   k_i^{ci} = capacity ratio with injection medium in the analytical column.
   In the case of precolumn concentration, the fraction can also be transferred via a precolumn in a two-step manner. In step 1, fraction B is compressed on top of the precolumn as in Equation 1. The switching valve is then turned to the inject posi-

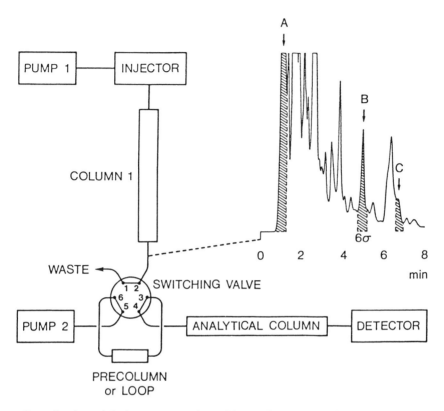

FIGURE 2.   A coupled column system used to explain some important operations and effects related
to the technique.

tion, and mobile phase 2 is used to elute the trapped compound (or compounds) by
reversing the flow on the precolumn (*backflush*). This last procedure has been examined theoretically and practically by Lankelma and Poppe (11), who derived an
expression for the concentration factor in the entire system.

Heart Cut

Heart cut refers to a selected part (B) of the eluate from column 1 that is taken
out and transferred to the analytical column.

Front Cut

Front cut is the first part (A) of the eluate from column 1 and is directed to the
analytical column. (This term might also refer to the first part of an ascend-
ing peak.)

End Cut

The *end cut* is usually a part (C) of a tailing peak eluting from column 1 and is transferred to the analytical column.

Backflush

Backflush refers to a process by which compounds trapped on the precolumn are transferred to the analytical column by reversing the flow on the precolumn.

Multidimensional Chromatography

Multidimensional chromatography refers to the combination of different *modes* of chromatography (7). In CCC, it is possible to make such combinations in an on-line manner. For example, as Figure 2 shows, column 1 could be an ion-exchange column and the analytical column of the reversed-phase type, thus allowing for different modes of chromatography. As will be shown here, such combinations can be very useful — especially for analysis of complex samples.

Trace Enrichment

The term *trace enrichment* refers to a process by which analytes at low concentrations in large sample volumes can be enriched (see *peak compression*) on a column if the sample solution is a weak eluent on that column (10). In principle, the sample could be transferred to the column in several ways — whether by syringe, with a pump, or as a fraction (*heart cut*) from another column.

SYSTEM SETUPS

As a first example, a simple CC system is shown that will be used here as a model (Figure 3). By extension of this system, other CC systems can be designed to solve other problems, as will be exemplified with references to other work.

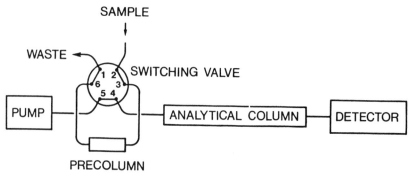

FIGURE 3.   Coupled column system 1.

Coupled Column System 1

The CC system in Figure 3 consists of two columns, one a precolumn and the other an analytical column. The precolumn is coupled in the loop position of a manually, pneumatically, or motor-actuated high-pressure valve. Samples can be introduced either manually with a syringe or can be delivered via an autoinjector with the aid of a pump. The sample and mobile phase should be totally miscible with each other. Chromatography using polar mobile phases such as water with organic modifiers — methanol or acetonitrile — as well as nonpolar mobile phases may be used. Several different types of packing materials can be used in the two columns. The precolumn can be used for concentration of diluted samples, direct injection of untreated samples, and on-line sample work-up. The procedures may be performed manually or may be totally automated, as will be exemplified below.

Terbutaline (T), as shown in Figure 4, is a selective β_2-receptor agonist widely used in the treatment of asthma. Plasma concentrations of unchanged T that are associated with effective therapy are in the range 10–30 pmol/mL. GC/MS was exclusively used for quantitation of T, but there was a need for a less expensive and more generally accessible method. It was found that LC with electrochemical detection could be used as an alternative (12). As much as 2 mL of plasma, however, had to be used, and it was therefore necessary to isolate T from the matrix and to eliminate the proteins.

It was earlier found that selective cleanup on manually operated ion-exchange columns was also a necessary step and could not be omitted. Quantitative elution of T from the ion-exchanger, however, required 3 mL of a buffer containing high salt concentrations. This solution could not be directly injected without creating a devastating effect on the detector signal. It was therefore necessary to enrich T from this solution to remove most of the salt.

The problem here was solved by using a CC system according to Figure 3. The ion-exchange eluate (3 mL) was applied to the precolumn manually with a sy-

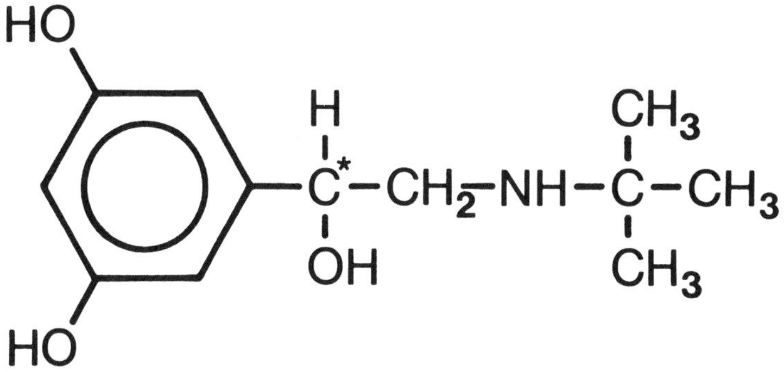

FIGURE 4.   Structural formula of terbutaline [1-(3,5-dihydroxyphenyl)-2-(t-butylamino)-ethanol]. Dissociation contants: pK_{a1} = 8.8 (phenol), pK_{a2} = 10.1 (amine), pK_{a3} = 11.2 (phenol).

TABLE I

TABLE I

**Experimental Details for Quantitative Analysis of Terbutaline
in Human Plasma**

Pump: Waters model M45 (Milford, Massachusetts)
Flow rate: 2.0 mL min^{-1}
Detector: Amperometric (glassy carbon, 0.9 V vs Ag/AgCl, 3 M NaCl)
Switching valve: Valco model CV-6-UHPa-N60 (Houston, Texas)
Injection volume: 3.1 mL

Column	Dimensions (mm)	Packing material	Mobile phase
Precolumn	23 × 3.9 (Waters guard column)	SepPak C_{18}	
Analytical	200 × 5	Nucleosil C_{18} (10 μm)	Methanol/ citric acid/ phosphate (pH 6.0)

ringe, so relatively large diameter packing was used in the precolumn. (For technical details see Table I.) Of the 3 mL, only about 200 μL — corresponding to the void volume of the precolumn — was left to be injected onto the analytical column. By backflushing the precolumn with the mobile phase, T and the internal standard (IS) were transferred to the analytical column for final separation.

Figure 5 shows that no significant band broadening was observed for T or the IS as compared with the injection of the same amount of substances in a small volume (20 μL) of T and the IS directly onto the analytical column. It should be observed that the on-line coupling of the precolumn to the analytical column will give these kind of results only if, during desorption from the precolumn, the retention of the enriched solute will be similar to or lower than retention on the analytical column; otherwise, too much peak broadening will occur (13).

A similar system configuration was proposed by Voelter and co-workers for investigation of drug metabolism to circumvent classical sample purification via extraction through use of direct purification and enrichment of the sample on the precolumn (14). Similar arrangements have also been used for precolumn sample enrichment for the quantitative analysis of serotonin and 5-hydroxyindolacetic acid in brain tissue (15) and methotrexate in human plasma (11) and by Johnsson and Bowers (16) for on-line hydrolysis of estriol conjugates.

Direct injection of biological fluids into conventional LC systems is possible, but it often leads to fast deterioration of column performance and/or clogging of the chromatographic system. Therefore, it can only be used for a limited number of injections of small sample volumes. This situation, however, can be greatly improved when CCC is used with on-line sample work-up.

Voelter and associates used a system similar to the one described above for direct manual injection of up to 1000 μL of plasma, with subsequent purification of the sample to remove proteins and other highly hydrophilic substances (14). After purification, the sample was transferred to the analytical column for final separation.

Roth and co-workers showed that the same principle, with on-line enrichment and sample work-up, could be used for direct injection with full automation (17,18). The principle suggested by Roth has also been adopted by other workers for automated analysis of antiepileptic drugs and their metabolites in serum (19,20).

Work by Nazareth and associates is a further extension of this approach and is an example of the inherent power of CCC (21). As Figure 6 demonstrates, a combination of on-column enrichment and boxcar chromatography could be used for direct injection of serum samples with a very high sample throughput (22). Primidone and phenobarbital, for instance, could be quantitated at a rate of 40 samples/h with a coefficient of variation of 1%. As can be seen from Figure 6, the incorporation of an automated injection system (A) and a boxcar system (B) in the basic CC system design of Figure 3 has extended the manually operated system with low sample throughput to a fully automated system with high sample throughput.

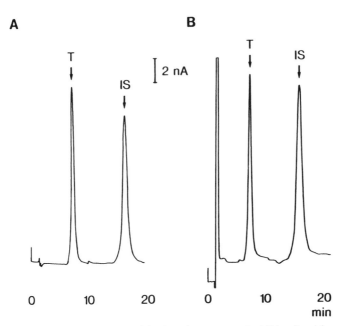

FIGURE 5.   Comparison of 20-μL loop injection of aqueous standard (A) and enrichment of 3-mL aqueous standard (1 *M* NaCl, pH 6) to simulate the properties of the eluate from the ion-exchange column (B). T = terbutaline, IS = internal standard. (Chromatographic conditions are listed in Table I.)

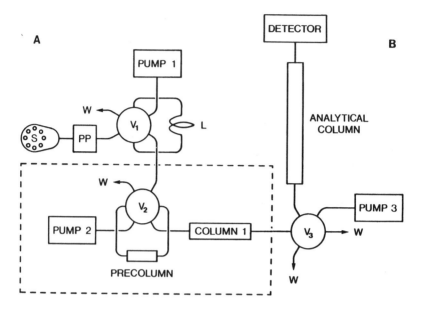

FIGURE 6.   Configuration of the LC system used by Nazareth and co-workers for automated analysis of directly injected serum samples (21). Samples are delivered from the autosampler (S) to fill the loop (L); at the same time the precolumn is equilibrated with water delivered by pump 2; switching valve V_1 is turned to deliver the sample for enrichment on the precolumn; switching valve V_2 is then turned to backflush (pump 2) the sample onto column 1; a heart cut (V_3) is then taken and delivered to the analytical column; W = waste.

Coupled Column System 2

By extending the CC system given in Figure 3 according to the illustration of Figure 7, further operations, such as preseparation, heart-cutting, and multidimensional chromatography, are possible. Added to the design in Figure 3 is a preseparation column (column 1), an automated injector, a pump, an extra detector, and a data handling system.

An untreated or partially pretreated sample is placed in the automated injector. After preseparation on column 1, a heart cut containing the analyte (or analytes) of interest is taken and directed to the precolumn (or a loop). This fraction is then concentrated on the precolumn, and the rest of the eluate from column 1 is directed to waste. The enriched compounds are then backflushed onto the analytical column.

The above system allows for total automation with direct injection, or simplification of sample pretreatment as a result of the high selectivity that can be obtained as different modes of chromatography are combined with selective detectors and heart-cutting. As in CCC system 1, the sample must be miscible in mobile phase 1, and the two mobile phases should be miscible. Polar as well as nonpolar mobile phases may be used. Mobile phase 1 should be chosen so that peak compression

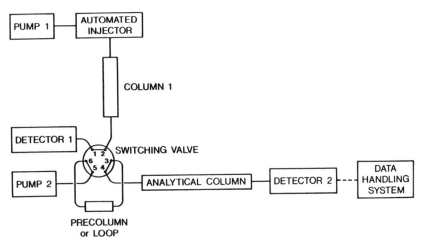

FIGURE 7.   Coupled column system 2.

and 100% recovery will be obtained on the precolumn. If a loop is used, peak compression should be obtained on the analytical column.

For determination of the volume to be trapped on the precolumn, column 1 is connected to a detector with the switching valve in the 1–2 position. The sample is then injected, and a window is chosen from the chromatogram (for example, fraction B in Figure 2). There are two different possibilities: either a part of a peak is trapped, or the complete peak is trapped. For quantitative analysis, the latter is preferred because it is less sensitive to retention time variations on column 1. Furthermore, better selectivity can be obtained if only a part of the eluate is taken.

The volume to be trapped on the precolumn may be chosen empirically, but to obtain 99.9% recovery a volume of 6τ is necessary for a Gaussian-shaped peak (Figure 2).

To illustrate quantitative analysis with CCC, studies of enprofylline will be described. Enprofylline (E), the structure of which is drawn in Figure 8, is a novel xanthine derivative with antiasthmatic properties and is now undergoing clinical trials. When administered orally or intravenously to humans, about 90% is recovered unchanged in the urine. The high concentrations (1–600 μg/mL) found in urine, together with its high molar absorptivity ($\varepsilon_{280} \cong 10^4$ Lmol^{-1}cm^{-1}), made it possible to separate and quantitate E with an automated isocratic LC system after direct injection of the sample onto a single analytical column (23). This approach, however, was not sufficient in all situations; selectivity problems became apparent.

With the use of the CC system of Figure 7 with preseparation on a strongly acidic ion-exchange column, trapping of E onto a precolumn, and final separation on a chemically bonded ODS column, a highly selective system was created. For technical details of this system see Table II. Direct injection was still possible, and

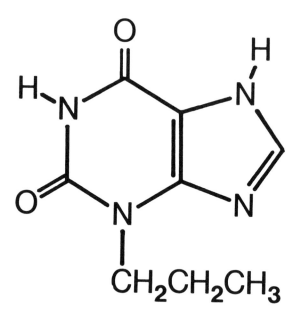

FIGURE 8.   Structural formula of enprofylline (3,7-dihydro-3-propyl-1*H*-purine-2,6-dione). Dissociation constant: $pK_a = 8.2$ (7-H).

TABLE II

Experimental Details for Quantitative Analysis of Enprofylline in Human Urine

Pumps: Waters model 45 or 6000A
Flow rates: 1–2 mL min^{-1}
Detector: Waters model 440 ($\lambda = 280$ nm)
Automated injector: Waters model 710B autosampler
Injection volume: 200 μL
Switching valve: Valco model CV-6-UHPa-N60 (pneumatically operated)
Data handling system: Hewlett-Packard 3388A (Avondale, Pennsylvania)*

Column	Dimensions (mm)	Packing material	Mobile phase
Column 1	50 × 5	Nucleosil SA (10 μm)	Sodium acetate (pH 4.6–4.7)
Precolumn	23 × 3.9 (Waters guard column)	SepPak C$_{18}$	
Analytical	150 × 5	Nucleosil C$_{18}$ (5 μm)	Methanol/ *o*-phosphoric acid (pH ~2.4)

* Data handling system also used to control valve switching.

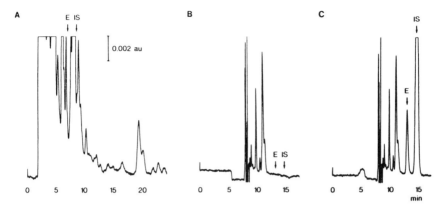

FIGURE 9. Quantitative analysis of urine samples. (A) Urine blank injected directly onto analytical column with analytical column coupled to the injector; (B) urine blank with the use of coupled columns according to Figure 7; (C) urine spiked with enprofylline (E) (8 µg/mL) and internal standard (IS) using coupled columns. (Chromatographic conditions are listed in Table II).

the system was fully automated with the use of an automated injector, a pneumatically actuated valve, and a reporting integrator with external events for actuation of the valve.

Figure 9A shows a chromatogram obtained after direct injection of urine onto the analytical column in accordance with the original method (23). For comparison, chromatograms in Figures 9B and 9C illustrate what is obtained using the CC system. In Table III, analytical characteristics obtained with spiked urine samples are presented. The system described has been running for several years 24 h/day without any interference problems attributable to other compounds.

Detailed description of the chromatographic behavior for E as well as other xanthine derivatives are described elsewhere (23,24). Because of the widespread consumption of caffeine (1,3,7-trimethylxanthine) containing food and beverages, human urine very frequently contains caffeine and its metabolites. Because they are closely related to E, they are also a potential source of interference.

TABLE III

Analytical Characteristics for Quantitative Analysis of Enprofylline in Human Urine

Amount added (µg/mL)	1.45	16.28	74.83
Amount found (µg/mL), mean	1.40	16.20	73.65
C_i (%, interassay)	5.0	1.7	1.5
n	33	73	46
C_i (%, intraassay)*	1.1–5.0	0.8–1.6	0–1.8
n	12	32	46
Total absolute recovery (%)	100	100	100

* Range found when determined on several occasions.

The pK$_a$ of enprofylline is 8.2 and the pK$_a$ values for many of the xanthines generated from caffeine metabolism are close to that value, so pH has little effect on retention and selectivity for these substances in the pH region most frequently used in reversed-phase chromatography. It was shown that it is possible to separate caffeine and all of its known metabolites in humans on the analytical column defined in Table II alone (23), but batch-to-batch variations in the packing materials made it necessary to test every new batch carefully.

On the ion-exchange column, sodium acetate buffer, pH 4.6, was used as mobile phase; in this system, pH has little effect on retention for the xanthines. The background, however, is dependent on pH (24). It was found that a pH of 4.6 was optimal. Figures 10A and 10B show chromatograms obtained with the ion-exchange column connected directly to the UV detector.

Generally, the ion-exchange material was chosen so that E and the internal standard eluted as close to each other as possible, with sufficient separation from caffeine. The ion-exchange material itself suffers from batch-to-batch variations, but this does not interfere with overall selectivity. In fact, batch-to-batch variations on the analytical column are of less importance given that, for example, caffeine and many other compounds are eliminated on column 1.

Another example will demonstrate how CCC can be useful in a bioanalytical situation in which a great deal of flexibility is needed and in which at the same time a

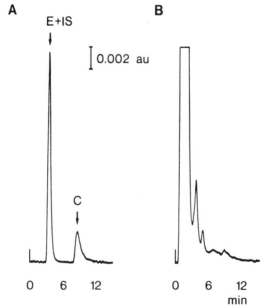

FIGURE 10.   Chromatograms obtained with the ion-exchange column connected to the detector in Figure 7. (A) Standard mixture containing enprofylline (E), internal standard (IS), and caffeine (C); (B) urine blank. (Chromatographic conditions are listed in Table II.)

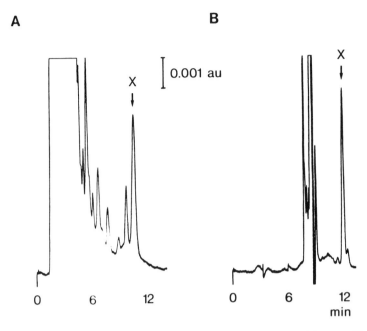

FIGURE 11. Chromatograms obtained after direct injection of rat urine containing a xanthine derivative (X). (A) Without coupled columns, samples injected onto the analytical column; (B) with coupled columns. (Chromatographic conditions are listed in Table II.)

lot of work can be saved using a rational system. In this example, it was necessary to measure the excretion of a number of unchanged xanthine derivatives in rat urine. Because of the analytical complexity, there was a need for a flexible, selective, and labor-saving screening procedure.

Preliminary studies showed that direct injection of samples and subsequent separation with a single isocratic system was impossible. A selective work-up step was thus necessary. Using the same setup as was used for enprofylline in urine, samples could be worked up on-line on the strongly acidic ion-exchange column. Only slight modifications of mobile phase 2 were needed: the amount and nature of the organic modifier had to be varied because of the difference in lipophilicity of the xanthine derivatives. Figures 11A and 11B show the efficiency of the system.

Method validation is another key concern in drug development. Bambuterol (B) is a prodrug (Figure 12) that has been developed to produce prolonged effect and reduced side effects of terbutaline (T), the active ingredient of the commercial products Bricanyl and Brethin.

For pharmacokinetic evaluation in humans, it was necessary to determine plasma concentrations of T. It was previously shown that a GC/MS procedure (25) and an LC method with a CC system (according to CC system 2) and electrochemical detection (LCEC) gave the same result on human plasma samples after T adminis-

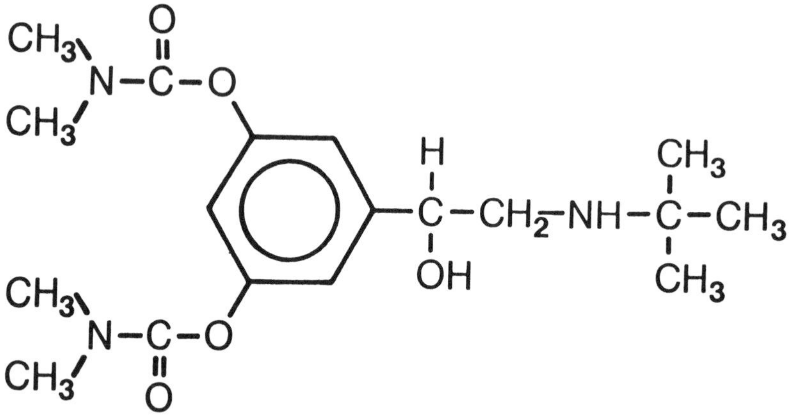

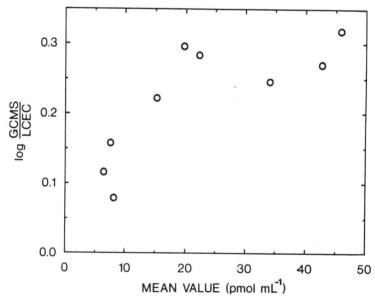

FIGURE 12.   Structural formula of bambuterol [1-[3,5-bis(*N,N*-dimethylcarbamoyloxy)phenyl]-2-*t*-butyl-aminoethanol].

tration (26). A comparative study of human plasma samples after administration of B to humans was undertaken using the two methods.

It was found that, in general, the GC/MS method gave higher results. From Figure 13, it is clear that in the region of 20–50 pmol/mL, GC/MS gave about twice the values obtained by LCEC. Investigations have shown that one possible explanation

FIGURE 13.   Comparison of GC/MS and LCEC for quantitative analysis of terbutaline (T) in human plasma after administration of bambuterol.

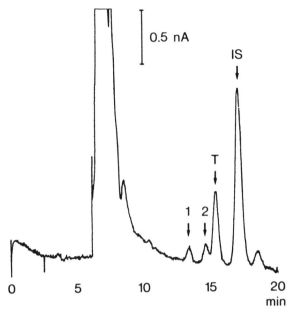

FIGURE 14.   Chromatogram of a plasma sample after administration of bambuterol. (Peaks 1 and 2 are of unknown identity, but are generally found in samples after administration of B.) Terbutaline (12 pmol mL^{-1}), internal standard (50 pmol mL^{-1}).

for the high values obtained by the GC/MS procedure is that hydroxylated metabolites of B are transformed into T in the derivatization step preceding GC/MS.

Figure 14 shows a chromatogram of a plasma sample after administration of B using the CC system. Peaks designated 1 and 2 are of unknown identity but are generally found in samples after administration of B. By combining the trapping of the analyte with two modes of chromatography (Table IV) — that is, reversed-phase chromatography and ion-exchange chromatography with electrochemical detection — a highly selective system was obtained. This is now the method of choice for quantitative analysis of T in human plasma after administration of B, and the method is now in use for pharmacokinetic evaluations of B.

Today there are few examples of the use of CCC in studies of metabolism. CCC could be a useful tool in many situations, however, in which conventional LC approaches would be insufficient or tedious. For example, the systems given for simplification of sample work-up by manual (14) or fully automated (17) procedures as described above have been useful in studies of metabolism. CC system 2 as designed for quantitative analysis of enprofylline in urine was, for instance, used in studies of the metabolism of enprofylline (E). About 90% of E is recovered unchanged in the urine when administered to humans, and preliminary studies indicate that E was partly metabolized to the parent uric acid (Figure 15).

TABLE IV

Experimental Details for Quantitative Analysis of Terbutaline
after Administration of the Prodrug Bambuterol

Pump: Waters model M45 or M6000A
Flow rates: 2.3 mL min^{-1} (both pumps)
Detector: Amperometric (glassy carbon, 0.9 V vs Ag/AgCl, 3 M NaCl)
Automated injector: Waters model 710B autosampler
Injection volume: 200 μL
Switching valve: Valco model CV-6-UHPa-N60 (motor-driven)
Data handling system: Hewlett-Packard 3388A*

Column	Dimensions (mm)	Packing material	Mobile phase
Column 1	150 × 4.6	Nucleosil phenyl (7 μm)	Sodium acetate (pH 4.6–4.7)
Precolumn	23 × 3.9	SepPak C$_{18}$	
Analytical	200 × 4.6	Nucleosil SA (10 μm)	Acetonitrile/ sodium acetate/ citric acid (pH 4.6)

* Data handling system also used to control valve switching.

Synthetic uric acid was used to determine the window in which a fraction should be trapped. Figures 16A and 16B show that, after administration of E to humans, urine contains a compound further identified by GC/MS to be the uric acid analogue to E. Again the selectivity obtained by combining two modes of chromatography with trapping of the analyte made it possible to run selective analyses with very little effort. In principle, the same approach can be used in other studies of metabolism for trapping of one or more metabolites for identification and quantitation. It is obvious that the fraction collected after cleanup in a multidimensional system would be very clean — which in turn could be helpful for subsequent structure elucidation with, for example, mass spectrometry.

The general elution problem (2) is a common one frequently encountered in studies of metabolism, because great polarity differences very often exist between

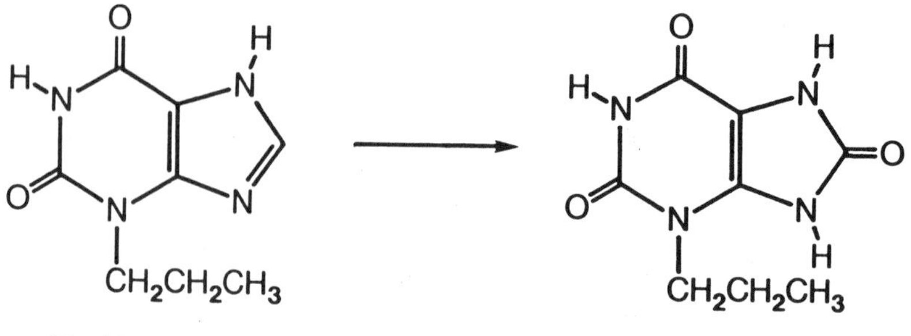

FIGURE 15.   Metabolism of enprofylline to the parent uric acid.

A

B

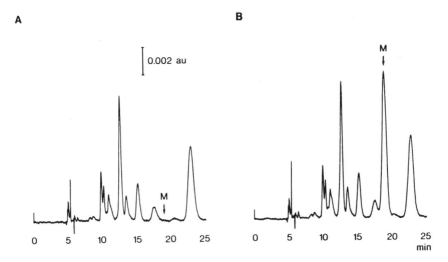

FIGURE 16.   Chromatograms of human urine using coupled column chromatography. (A) Blank urine; (B) urine, obtained after administration of enprofylline, containing the corresponding uric acid (M).

a drug and its metabolites. One way to solve this problem is to use gradient elution, but CCC offers an alternative to this approach, giving the advantage of optimizing the speed of the analysis and enhancing system selectivity (27,28). In other work, a bimodal system was used by Weidolf (29) to optimize the separation of six urinary metabolites of [^{14}C]felodipine; the system allowed for their quantification using an on-line radioactivity detector.

The use of LC for separation of enantiomers is well established and can be performed in several ways (30). When applied to samples such as biosamples, it is necessary in a chiral system to precede final separation with a selective cleanup step. CCC should have a great potential for use in such applications through use of the techniques of preseparation and subsequent transfer of the analyte for final separation in, for example, a chiral system. To the author's knowledge, however, no such work has yet been published. Nonetheless, it was shown by Tapuhi and co-workers that a complex mixture of dansyl-amino acids of nonbiological origin could be separated using this technique (31).

As an example of enantiomeric separations, a study of terbutaline was chosen. Terbutaline, as was shown in Figure 4, has an asymmetric center at the α-carbon and is usually administered as a racemate. Effect studies have shown that only the L- form has the desired pharmacological effect (32). The pharmacokinetics of the two enantiomers have been studied after administration of the respective forms, but further studies are needed to evaluate the kinetics of the two enantiomers when the drug is administered as a racemate. For this reason, a chiral system with the ability to separate the two enantiomers of T without the need for derivatization was sought.

It was found that the enantiomers of T could be separated on a chemically bonded cyclodextrin phase (33), with a mobile phase very similar to the one used on the analytical column mentioned above for bambuterol. Amperometric detection was thus possible, and the preseparation system could be used intact. Because T is less retarded on the cyclodextrin phase than on the ion-exchange column discussed in the previous example of bambuterol, less organic solvent was needed.

This arrangement precluded the use of a precolumn, and a loop was used instead to obtain peak compression. Application of this configuration made it possible to isolate, separate, and detect the enantiomers of T in human plasma at relevant therapeutic levels. Figure 17 shows a chromatogram containing ∼ 30 pmol/mL of the racemate. The procedure for determination of volume to be trapped was the same as described above.

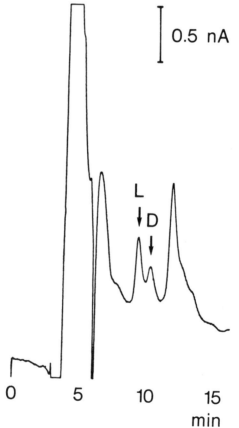

FIGURE 17.   Separation of the entantiomers of terbutaline (L and D). Analytical column: Cyclobond I β-cyclodextrin, 250 mm × 4.6 mm. Mobile phase 2: methanol/citric acid-phosphate buffer, pH 6.0, 5:95 (v/v). Other chromatographic conditions are listed in Table IV.

It should be noted that, if the ratio of the enantiomers alone is to be determined, the cut could be very close and thus the selectivity could be further enhanced. The result shown in Figure 17 was obtained after sampling of the complete peak.

The principle of preseparation with subsequent heart-cutting and transfer of the analyte (or analytes) to an analytical column for final separation and detection has been used by several workers to solve bioanalytical problems. In Table V, a summary is presented of some of this published work in which CC designs along the lines of CC system 2 have been used.

As was discussed at the outset, rate-limiting factors for sample throughput in most bioanalytical work using LC include sample work-up, run times for each sample, and the total time the system is working. In an automated LC system working on a 24-h basis, the sample work-up is often the rate-limiting factor, and rationalization of this procedure would be one way to increase sample throughput. It was shown above that CCC could be very useful for this purpose by allowing for either simplified sample work-up or direct injection of the sample with on-line sample work-up. With the use of boxcar chromatography (22), the sample throughput could be made very high (21).

TABLE V

Bioanalytical Applications Using the Principles of Preseparation, Heart Cut, and Transfer of the Analyte to an Analytical Column (According to Coupled Column System 2)

Application, objective, and reference	Column 1	Precolumn	Analytical column	Comments
Quantitative analysis of theophylline in plasma; labor saving (34).	Gel-permeation	Direct transfer of analyte to analytical column	Reversed-phase	Direct injection Different mobile phases
Quantitative analysis of cyclosporine in plasma and blood. Labor saving, improved analysis time and detection (35).	Reversed-phase	Direct transfer as above	Reversed-phase	Automated sample preparation preceding LC 1000 samples could be analyzed without deterioration of method parameters Different mobile phases
Quantitative analysis of riboxamide in plasma. Selective analysis, providing a low detection limit (36).	Silica gel with ion-pairing agent in neat aqueous eluent	Direct transfer as above	Reversed-phase with ion-pairing agent in neat aqueous eluent	Same mobile phase

(CONTINUED)

TABLE V (*Continued*)

Quantitative analysis of cisplatin in urine. Selective analysis with minimum of sample preparation (37).	Silica gel with ion-pairing agent in neat aqueous eluent	Direct transfer as above	Reversed-phase with ion-pairing agent in neat aqueous eluent	Direct injection Off-line detection Same mobile phase
Several applications to show the inherent potential of coupled column chromatography. Examples are xanthine and catecholamines in biological material (38).	Gel-permeation	Loop	Reversed-phase Normal-phase* Ion-exchange	Direct injection Single pump operation Qualitative analysis
Quantitative analysis of prostaglandins in plasma to obtain a highly selective and flexible system (39).	Normal-phase	Loop (2.2 mL)	Normal-phase	One of the few applications showing that the technique is also useful for normal-phase LC. Different mobile phases

* Chemically bonded NH_2, aqueous mobile phase.

For quantitative analysis of enprofylline (E) in plasma, single column LC was used (23). To obtain sufficient selectivity, the sample had to be worked up with an extraction procedure and subsequent protein precipitation before LC. Large-scale clinical studies required a new method with higher sample throughput. In this case, a simplified work-up would suffice, but that would also call for an LC system with much higher selectivity for E. It was found that, by using CCC with on-line sample work-up, this could be obtained (40).

In Figure 18, the CCC system used is shown; Table VI displays technical details. The system is an extension of CC system 2. Added to this design is equipment for selective elution of E and the internal standard (IS) and for acidification of the eluate from column 1.

The procedure moves as follows. First, the protein-precipitated sample, with IS added, is injected onto column 1, which is packed with a poly(styrene-divinyl-benzene) material (pH range 1–13) with hydrophobic properties. Column 1 is then washed with a mobile phase containing methanol in a buffer solution at pH 7.0. At this pH, E and IS are neutral ($pK_a = 8.2$) and strongly retarded on column 1 (Figure 19A). An alkaline solution (BASE) at pH~12 is then used to elute E and IS from the column, and a peristaltic pump (PP) is used to fill the alkaline solution into loop 1. At pH~12, E and the IS are completely ionized and are eluted close to the front (Figure 19B).

The analytical column contains a silica-based packing material, and therefore the sample must be acidified, which is done by mixing dilute phosphoric acid (de-

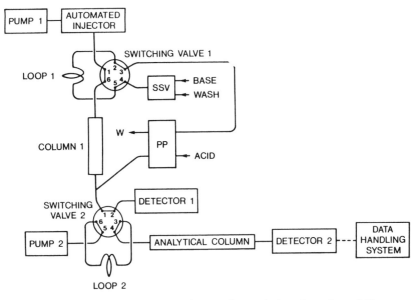

FIGURE 18.   Coupled column system used to increase the sample throughput of enprofylline.

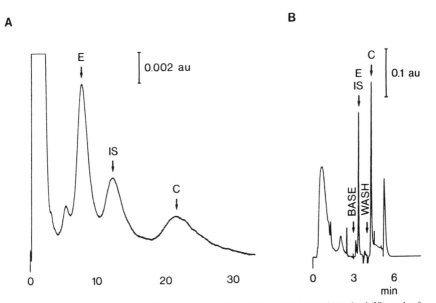

FIGURE 19.   Chromatogram showing the elution of enprofylline (E), internal standard (IS), and caffeine (C) from column 1 after injection of a spiked plasma sample (10 μg/mL). (A) Mobile phase: methanol/buffer (pH 7.0); (B) *first*, selective elution of E + IS using an alkaline solution (BASE, pH 12), *second*, a washing solution (WASH) is used to clean column 1 whereby C is eluted.

TABLE VI

Experimental Details for the Coupled Column System Used to Increase the Sample Throughput of Enprofylline

Pumps (high pressure): Waters model 590 (with external event control, used for valve switching) and model 510

Flow rates: 0.4 mL min^{-1} (both pumps)

Pumps (low pressure): ALITEA C-4V (Ventur, Sweden) peristaltic pump

Detector: Waters model 440 dual channel ($\lambda = 254$ nm)

Automated injector: Waters model 710B autosampler

Injection volume: 150 μL

Loop 1: 500 μL; loop 2: 200 μL

Switching valve 1: Rheodyne 7000 with pneumatic actuator 7001 (Cotati, California)

Switching valve 2: Rheodyne 7000 with pneumatic actuator 7001 and pilot valves (Reference 41) for faster switching

Solvent select valve: Rheodyne 5301 with pneumatic actuator 5300

Data handling system: Hewlett-Packard 3388A

Column	Dimensions (mm)	Packing material	Mobile phase/ solutions
Column 1	32 × 2.1	Hamilton (Reno, Nevada) PRP-1 (10 μm)	Methanol/phosphate buffer (pH 7.0)
			BASE: acetonitrile/sodium hydroxide (pH ~12)
			WASH: acetonitrile/ phosphate buffer (pH 7.0)
			ACID: o-phosphoric acid (pH 1.7)
Analytical	100 × 3.0	Nucleosil C$_{18}$ (5 μm)	Methanol/o-phosphoric acid (pH 2.4)

livered with the peristaltic pump) with the fraction after column 1. The acidified fraction with E and the IS is then filled into loop 2 and thereafter injected onto the analytical column. While E and the IS are chromatographed on the analytical column, the following procedures are performed: column 1 is washed with the washing solution (WASH) injected from loop 1 (Figure 19B), and a new sample is injected onto column 1.

This CC system has the following advantages compared to the old method (23): First, sample throughput is increased by a factor of at least two. Second, there is less interruption attributable to batch-to-batch variations in the packing material used in the analytical column. Finally, the selectivity of the system partly eliminates the problem of separating xanthines emanating from caffeine metabolism. The efficiency of the CC system in cleaning up the sample can be seen in Figures 20A and 20B.

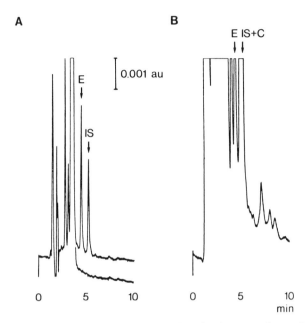

FIGURE 20.   Chromatograms obtained after injection of a plasma sample containing xanthines resulting from food intake. The sample was further spiked with enprofylline (E) and internal standard
(IS) at 0.3 μg/mL each. (A) Using the CC system; lower trace, plasma blank; (B) direct injection onto
the analytical column. (Chromatographic conditions are listed in Table VI.)

PRACTICAL CONSIDERATIONS

For proper use of CCC, some practical details must be considered, among them
band broadening, valve switching, valve control, precolumn use, and a range of
others discussed in turn below.

Band Broadening

Band broadening (2), which occurs during the passage of a solute through the
chromatographic system, is even more pronounced in CCC as compared to ordinary LC in which single columns are used. Apart from the extra band broadening
that occurs through the use of multiple columns, extracolumn effects (2) play an
important role. Void volumes attributable to connecting tubings or valves must be
kept to a minimum so that efficiency and sensitivity are not impaired, and in cases
in which peak compression can be obtained extracolumn effects can be partly
or totally eliminated. Band broadening can also occur if the injected solute is dissolved in a strong eluting solvent.

Valve Switching

For automation, either pneumatically or motor-driven valves can be used. In the applications described above, six-port valves were used. A motor-driven valve might be preferred in cases in which pressurized gases are not available; however, pneumatically driven valves can operate faster, which might be a plus for extending the lifetime of columns (41). In the study of enprofylline in plasma, the valves could be moved about ten times faster by adding pneumatically operated pilot valves to the original configuration (Table VI).

Control of Valve Switching

For automation, it is necessary to use a system controller for activation of valves at preprogrammed times. The system controller can be either a stand-alone device or a part of the LC system, as is the case with an integrator with external events. Stand-alone systems are preferred, however, because they allow for increased sample throughputs in that new samples can be injected into the CC system while another sample is under analysis (Table VI).

Use of a Precolumn

The use of a precolumn in CCC has several advantages and allows for simplified work-up, higher total absolute recovery, less risk of losses to adsorption or sorption during sample work-up, automation, preseparation and further cleanup of the sample, increased detectability through handling of large sample volumes, and, finally, transfer of a minor volume of sample medium to the analytical column with less disturbance of the detector signal and the separation.

The precolumn should contain an amount of packing material sufficient to allow trace enrichment of the desired amount of analyte from a given volume of sample solution. The void volume should be low to prevent extracolumn band broadening and to minimize the trapping of constituents that may contaminate the analytical column or interfere with the detection. Back pressure limitations should not be set by the precolumn during sampling at high flow rates.

In addition, the precolumn should be easy to replace and repack. The packing material should display high retention for the analyte during the sampling step and have high loadability — that is, pellicular packings are not recommended. High retention during sampling must be combined with negligible retention during desorption in order to minimize extracolumn band broadening during elution.

In an investigation of Goewie and co-workers (42), it was shown that, for most practical purposes, precolumns having an inner diameter of 2–4.6 mm and a length of 2–10 mm are recommended. For very polar compounds, longer columns might be needed as in the examples above.

Chromatography

It was mentioned above that polar as well as nonpolar mobile phases could be used but that they must be miscible. Another necessity is that, if the collected fraction from one column is to be transferred to another column, the mobile phase in the first column must not be a strong solvent in the second column to avoid band broadening. If large volumes are to be injected, this criterion becomes even more critcal. In all but one (39) of the examples given above, polar mobile phases have been used (Table V). The possibilities of combining different modes of chromatography for on-line coupling have been further discussed by Majors (6).

Column Packing Material

In most LC applications, silica-based column packing materials have been used. Because they are sensitive to both chemical (pH must be between 2–7) and mechanical degradation, their use in a laboratory in which uninterrupted work is needed may sometimes be troublesome.

For some applications an alternative might be to use the new type of packing materials based on polymers (43) — for example, poly(styrene-divinylbenzene). In CCC, where on-line sample work-up may be performed, this type of packing material may be of special value because the materials are both chemically (pH can range between 1–13) and mechanically stable. As was shown in the discussion of enprofylline in plasma, a poly(styrene-divinylbenzene) based material may be used for selective cleanup of the sample at pH 12.

Batch-to-batch variation is another problem with chemically bonded packing materials (2). This drawback might be partly obviated in CCC, where the sample is preseparated and a cut is transferred to an analytical column for final separation, because only a limited number of compounds have to be separated.

Miniaturization

The use of miniaturized columns with interal diameters in the range 10–3000 μm is increasing (44–47). Miniaturization offers some advantages as compared to conventional LC: for example, reduction of cost for packing material and solvent, the possibility of direct coupling to a mass spectrometer, generation of high plate numbers, and lowering of detection limits are advantages gained with miniaturized columns.

For the smallest-diameter columns, very stringent requirements are placed on the equipment used (44), so they are not readily useful in practical work. But even a small reduction of the diameter from, for instance, 5 mm to 2 mm is attractive as it can reduce the consumption of solvent and packing material by a factor of about six. Additionally, 2–3 mm columns can be used with conventional LC equipment.

Although miniaturized columns have not as yet been used in CCC applications to any extent, their use should be attractive for the same reasons given above. As was shown by Scott (47), a CC design using CC system 1 could be used as an injection device for large volumes (400 μL) of serum onto a miniaturized column, thereby taking advantage of the high mass sensitivity of that column.

Internal Standards

To find an appropriate internal standard in CCC might be a more difficult problem than in conventional LC approaches, especially if multidimensional chromatography is used. Bioanalysts working in drug development, however, might find it easier because a large number of structures similar to an analyte are available.

In the examples shown above, internal standards were used — although it could be argued whether this is necessary when the samples could be directly injected with 100% recovery (17). But even in such cases, the use of an internal standard might be profitable when direct injection is used (with 100% recovery); in an earlier discussion of enprofylline, it was found that the use of an internal standard gave much better precision.

CONCLUSION

This chapter has summarized some of the coupled column approaches taken to date in the analysis of substances in drug development. Continuing work in this area will seek to exploit the key advantages displayed by such coupled systems, namely, speed, rationalization of approaches, automation, and general efficiency.

REFERENCES

(1) B.L. Karger, R.W. Giese, and L.R. Snyder, *Trends in Anal. Chem.* **2** (5), 106 (1983).
(2) L.R. Snyder and J.J. Kirkland, *Introduction to Modern Liquid Chromatography*, 2nd Ed. (John Wiley & Sons, New York, 1979).
(3) G. Schill, H. Ehrsson, J. Vessman, and D. Westerlund, *Separation Methods for Drugs and Related Organic Compounds*, 2nd Ed. (Swedish Pharmaceutical Press, Stockholm, 1983).
(4) J. Lawrence and R.W. Frei, *Chemical Derivatization in Liquid Chromatography*, Volume 7, J. Chromatogr. Library (Elsevier, Amsterdam, 1976).
(5) C.J. Little, O. Stahel, and K. Hales, *Intern. J. Environ. Anal. Chem.* **18**, 11 (1984).
(6) R.E. Majors, *J. Chromatogr. Sci.* **18**, 571 (1980).
(7) D.H. Freeman, *Anal. Chem.* **53**, 2 (1981).
(8) R.W. Frei, *Swiss Chem.* **6** (11), 55 (1984).
(9) C.J. Little, D.J. Tompkins, O. Stahel, R.W. Frei, and C.E. Werkhoven-Goewie, *J. Chromatogr.* **264**, 183 (1983).
(10) M. Broquarie and P.R. Guinebault, *J. Liq. Chromatogr.* **4** (11), 2039 (1981).
(11) J. Lankelma and H. Poppe, *J. Chromatogr.* **149**, 587 (1978).
(12) S. Bergqvist and L.-E. Edholm, *J. Liq. Chromatogr.* **6** (3), 559 (1983).

(13) C.E. Werkhoven-Goewie, U.A.Th. Brinkman, and R.W. Frei, *Anal. Chem.* **53**, 2072 (1981).

(14) W.V. Voelter, T. Kronbach, K. Zech, and R. Huber, *J. Chromatogr.* **239**, 475 (1982).

(15) D.D. Koch and P.T. Kissinger, *Life Sci.* **26**, 1099 (1980).

(16) P.R. Johnsson and L.D. Bowers, *Anal. Chem.* **54**, 2247 (1982).

(17) W. Roth, K. Beschke, R. Jauch, A. Zimmer, and F.W. Koss, *J. Chromatogr.* **222**, 13 (1981).

(18) W. Roth and K. Beschke, *J. Pharm. Biomed. Anal.* **2** (2), 289 (1984).

(19) W. Kuhnz and H. Nau, *Ther. Drug Monit.* **6**, 478 (1984).

(20) U. Juergens, *J. Chromatogr.* **310**, 97 (1984).

(21) A. Nazareth, L. Jaramillo, B.L. Karger, R.W. Giese, and L.R. Snyder, *J. Chromatogr.* **309**, 357 (1984).

(22) L.R. Snyder, J.W. Dolan, and Sj. van der Wal, *J. Chromatogr.* **203**, 3 (1981).

(23) L.-E. Edholm, L. Heintz, and L. Ögren, manuscript in preparation.

(24) L.-E. Edholm and C. Bodenäs, manuscript in preparation.

(25) S.-E. Jacobsson, S. Jönsson, C. Lindberg, and L.-Å. Svensson, *Biomed. Mass. Spectrom.* **7**, 265 (1980).

(26) L.-E. Edholm, B.-M. Kennedy, and S. Bergqvist, *Chromatographia* **16**, 341 (1982).

(27) J.W. Cox and R.H. Pullen, *J. Chromatogr.* **307**, 155 (1984).

(28) M. Johansson and C. Svensson, poster presented to the 15th International Symposium on Pharmaceutical Sciences, 18–21 June 1984, Ronneby, Sweden.

(29) L. Weidolf, *J. Chromatogr.* **343**, 85–97 (1985).

(30) W. Lindner and C. Pettersson, in *Liquid Chromatography in Pharmaceutical Development: An Introduction*, I.W. Wainer, ed. (Aster Publishing Corp., Springfield, Oregon, 1985).

(31) Y. Tapuhi, N. Miller, and B.L. Karger, *J. Chromatogr.* **205**, 325 (1981).

(32) A.-B. Jeppsson, U. Johansson, and B. Waldeck, *Acta Pharmacol. et Toxicol.* **54**, 285 (1984).

(33) D.W. Armstrong and W. De Mond, *J. Chromatogr. Sci.* **22**, 411 (1984).

(34) S.F. Chang, T.M. Welscher, and R.E. Ober, *J. Pharm. Sci.* **72**, 353 (1984).

(35) H.T. Smith and W.T. Robinsson, *J. Chromatogr.* **305**, 353 (1984).

(36) C.M. Riley, L.A. Sternsson, and A.J. Repta, *J. Chromatogr.* **276**, 93 (1983).

(37) C.M. Riley, L.A. Sternsson, and A.J. Repta, *J. Chromatogr.* **229**, 373 (1982).

(38) J.A. Apfel, T.V. Alfredsson, and R.E. Majors, *J. Chromatogr.* **206**, 43 (1981).

(39) J.W. Cox and R.H. Pullen, *Anal. Chem.* **56**, 1866 (1984).

(40) L. Ögren, E. Albertsson, and L.-E. Edholm, manuscript in preparation.

(41) M.C. Harvey and S.D. Stearns, *Anal. Chem.* **56**, 837 (1984).

(42) C.E. Goewie, M.W.F. Nielen, R.W. Frei, and U.A.Th. Brinkman, *J. Chromatogr.* **301**, 325 (1984).

(43) J.R. Benson and D.J. Woo, *J. Chromatogr. Sci.* **22**, 386 (1984).

(44) F.J. Yang, *HRC&CC, J. High Resolut. Chromatogr. Chromatogr. Commun.* **6**, 348 (1983).

(45) M. Novotny, *Anal. Chem.* **53**, 1294A (1981).

(46) M. Novotny, *Clin. Chem.* **26**, 1474 (1980).

(47) R.P.W. Scott, *J. Chromatogr. Sci.* **18**, 49 (1980).

PART FOUR

LC IN THE QUALITY CONTROL OF MARKETED PHARMACEUTICALS

TLC AND HPTLC IN THE QUALITY CONTROL OF PHARMACEUTICALS

Robert A. Egli

Cilag AG
CH 8201 Schaffhausen, Switzerland

Recent developments in thin-layer chromatography (TLC) have increased the utility of the technique for quality control of pharmaceuticals. This chapter will survey some of these developments in an attempt to encourage HPLC specialists to use TLC in addition to HPLC in cases in which TLC may provide additional information or may be a more practical approach to solving a particular problem. Chromatographic behavior on silica gel as well as reversed-phase (RP) C_{18} phases will be compared in TLC and HPLC methods, and versatile and efficient acidic, basic, and neutral mobile phases of varying elution power for silica gel and RP C_{18} phases will be recommended. In this discourse, it will become apparent that TLC is a simple working technique that can be used to improve productivity in quality control departments of pharmaceutical companies.

When compared with HPLC, TLC is advantageous because, from the starting line of sample application to the solvent front, nearly all components are detectable; versatile detection schemes, such as iodine chambers and a great number of specific detection reagents are available, and substances that are not UV-absorbing can easily be detected; many substances or many samples, including reference materials, can be chromatographed in one run; and dirty or turbid samples can be analyzed. TLC also is simple and inexpensive, methods development and elaboration are fast and easy, and UV-absorbing or corrosive mobile phases can be used to develop a chromatogram. The latter point is especially important because some of the most efficient mobile phases used in silica-gel TLC of pharmaceuticals usually cannot be used in HPLC because of the UV-absorbing properties of the solvents.

HPLC does have its advantages over TLC, especially in routine assays, because HPLC assays can be fully automated and are more accurate than those of quantitative TLC; HPLC calibration curves are usually linear, in contrast to TLC (1); resolution and UV sensitivity are better; substances sensitive to light or oxygen are less of a problem to analyze by HPLC than with TLC; and volatile substances can be analyzed. Nonetheless, TLC is valuable in the quality control of pharmaceuticals when it is used for identification, purity testing, and stability testing. These points will be considered in more detail in the sections that follow.

FUNDAMENTAL PRINCIPLES OF TLC

Thin-layer chromatography uses open layers of sorbents on plates or foils to separate components of a sample. Solvents or solvent mixtures (solvent systems) move through the sorbent by capillary forces, enabling the separation of mixtures applied at the starting point. Along with paper chromatography, TLC is considered to be planar or flat-bed chromatography, which in turn is a type of liquid chromatography. Although the principles and the theory of TLC are similar to column liquid chromatography, there are some major differences between the two techniques.

In TLC, the sorbent is usually dry and not in equilibrium with the mobile phase, so the composition of the mobile phase and with it the activity of the sorbent usually vary over the distance from the starting point. In addition, detection in TLC is performed after the chromatography is finished. Instead of capacity factors (k'), TLC uses R_F values to express the position of a spot on a developed chromatogram. The R_F values are the ratio of the distance of the spot center from start to the distance of the solvent front from start. In a thin-layer chromatographic system, R_F values between 0.25–0.6 are preferred. For convenience, the quantity hR_F — defined to be 100 times R_F — is often used.

Other basic principles of TLC involve resolution, efficiency, and selectivity. The *resolution, R,* is the difference of the moving distances of two spot centers divided by the average spot diameter. According to Delley and Székely (2), the spot diameters must be measured at spot sizes about eight times the detection limit either by eye or preferably by densitometry at one-eighth of the peak height.

The *efficiency, E,* in TLC is, according to Delley and Székely (2), the spot distance from the start divided by the spot diameter. For comparisons of efficiency, R_F values of around 0.5 should be chosen. The efficiency can be used to calculate the number of theoretical plates, N. In TLC theoretical plate numbers from 1000 to 5000 can be obtained using the formula:

$$N = 16 \cdot E^2 \qquad [1]$$

The *selectivity, S,* in TLC is the difference of the moving distances of two spots divided by the average moving distance as measured from the start. The selectivity relates to the resolution as follows:

$$R = E \cdot S \qquad [2]$$

A large number of monographs (3–16) and reviews (17–19) have discussed the principles, techniques, and aspects of TLC, and a current listing of papers is available in the bibliography section of the *Journal of Chromatography* (20). This chapter, therefore, will cover only the most important practical aspects regarding the TLC of pharmaceuticals.

Sorbents for TLC

Most drug substances have polar functional groups and are, therefore, well suited for adsorption chromatography on silica gel. Silica gel is, by far, the most widely used sorbent: The majority of drug substances are similarly well suited for reversed-phase TLC on bonded-phase plates, but these plates are less known and more expensive. Other TLC sorbents are very seldom needed. This discussion will include only practical details regarding silica gel and RP C_{18} sorbents.

Silica gel separates mixtures according to differences in polarity — that is, polar substances need polar solvent systems. Functional groups in a molecule provide retention in an increasing order as follows:

$$\overset{|}{\underset{|}{C}} = \overset{|}{\underset{|}{C}} \; < -Cl < -OCH_3 < -NO_2 < -N(CH_3)_2 < -COCH_3$$

$$< -OCOCH_3 < -NH_2 < -NHCOCH_3 < -OH < -CONH_2 < -COOH$$

When the molecule contains two or more functional groups, an especially high retention can be expected. When two molecules have differences only in the nonpolar part of the molecule, differences in number of $-CH_3$ or $-CH_2-$ groups, for example, separation often is not possible.

Reversed-phase C_{18} sorbents usually have an opposite retention behavior — a low retention for polar substances and a high retention when the molecule contains a nonpolar part. The author's experience has been that there is usually no advantage in using RP C_8 or RP C_2 instead of RP C_{18} sorbents for the TLC of pharmaceuticals.

Solvent Systems

The solvent systems are the most important variables in silica-gel TLC. In the development of TLC quality control (QC) methods for pharmaceuticals, an approach that uses some versatile solvent systems of different selectivity is much more successful than one that varies the types of chambers, activity of sorbent, humidity, saturation of chamber, temperature, or brand of plate. Versatile solvent systems in silica-gel TLC usually consist of at least two components, one of which is often chloroform or dichloromethane. These solvents, however, present an environmental and health problem and can, in most cases, be replaced by superior solvent systems (21). One of the consequences of a two-component system is the phase gradient over the total developing distance: the greatest polarity is found near the start and the lowest near the solvent front.

Spotting the Plates and Development

The best resolution in TLC is obtained using small sample amounts in small volumes (0.5–1 μL, for example), and small starting spots are desired. These conditions are necessary for identification and for densitometric (quantitative) assays. For purity testing, however, relatively large sample amounts are needed to enable the detection of low levels of impurities. In such cases it would not be advantageous to apply a sample smaller than 2 μL, because big main spots result from the larger loading regardless of the volume as a result of sorbent capacity.

Developing distances farther than 10 cm (more than 6 cm for HPTLC plates) are seldom favorable because detection limits increase and a longer developing time is needed. The spot sizes also increase with the distance traveled because of diffusion, so there is usually not much gain in resolution afforded by developing for distances more than 10 cm.

Reproducibility

To obtain reproducible R_F values and separating power, TLC conditions must be kept constant. This constancy is only partly possible, so reference standards are usually run concomitantly.

In the author's experience, best reproducibilities are obtained using normal chambers without saturation (no paper wicks) and with commercially available TLC plates put diagonally into the tank with the coating downwards. The starting line on normal plates should be 10 mm above the immersion height of the solvent system. Sunlight, drafts, or changes in room temperature should be avoided.

TLC AND HPTLC

There is no clear definition for high performance TLC (HPTLC). According to the proponents of HPTLC (7,12,22–24) the following conditions are the most important in HPTLC:

- use of HPTLC plates, which have a slightly lower mean particle size and a narrower sorbent particle size distribution as well as a slightly reduced thickness compared with normal TLC plates
- starting spot diameters of not more than 1 mm and small sample amounts between 0.005 μg and 0.5 μg (5–500 ng)
- development distances between 2 cm and 6 cm
- use of special development chambers (for example, a "U" chamber for circular or anticircular chromatography).

From comparisons of commercially available precoated HPTLC and TLC plates done by Brinkman and co-workers (26,27) and the author's experiences with silica-gel plates from Merck, it can be concluded that HPTLC plates have their highest efficiency between 4 cm and 6 cm development distance (2–2.5 cm using the "U" chamber). Resolution can be increased by about 10%–30% on HPTLC plates, but with longer development distances the efficiency decreases rapidly because of the lower mobile phase velocities in HPTLC. The decrease is less pronounced in the HPTLC of relatively big molecules such as the usual test dyes. Normal TLC plates have their highest efficiencies around a 10-cm development distance. Delley and Székeley (2) obtained similar efficiencies with both kinds of plates with an 8-cm linear development distance.

It seems that the best approach for routine TLC work is to apply part of the HPTLC working technique — that is, small starting spots and development distances as short as possible — on normal 10 cm × 10 cm TLC plates. In this manner, an 8-cm development distance on normal plates usually gives better resolution than does 4-cm development using HPTLC plates. The longer development time in normal TLC is more than compensated for by its much easier working technique. In fact, for purity testing, HPTLC is definitely less suited than TLC for sorbent capacity reasons.

Because there is no general advantage and no clear definition of HPTLC, the author prefers to avoid the expression HPTLC and call both kinds of TLC simply TLC. The commercially designated HPTLC precoated plates may just be regarded as higher-priced TLC plates. Nevertheless, HPTLC plates are a valuable additional tool for special purposes.

SORBENTS AND LAYERS

Ready-to-use TLC plates usually give superior and more reproducible results than do laboratory-prepared plates. Plates with the following sorbents on glass, aluminum, or plastic foils — and with and without binders and fluorescent indicators — are available from several suppliers:

- Silica-gel 60 (pore size 6 nm)
- Silica-gel 40 (pore size 4 nm)
- Silica-gel 100 (pore size 10 nm)
- Silica-gel 150 (pore size 15 nm)
- SiO_2 50,000
- Kieselguhr (diatomaceous earth)
- Aluminum oxide
- Magnesium silicate
- Cellulose
- Cellulose (acetylated)
- Ion-exchange cellulose
- Polyamide
- RP C_{18} silica gel
- RP C_8 silica gel
- RP C_2 silica gel
- NH_2 silica gel
- Ion-exchange resins
- Mixtures of silica gel and cellulose (silcel)
- Mixtures of Al_2O_3 and acetylated cellulose (alox/cel).

Most plates are available as standard or HPTLC plates (nano plates) with these sorbents. Many kinds of plates can also be obtained with a low adsorptive concentrating zone at the start. If iodine detection is to be used, glass plates have the lightest background and therefore the highest sensitivity.

Despite the variety of available tools, in the quality control of pharmaceuticals more than 95% of the problems encountered can be solved on standard silica-gel 60 or RP C_{18} silica-gel plates. Humidity, however, can have a considerable influence on the TLC result when using silica gel plates. Drying the plate before application does not make much sense because the adsorption of water from the air is very rapid. According to Geiss (4), one-half of the possible uptake of water occurs within 90 seconds. Plates in equilibrium with air of around 50% relative humidity are well suited in most cases. If drying is needed, the author recommends drying the plates after sample application for 30 min in an oven at 50 °C, cooling them rapidly on a thick plate, and putting the subject plate into the developing chamber immediately. The samples, of course, must be stable enough to withstand heating. RP C_{18} plates do not need drying before use, but care must be taken that they do not take up solvent vapors from laboratory air during storage.

SAMPLE PREPARATION AND APPLICATION

Water and viscous and high-boiling-point solvents should be avoided if possible when the sample is dissolved. Appropriately dissolved sample solutions then can be applied as spots or streaks. It is preferable to use the simpler spot application onto points that are marked with a soft pencil. To obtain as small starting spots as is possible, the solvent for the sample should have an elution power as low as possible. This criterion ensures that the sample remains near the capillary outlet. Small sample volumes are even more effective for obtaining small starting spots and good resolution. The smallest amount for easy and reproducible application is approximately 0.5 μL. The author prefers to use 1-μL Microcaps (Drummond Scientific, Broomall, Pennsylvania), cut in half with the even side down, in a Scilab device (Scilab, Therwil, Switzerland).

In an extreme case, when the sample moves with the front of the application solvent, the following starting spot sizes are obtained on 0.25-mm silica-gel plates:

sample volumes:	0.2 μL	0.5 μL	1 μL	2 μL	10 μL
starting spot diameters:	1 mm	2 mm	3 mm	4 mm	10 mm

Volumes greater than 2 μL are usually applied stepwise with drying between steps or by drying with air during application. Robots that deliver small sample increments between selected time intervals are available, as is an instrument for contact spotting (12) and a fully automatic TLC spotter (Camag, Muttenz, Switzerland). Too high a concentration of sample on too small a starting spot or sample amounts that are too large may cause tailing if the substance is not soluble enough in the solvent system. Insufficient drying of applied aqueous solutions may also cause tailing. To avoid this, dry the plate at room temperature for at least 30 min after sample application.

Because the mobile phase velocity is extremely high at the beginning of the chromatographic process, the samples should be spotted about 10 mm above immersion level in the tank (15 mm above the edge at a 5-mm solvent system level and at least 15 mm from the side edge).

Usual sample amounts used in spotting are:

- densitometry, 0.1 μg–1 μg (0.5 μL 0.02%–0.2% solutions)
- identifications, 1 μg–20 μg (1 μL 0.1%–1% solutions)
- purity testing, about 100 μg (2 μL 5% solutions)

CHOICE OF DEVELOPING CHAMBERS

For routine work with 20 $\times$ 20 cm plates one can use the normal 6 $\times$ 21 $\times$ 21 cm flat-bottom chamber (tank) — which requires 50 mL of solvent to obtain a solvent height of 5 mm — or the twin-trough chamber from Camag, which needs about 15 mL of solvent. For 10 $\times$ 10 cm plates, the same two types of chambers in the corresponding smaller size are available. Each needs about 20 mL or 6 mL of solvent, respectively, for a solvent height of 5 mm. The best reproducibilities and nearly even front lines are obtained when normal flat-bottom chambers are used without a paper wick and the plates are inserted diagonally with the coating downwards. Sandwich chambers also give good results, but for the reason of rationalization one should avoid them as well as other kinds of chambers in routine work.

"U" chambers are recommended (7,12) for development distances of either 20–25 mm or 30–40 mm. In combination with HPTLC plates, these chambers can be used for the circular and, more infrequently, the anticircular technique to solve special problems that might not be approached as well using linear development, such as using TLC as a pilot technique for LC. The disadvantages are that a great deal of skill is needed to use these techniques and that only a few samples can be applied. Furthermore, two-dimensional TLC is not possible when a "U" chamber is used for development.

DEVELOPMENT OF THE TLC PLATES

In silica-gel TLC, vapor saturation in the chambers usually creates poorer separations and should be avoided. When changing from a method with a paper wick to one without, the solvent system may require adjustment to a lower polarity. RP C_{18} plates also do not need chamber saturation.

Because mobile phase velocity decreases with time, short developing distances can save time (about one-fourth of the time is needed for half the distance) and also gain sensitivity. Delley and Székely obtained similar efficiencies at 8-cm and 14-cm development distances on silica-gel TLC plates, but for 14 cm they needed 5 μL instead of 1 μL application solution to obtain the same detection limit as for 8 cm (2). Values of R_F, however, may differ considerably for different developing distances, and a constant distance should be maintained to gain reproducible results.

When TLC is used to monitor organic syntheses for starting materials, products, and decomposition products, it may be of special value to use development distances as short as possible to obtain quick results while the reaction is still running. The preferred developing distances in quality control for silica-gel TLC are 7–10 cm. For RP C_{18} TLC, it is preferable to use developing distances of 6–8 cm (using normal plates in both cases). In addition, if very sharp starting spots or lines are needed, the author uses a short prerun with a solvent of a high elution power.

SOLVENT SYSTEMS FOR SILICA-GEL TLC

Obviously, most drug substances are either acids or bases and have different R_F values in acidic and basic solvent systems of similar polarity. Using both kinds of solvent systems enhances the probability of finding impurities as well as the reliability of identifications. Acidic, basic, and neutral solvent systems all have the potential to be suitable for acidic, basic, or neutral substances, but the acidic solvent sys-

TABLE I

Properties of Solvents for LC and TLC[*]

Solvent	UV cutoff (nm)	Boiling point (°C)	Activity coefficient	Eluotropic strength (normal phase)[‡]	Eluotropic strength (reversed phase)
n-Hexane	195	69	–	0.0	–
Cyclohexane	205	81	–	0.4	–
Toluene	285	111	0.8	2.9	–
Diethylether	210	35	1.2	3.8	–
Chloroform	245	61	1.0	4.0	–
Dichloromethane	232	40	–	4.2	–
Tetrahydrofuran	220[†]	66	–	4.5	4.4
Dioxane	215	101	3.2	5.6	3.5
Acetone	330	56	5.6	5.6	–
Ethylacetate	260	77	2.9	5.8	–
Acetonitrile	191	82	–	6.5	3.1
2-Propanol	207[†]	82	5.2	8.2	4.2
Ethanol	200	78	6.9	8.8	3.6
Methanol	204[†]	65	10.0	9.5	2.6
Water	<190	100	18.6	high	0.0

[*] Data from References 9, 10, 24, and 33 and the author's unpublished observations.
[†] UV cutoff point at absorbance, $A = 1.0/1$ cm.
[‡] Elution power on Al_2O_3, $E^0 \cdot 10$.
[§] Defined in Reference 33.

tems are usually the best. One limitation is that the samples must have enough solubility in the solvent system. Properties of common solvents are listed in Table I.

From the author's experience with drug substances, five acidic, five basic, and five neutral solvent systems with increasing elution power can be recommended (Table II). Numbers 2 and 3 of each group are of similar elution power but have different selectivity. The most universal systems are marked with asterisks. Elution powers can be varied simply by varying the percentage of toluene in the mixtures, and, for most routine work, technical-grade solvents can be used. Solvent systems B3 and B4 (with NH$_3$) are unstable and must be prepared each time they are needed, but the other solvents can be stored in bottles and also can be used several times. The mixtures containing NH$_3$ should be used only once and in an

TABLE II

Solvent Systems for Silica-gel TLC Listed by Increasing Elution Power

Acidic solvent systems	Volumes
(A1) Toluene + 100% formic acid	98 + 2
(A2*) Toluene + ethyl acetate + 85% formic acid	50 + 45 + 5 or
(A3) Toluene + methyl ethyl ketone + 85% formic acid	45 + 50 + 5
(A4) Toluene + acetone + 2 N formic acid	30 + 65 + 5
(A5*) Toluene + 2-propanol + ethyl acetate + 2 N acetic acid	10 + 35 + 35 + 20
Basic solvent systems	
(B1) Toluene + 2-propanol + diethylamine	85 + 10 + 5
(B2*) Toluene + 2-propanol + NH$_3$ (concentrated)	70 + 29 + 1 or
(B3) Toluene + acetone + NH$_3$ (concentrated)[†]	40 + 58 + 2
(B4) Ethyl acetate + methanol + NH$_3$ (concentrated)[†]	85 + 10 + 5
(B5*) Toluene + dioxane + methanol + NH$_3$ (concentrated)	20 + 50 + 20 + 10
Neutral solvent systems	
(N1) Toluene + acetone	95 + 5
(N2*) Toluene + ethyl acetate	50 + 50 or
(N3) Toluene + 2-propanol	90 + 10
(N4) Toluene + ethyl acetate + ethanol	60 + 20 + 20
(N5*) Toluene + acetone + water	25 + 70 + 5

[*] most versatile
[†] freshly prepared

amount sufficient to enable NH_3 to permeate the atmosphere in the chamber (flat-bottom). If diethylamine is used instead of NH_3, it has the disadvantage of interfering with iodine detection. Whenever possible, one should avoid the use of chloroform or dichloromethane because of their toxicity and the disposal problem. Furthermore, these solvents can produce explosive mixtures with acetone in the presence of amines or alkali (25).

Analysis of TLC data in an earlier study showed that the acidic mobile phase A2 and the basic mobile phase B5 were the most efficient for 35 acidic and 47 neutral drug substances; A5 and B2 were best for 69 basic drug substances (28). In another study, 22 of 28 drug substances investigated were acidic or neutral (21). Superior TLC results were obtained for the 22 substances with mobile phases A2 and B5 compared with results obtained using four solvent systems suggested by Stead and associates (29); the number of acceptable spots with R_F values between 0.10 and 0.90 using A2 and B5 combined was 42 of 44.

Solvent System for RP C_{18} TLC

The solvent systems for RP C_{18} TLC consist of mixtures of an aqueous 3% sodium chloride solution with water-miscible organic solvents. The usual solvents, such as methanol, ethanol, 2-propanol, acetonitrile, acetone, dioxane, and tetrahydrofuran (THF), demonstrate much lesser selectivity differences than are seen with silica-gel TLC. For method elaborations, therefore, it is even more important than in silica-gel TLC to try an acidic and a basic solvent system, especially when chromatographing basic substances. In purity testing of 22 acidic and neutral substances (21) the greatest number of acceptable spots with R_F values between 0.10 and 0.90 (40 of 44) were obtained when the following two solvent systems were used:

(1) tetrahydrofuran + methanol + 0.5% phosphoric acid + 3% sodium chloride
   28   +   28   +   10   +   34 (volumes)

(2) methanol + 2 *N* ammonia + 3% sodium chloride
   60   +   10   +   30 (volumes).

COMPARISON OF REVERSED-PHASE AND SILICA-GEL TLC

The strong retention of ions on silica gel means that, in general, a strongly polar, basic solvent system and a relatively weakly polar, acidic mobile phase are needed for the TLC of acids and that a strongly polar, acidic mobile phase and a relatively weakly polar, basic mobile phase are needed for the TLC of bases. In RP TLC, the ionization effect is reversed and much weaker. Furthermore, the differences between the R_F values obtained with solvent systems of similar eluotropic strength are much lower than on silica gel. In identity testing by RP TLC, therefore, the use of more than two solvent systems does not bring out very much additional information. With RP TLC it is very easy to find a good mobile phase because there are only a few volatile water-miscible solvents available, and most of these are generally suitable; the wider variety of good mobile phases available in silica-gel TLC, however, increases the separation potential of this technique.

During purity testing, 31 impurities in 22 acidic and neutral drug substances were found on silica gel with the mobile phases A2 and B5, and 36 impurities were found using the RP solvent systems 1 and 2. Only 21 of the total of 44 impurities revealed by using all four solvent systems were found on silica gel as well as on the RP bonded phase. It must be considered that separation on silica gel is attributable mainly to differences in the polarities of the substances; on RP plates, solubility in the solvent system is influential, as are structural differences in the nonpolar part of the molecule. It is therefore advisable in purity testing to use both TLC methods.

DISCUSSION OF RP TLC OPTIMIZATIONS IN IDENTITY AND PURITY TESTING

For unknown substances, the author recommends starting with the two solvent systems mentioned above, even though Brinkman and de Vries (30) suggest that mixtures with acetone be used as the first choice. To solve selectivity problems other solvents or mixtures may be used, but acetonitrile is to be avoided because of its toxicity. Sulfuric acid may be used instead of phosphoric acid, and for neutral substances neutral solvent systems can be used. The samples must, however, have adequate solubility in the solvent system. If the R_F values are too low or too high, the concentration of the organic solvent must be adjusted according to the empirical rule that a reduction of 10 mL of methanol per 100 mL of solvent decreases the R_F value by about 0.10–0.20. Sodium chloride in the solvent systems enhances the wettability of the plates and is especially necessary when the water content exceeds about 20%. Considerable differences in separation may be obtained if sodium chloride is not added. A water content exceeding about 60% usually produces poor spot shapes.

Brinkman and de Vries compared eight types of commercially available RP TLC plates (26). One of their conclusions was that the different behavior of the eight types increases the potential of the technique as a tool for separation. The author has used Merck RP-18 and Whatman KC 18 plates, and each has its advantages. Using spray reagents as well as UV detection and an iodine chamber, better sensitivities were obtained by the author with Whatman KC 18 plates because of the lighter plate background (especially, for spots near the front, in UV light). Furthermore, Whatman KC 18 provide faster runs and as a consequence more compact spots.

COMPARISON OF TLC AND LC ON SILICA GEL

Most of the versatile and efficient TLC mobile phases containing toluene, acetone, or ethyl acetate for use with silica-gel plates cannot in most cases be used in silica-gel LC because of their UV-absorbing properties. Thus, the solvents n-hexane or dichloromethane are mostly used as the main component along with methanol, 2-propanol, or THF. The resulting TLC solvent systems are relatively inefficient, and their versatility is limited. Four LC solvent systems were used for a TLC study:

- dichloromethane/methanol/acetic acid (90:8:2)
- n-hexane/THF/acetic acid (25:73:2)
- dichloromethane/THF/ammonia (60:39:1)
- n-hexane/THF/ammonia (25:74:1).

The number of spots in the author's study that had acceptable R_F values (ranging from 0.10 to 0.90 with the four solvent systems) was six times less than with the four best TLC solvent systems in Table II. The use of silica-gel TLC as a pilot technique for finding suitable silica-gel LC solvent systems thus can have only restricted applications.

COMPARISON OF RP TLC AND RP LC

Table III compares RP TLC and RP LC data for neutral substances; the data were obtained using methanol and acetonitrile mobile phases. The table illustrates the well-known fact that C_{18} LC columns from different suppliers have different selectivities; the same holds true about different brands of RP TLC plates.

Table IV gives an approximate idea of what k' value may be expected for a known R_F value when the same neutral solvent system is used. The values show good linearity using a value of 0.5 for k_f and the equation of Geiss (4),

$$k' = k_f \left[\frac{1}{R_F} - 1 \right] \qquad [3]$$

where k_f is the transference factor to convert from TLC to LC.

To obtain k' values between 1 and 4 in LC, it is necessary to use a solvent system that gives hR_F values between 10 and 30 in TLC. As an empirical rule in practical work, it was found that decreasing the methanol concentration by 10 mL per 100 mL of methanol/water mixture roughly doubles k' in RP LC and reduces the hR_F values in RP TLC by about 10–20.

Jost and co-workers (31) as well as Jork and associates (32) studied possibilities and limitations of linear TLC for the prediction of LC solvent systems. Alternatively, the "U" chamber (7,12) may have a special advantage when using TLC as a pilot technique for LC because the possibility of in-flow sample injection produces conditions that are more similar to LC than linear TLC. There may also be a great deal of difference in resolution between RP TLC and LC. It is certainly possible that two substances with the same hR_F value could be separated by LC because in LC selectivity may be slightly different or resolution may be better.

An advantage of RP TLC in purity or stability testing stems from the fact that low levels of impurities with high retention may not be detectable in LC because of peak broadening. RP TLC also gives the most sensitive or sharpest spots with highly retained substances. Furthermore, impurities in LC that have very low retention may appear unresolved near $k' = 0$, but in TLC they may separate between hR_F values of 50 and 90. Even impurities at the start line or the solvent front can be detected in RP TLC.

TABLE III

Comparison of Retention Capacity Factors (k') in LC with hR_F Values in TLC[*]

| | Methanol 60% (v/v) | | | | Acetonitrile 50% (v/v) | | |
| | RP LC | | RP TLC | | RP LC | RP TLC | |
Substance	k' SPH[†]	k' Bond[‡]	hR_F Whatman[§]	hR_F Merck[‖]	k' SPH	hR_F Whatman	hR_F Merck
Sulfanilamide	0.1	0.0	93	88	0.2	70	62
Paracetamol	0.2	0.1	80	72	0.2	71	62
Triamcinolone	0.6	0.6	44	38	0.3	65	58
Methylparaben	0.6	0.6	42	35	0.6	47	42
Chlorzoxazone	1.1	1.0	32	23	0.9	42	34
Estriol	1.2	1.0	29	22	0.5	58	47
Propylparaben	1.6	1.8	20	16	1.4	32	25
Triamcinolone acetonide	1.8	2.0	19	13	1.1	43	33
Dienestrol	3.5	4.4	11	7	2.6	22	17
Ethinyl estradiol	4.1	4.1	11	7	2.2	26	17
Norethisterone	4.3	3.3	12	7	2.6	22	14
Mestranol	16.5	19.5	0	0	7.9	8	4

[*] k' values are calulated using the methanol peak as zero retention ($k' = 0$); retention times in minutes for SPH: $1.3 + 1.3 k'$; for Bond: $2.8 + 2.8 k'$.

[†] SPH = Spherisorb ODS 5 μm, 125 mm × 4.6 mm.

[‡] Bond = Micro-Bondapak C_{18}, 300 mm × 4 mm.

[§] Whatman = KC 18 F RP-TLC plates, 20 cm × 20 cm, 0.20 mm, No. 4803-800.

[‖] Merck = RP-18 F_{254} S plates, 10 cm × 20 cm, 0.25 mm, No. 15423.

TABLE IV

RP LC Predictions from RP TLC Values, Tested with Neutral Substances according to Table III

TLC (hR_F)	83	66	50	32	18	10	5	0
LC (k')	0.1	0.25	0.5	1	2	4	8	16

DETECTION METHODS IN TLC

When using a detection method with low sensitivity, often only the center of a spot (the tip of the peak) can be seen. This results in a spot diameter too small and in a resolution too good (18). Thus, an important aspect of TLC is an accurate and sensitive method of detection.

The most important TLC detection methods are fluorescence quenching of UV light, exposure to iodine vapor, and chemical reactions for specific compounds, in addition to a few special techniques. These techniques will be discussed briefly.

UV Detection

Usually TLC plates are prepared with fluorophores added to the sorbent. Upon irradiation by UV light at 254 nm (366-nm light is used infrequently) sample spots on the plate absorb some of this UV light and cause fluorescence quenching (dark spots on the fluorescing background). For substances that absorb light at other wavelengths, specialized TLC spectrophotometers (densitometers) are needed for detection. Densitometers also enable the recording of UV spectra on the plate.

Iodine Chamber

In an iodine chamber, a layer of a few millimeters of iodine crystals is put into a flat bottom development chamber with a ground cover, and the TLC plates are usually placed in it for 1 h. Iodine detection is more generally applicable and gives a better estimate of the amounts of unknown substances than UV detection. Compounds with unsaturated bonds react more quickly than saturated molecules in the iodine chamber.

Chemical Detection Reagents

These reagents are usually applied by spraying the plate, but sometimes the plate is dipped into the reagent. Several hundred of these reagents are known, but the most important ones and the types of compounds they detect are:
- iodoplateate (N-containing heterocycles)
- Dragendorff (alkaloids and bases)
- molybdatophosphoric acid with 10-min heating to 100 °C–110 °C (many pharmaceutical substances)
- sulfuric acid/anisaldehyde with heating (many naturally occurring drugs)
- derivatization with several fluorescent reagents — for example, fluorescamine — permits detection of substances with amino and phenolic hydroxyl groups with very high sensitivity (18).

Universal Special Methods

For silica-gel TLC, diphenylamine (0.1%) in acidic solvent systems and, after drying, irradiation with 254-nm UV light gives colored spots for many substances (34). Another reagent for silica-gel TLC is phenothiazine (0.05%) added to the solvent systems. The plates are dried and put into an iodine chamber for 1 h, into a hood for 10 min, and then into a double-trough chamber with NH_3 vapors. This method gives different colored spots for most pharmaceutical substances (35). The method is not very sensitive and usually needs 1 μL of 0.2%–2% solutions, but it greatly enhances the reliability of identifications.

Water soaking by spraying until the plate is completely wet (or immersion into water for 1 s) followed by partial drying at room temperature until the spots are just of an optimal visibility is a very sensitive and universal detection method. It gives the most consistent sensitivities for different kinds of substances (that is, similar amounts produce similar intensities).

QUANTITATIVE TLC

The methodology and theory of quantitative TLC were described by Ebel (13), and the literature has been reviewed by Jork and Wimmer (38). Determinations of drugs and metabolities in biological samples are described in the books of Heftman (16) in part B and Götz and associates (39). Important points will be considered here.

Most quantitative analyses are carried out in situ, and the UV absorption of the spots on the plate is measured with a densitometer. Reference standards must be run on the same plate as the samples. The plate is moved automatically past a slit with a light beam of a chosen wavelength, the remitted (reflected) light is measured, and the signal is recorded. Transmission measurements are also possible, but not in the short UV range because of the absorptivity of the sorbent. Computerized densitometric systems are available (12).

Peak heights or peak areas are not directly proportional to the amounts of substance in the TLC spots; Figure 1 is a typical example. A good linearization of the calibration curves by a quadratic-logarithmic approximation with a computer program is given by Müller (36). TLC reflectance calibration curves can best be linearized (1) using a pocket calculator such as a Hewlett-Packard HP-97 or Texas Instruments SR-52 using the equations:

$$\ln S = a \cdot F^b \tag{4}$$

or

$$\ln S = a \cdot H^b \tag{5}$$

where
S = the amount of substance

F = the peak area
H = the peak height
a,b = adequate scale figures for the concentration.

A good correlation of the calculator fit with the measured data was obtained (within the range of the TLC error) for 10 sampled drugs. For the amount of substances, S, a suitable scale must be chosen empirically using a calculator; in most cases 10–100 or 100–1000 and seldom 1000–10,000 gives the best curve fitting. For calibration ranges as narrow as 80%–100% (such as in assays of drug substances) or 75%–

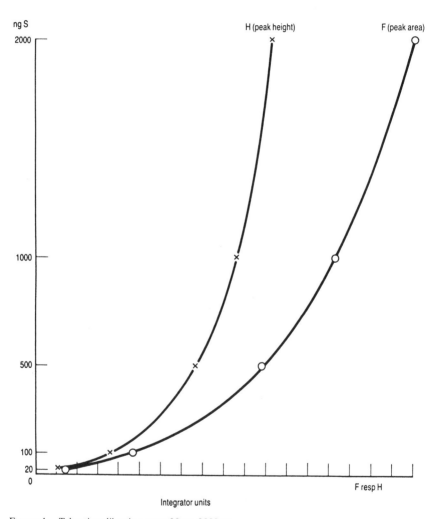

FIGURE 1.   Tolmetin calibration curve 20 ng–2000 ng

125% (for content uniformity testing) the function $S = a \cdot F^b$ (or $S = a \cdot H^b$) may be sufficient without taking the logarithms of S. Relative standard deviations of the results between 1%–3% can be obtained if a good working technique and enough skills are applied. The following procedure is recommended for calibration.

Plate Preparation

In plate preparation, the plate is first cleaned with a soft brush and checked by eye in visible and UV light for possible irregularities or contaminants. Next, the starting points are marked, using a very soft pencil and without damaging the layer, 15 mm above the lower edge (10 mm above the solvent system). Then, a 2-mm wide front line is scratched out 40–80 mm above the starting line, and again the plate is cleaned with a soft brush.

Application

To create starting points as small as possible, nonpolar solvents of a rather low volatility (such as toluene) are preferred — if the sample is soluble in that solvent. Sample size is preferably 0.5 μL (about 0.1–1 μg); this volume can be applied with a reproducibility of less than 0.5% relative standard deviation. Samples and reference standards are applied as data pairs (37). The same capillary is used to apply all the spots on one plate, and it is rinsed three times between each spot with the solution used for application. Finally, on both sides of the plate, additional edge spots are applied. These are never used for calculations and they serve to eliminate the possibility of an edge effect.

Conditioning

After sample application, the plates are dried at 50 °C to obtain a constant humidity. The plates are cooled to room temperature as quickly as possible on a thick plate immediately before development.

Development

For development, a mobile phase is chosen that gives a smooth baseline in measuring the R_F values, which preferably are between 0.25 and 0.60, produce regular round main spots, and separate side components well. After the solvent system has reached the carved front line, one should wait a few minutes until the layer shows a homogeneous transparency before removing the plate.

Measuring

The remitted UV light is usually measured using a slit width of 1 mm (a smaller width gives baselines that are not as good) and a slit length 1–2 mm larger than the diameter of the spot. For the estimation of unknown spots in purity testing, 1:1000 to 1:200 diluted standard solutions are applied for the comparison of peak heights or peak areas. The instrument is capable of measuring 26 spots (four times each) in the development direction within about 30 min. Each spot must be positioned to the minimum remission, and rectangular measuring takes only about 15 min and is therefore preferred despite the somewhat lower accuracy of that technique.

Calculations

The average of four measurements (twice back and forth) is usually used in calculations; if the standard deviation of the four is more than 5%, the experiment is repeated. Furthermore, in this approach the responses are averaged and all the data pairs are summed. In this way, differences caused by position (side or middle) are nearly eliminated and erroneous results (differences of more than 10% between the results of one data pair) can be recognized as outliers. Using elementary programming, a personal computer can do the curve fitting and calculation work (12,13).

COMPARISION OF QUANTITATIVE TLC WITH LC AND GLC

Table V summarizes the author's experiences in developing quantitative methods for the purity of drug substances using three chromatographic techniques: TLC, LC, and gas-liquid chromatography (GLC). TLC has the smallest number of unfavorable parameters (accuracy and reproducibility). Although TLC does

TABLE V

Comparison of Quantitative TLC with LC and GLC

Parameter	TLC	GLC	LC
Method development time	+	+ +	0
Analysis time	+	+ +	+ +
Costs	+ +	+	0
Accuracy and reproducibility	0	+	+ +
Separation power for impurities	+	+ +	+
Detection reliability for unknowns	+	+ +	0
Detection sensitivity	+	+ +	+
Analysis of low volatile substances	+ +	0	+ +
Analysis of temperature-sensitive substances	+ +	0	+ +

0 = unfavorable
+ = favorable
+ + = very favorable

not have as many *very* favorable parameters as the other two techniques, it is an effective method to use in testing the purity of drug substances.

PROCEDURE FOR CONTENT UNIFORMITY TESTING OF TABLETS

The following method illustrates how to use TLC to measure tablet content uniformity. Ten Luvatrene tablets (5 mg moperone · HCl) are each disintegrated in 1.00 mL water/methanol (v/v) for 10 min in an ultrasonic bath; 4.00 mL methyl ethyl ketone is then added to each sample; the mixture is shaken for 15 min and then is centrifuged. Next, 0.5 μL of the supernatant solution is applied on 20 cm × 20 cm Merck Kieselgel 60 F_{254} (0.25-mm) plates according to the following scheme:

$$R_E \ R_I \ R_{II} \ 1 \ 2 \ 3 \ 4 \ 5 \ R_{III} \ 6 \ 7 \ 8 \ 9 \ 10 \ R_I \ R_{II} \ 1 \ 2 \ 3 \ 4 \ 5 \ R_{III} \ 6 \ 7 \ 8 \ 9 \ 10 \ 10_E$$

where

E = edge spot, not to be used for calculations
R = reference spots (R_I = 75%, R_{II} = 125%, R_{III} = 100% of declared value)
1–10 = sample spots (tablet extract 1–10).

The spots are applied at a distance of 19 mm from the side edge with 6 mm between each pair of spots. The plate is dried at 50 °C for 30 min and, immediately after cooling down, is developed over a distance of 5 cm using toluene/2-propanol/NH_3 (concentrated) (84.5 + 15 + 0.5 volumes).

After air drying at room temperature for 30 min, the remission of the plate is measured perpendicular to the development direction at 252 nm using a densitometer with a 1 mm × 8 mm slit at a speed of 100 mm/min.

SPECIAL TECHNIQUES

Occasionally, some special techniques are used for pharmaceutical analyses. (The reader is referred to the original literature for more details.)

Impregnation

Sorbent impregnations are widely used, primarily to modify sorption properties for particular applications (16). Liquid stationary phases (paraffin oil, for example), ion-exchange materials, buffers, acids, bases, silver nitrate, or detergents are possible compounds for use in sorbent impregnation.

Ion-pair Chromatography

In TLC, ion-pairing is used much less than in LC. Studies with reversed-phase plates are described by Jost and associates (40) and Ruijten and co-workers (41). Basic drugs on silica gel were investigated by De Zeenw and associates (42).

Two-dimensional TLC

In two-dimensional TLC, samples are spotted in two corners of a TLC plate, developed first in one direction and then at 90° using a different solvent system. This technique in general surpasses the separation power of LC. Wilson used normal-phase TLC on silica gel with simultaneous paraffin impregnation to enable subsequent RP TLC in the second dimension (43).

Multiple Development

In multiple development, the plates are repeatedly developed in the same direction with the same or different solvent systems with drying in between. Programmed multiple development does this automatically (44). Burger made comparisons of multiple development schemes with LC gradient elution (45).

Continuous Development

An improved resolution for strongly retained substances can be provided by continuous development because the solvent is allowed to evaporate during development (8,46).

Sequential TLC

Sequential TLC is especially suited for preparative TLC (47), but it needs special instrumentation available from Scilab.

Overpressured TLC

The application of pressure in a special chamber (known as overpressured TLC) shortens migration times and enables long development distances without much spread of spots (48).

Enantiomer Resolution

Direct resolutions of enantiomers by TLC using a chiral sorbent (ionically bonded to aminopropyl silanized silica gel) were described by Wainer and coworkers (49). A process using RP plates [pretreated with a Cu(II) complex of N,N-di-n-propyl-L-alanine] was used by Weinstein for separating dansyl amino acids (50). Another method developed by Lepri et al. uses impregnated layers of silanized silica gel for resolution of enantiomers (51).

CONCLUSIONS

It is beyond the scope of this contribution to give a comprehensive overview of the TLC of pharmaceuticals, and the reader is referred to literature, especially References 3–21 and 52. In sum, however, TLC is a useful technique for quality control in pharmaceutical operations. Both silica-gel and reversed-phase plates can be used with a variety of solvent systems to monitor identity and purity testing. TLC methods supplant LC procedures in the pharmaceutical laboratory, and the relative simplicity of the methods and inexpensive cost of operation make it the method of choice in many applications.

REFERENCES

(1) R.A. Egli, H. Müller, and S. Tanner, *Fresenius' Z. Anal. Chem.* **305,** 267 (1981).

(2) R. Delley and G. Székely, *Chimia* **32,** 261 (1978).

(3) E. Stahl, ed., *Thin-Layer Chromatography,* 2nd Ed. (Springer, Berlin, 1969).

(4) F. Geiss, *Parameter der Dünnschichtchromatographie* (Vieweg, Braunschweig, 1972) and *Parameters of Thin-Layer Chromatography* (Hüthig, Heidelberg, Revised Edition, 1985).

(5) I.M. Hais, M. Lederer, and K. Macek, eds., *Identification of Substances by Paper and Thin-Layer Chromatography* (Elsevier, Amsterdam, 1970).

(6) G. Zweig and J. Sherma, eds., *Handbook of Chromatography,* Volumes 1 and 2 (CRC Press, Cleveland, 1972).

(7) A. Zlatkis and R.E. Kaiser, eds., *HPTLC: High Performance Thin-Layer Chromatography,* J. Chromatogr. Library Series, Volume 9 (Elsevier, Amsterdam, 1977).

(8) J. Kirchner and E.S. Perry, *Thin-Layer Chromatography* 2nd Ed. (John Wiley & Sons, New York, 1978).

(9) J.C. Touchstone and M.F. Dobbins, *Practice of Thin-Layer Chromatography* (John Wiley & Sons, New York, 1978).

(10) J. Gasparic and J. Churacek, *Laboratory Handbook of Paper and Thin-Layer Chromatography* (John Wiley & Sons, Chichester, U.K., 1978).

(11) J.C. Touchstone and J. Sherma, eds., *Densitometry in TLC, Practice and Applications* (John Wiley & Sons, New York, 1979).

(12) W. Bertsch and R.E. Kaiser, *Instrumental HPTLC* (Hüthig, Heidelberg, 1980).

(13) S. Ebel, in *Ullmanns Encyklopädie der Technologie Chemie,* Volume 5, H. Kelker, ed. (Verlag Chemie, Weinheim, 1980), pp. 205–213.

(14) J. Cazes, ed., *Thin-Layer Chromatography,* Chromatographic Science, Volume 17 (Marcel Dekker, New York, 1982).

(15) H. Wagner, S. Bladt, and E.-M. Zgainski, *Drogenanalyse* (Springer, Berlin, 1983).

(16) E. Heftman, ed., *Chromatography, Part A: Fundamentals and Techniques; Part B: Applications* (Elsevier, Amsterdam, 1983).

(17) E. Stahl, *J. Chromatogr.* **165,** 59 (1979).

(18) E. Stahl, *Angew. Chem.* **95,** 515 (1983).

(19) U.A.Th. Brinkman and G. de Vries, *HRC&CC, J. High Resolut. Chromatogr. Chromatogr. Comm.* **5,** 476 (1982).

(20) M. Lederer, ed., *Chromatographic Reviews,* Volumes 1–28 (Elsevier, Amsterdam, 1959–1984).

(21) R.A. Egli and S. Keller, *J. Chromatogr.* **291,** 249 (1984).

(22) H. Halpaap and J. Ripphahn, *Chromatographia* **10,** 613 and 643 (1977).

(23) H.E. Hauck and H. Halpaap, *Kontakte* (Merck, Darmstadt) **3**, 29 (1979).

(24) H. Wimmer and H.E. Hauck, *GIT Fachz. Lab.*, *Supplement Chromatographie* (Mannheim) **26**, 45 (1982).

(25) E.L. Grew, *Chemistry and Industry*, 491 (1970).

(26) U.A.Th. Brinkman and G. de Vries, *J. Chromatogr.* **258**, 43 (1983).

(27) U.A.Th. Brinkman and G. de Vries, *J. Chromatogr.* **198**, 421 (1980).

(28) R.A. Egli and S. Tanner, *Fresenius' Z. Anal. Chem.* **295**, 398 (1979).

(29) A.H. Stead, R. Gill, T. Wright, J.P. Gibbs, and A.C. Moffat, *Analyst* **107**, 1106 (1982).

(30) U.A.Th. Brinkman and G. de Vries, *J. Chromatogr.* **265**, 105 (1983).

(31) W. Jost, H.E. Hauck, and F. Eisenbeiss, *Fresenius' Z. Anal. Chem.* **318**, 300 (1984).

(32) H. Jork, E. Reh, and H. Wimmer, *GIT Fachz. Lab.* **25**, 566 (1981).

(33) T.I. Bulenkov, *Zh. Anal. Khim.* **23**, 848 (1968).

(34) R.A. Egli and S. Tanner, *Fresenius' Z. Anal. Chem.* **296**, 45 (1979).

(35) R.A. Egli, *Fresenius' Z. Anal. Chem.* **259**, 277 (1972).

(36) M. Müller, *Chromatographia* **13**, 557 (1980).

(37) H. Bethke, W. Santi, and R.W. Frei, *J. Chromatogr. Sci.* **12**, 392 (1974).

(38) H. Jork and H. Wimmer, *Literatursammlung Quant. Auswertung von Dünnschicht-Chromatogrammen*, Parts 1–4 (GIT-Verlag, Mannheim, 1984).

(39) W. Götz, A. Sachs, and H. Wimmer, *Dünnschichtchromatographie* (Gustav Fischer, Stuttgart/New York, 1978).

(40) W. Jost, H.E. Hauck, and H. Herbert, *Chromatographia* **18**, 512 (1984).

(41) H.M. Ruijten, P.H. Van Amsterdam, and H. De Bree, *J. Chromatogr.* **252**, 193 (1982).

(42) R.A. De Zeenw, F.J.W. Van Mansvelt, and Jan. E. Greving, *J. Chromatogr.* **148**, 255 (1978).

(43) I.D. Wilson, *J. Chromatogr.* **287**, 183 (1984).

(44) J.A. Perry, *J. Chromatogr.* **113**, 267 (1975).

(45) K. Burger, *Fresenius' Z. Anal. Chem.* **318**, 228 (1984).

(46) L. Zhou, C.F. Poole, J. Triska, and A. Zlatkis, *HRC&CC, J. High Resolut. Chromatogr. Chromatogr. Comm.* **3**, 440 (1980).

(47) P. Buncak, *Frsenius' Z. Anal. Chem.* **318**, 289 (1984).

(48) H.E. Hauck and W. Jost, *J. Chromatogr.* **262**, 113 (1983).

(49) I.W. Wainer, C.A. Brunner, and T.D. Doyle, *J. Chromatogr.* **264**, 154 (1983).

(50) S. Weinstein, *Tetrahedron Lett.* **25**, 985 (1984).

(51) L. Lepri, P.G. Desideri, D. Heimler, and S. Giannessi, *J. Chromatogr.* **265**, 328 (1983).

(52) K. Macek, ed., *Pharmaceutical Applications of TLC and LC* (Elsevier, Amsterdam, 1972).

COMPARISON OF INTERLABORATORY RESULTS OF DRUG ASSAYS: WHAT IS A VIABLE HPLC ASSAY?

Richard M. Venable and William Horwitz

U.S. Food and Drug Administration
Washington, DC 20204

The utility of any analytical method is best demonstrated when it is applied in a laboratory that has had no previous experience with the procedure. Most practicing chemists are aware that a method taken from the literature or obtained from another laboratory may require practice or modifications before it produces acceptable results in their laboratory. Differences in reagent impurities, column packing, instrument controls, and data-reduction algorithms are examples of conditions that may lead to a requirement for method changes. These environmental differences, sometimes deliberate but more often unconscious, result in variability between laboratories.

Thus, it is clear that *interlaboratory testing* gives more information about the viability of a method than any evaluation in a single laboratory. The collaborative study applied to several homogeneous materials permits a quantitative assessment of between-laboratory variability of a specific method. The most important statistical parameter used in assessing this variability is the *reproducibility relative standard deviation* (RSD_x), also called *reproducibility coefficient of variation* (CV_x). Because it is a relative parameter with the mean normalized out — (standard deviation/mean) $\times$ 100 — it may be used to compare precision results from different methods on different analytes at different concentrations. The RSD_x is intended to estimate the maximum variability to be expected in the application of a method by various laboratories.

The *repeatability relative standard deviation* (RSD_o), or the equivalent coefficient of variation (CV_o), is the corresponding parameter used to characterize precision within a single laboratory. The most reliable indicator of repeatability is obtained from analyses on blind replicates, that is, analyses of samples unknown to the analyst. Blind replication eliminates the problem of well-intentioned censoring

399

of data by the analyst. RSD_o is not the same as the relative standard deviation for replicate injections (RSD_e), which is a measure of the variability of the instrument alone. RSD_e does not include the variability of the chemistry involved in the preparation of the final test solution presented to the instrument.

These precision parameters form a hierarchy: RSD_x is the most general, containing the contribution of all the others; RSD_o contains the instrument error measured by RSD_e. Other intermediate precision parameters can exist, such as variability between days by a single analyst, between different analysts within a single laboratory, between instruments within a single laboratory, and so forth, but these are generally not stressed because, as comparative parameters, they are not as useful as the reproducibility, repeatability, and instrumental parameters.

The three most useful parameters are measured by performing a collaborative study. A collaborative study is an interlaboratory investigation requiring blind analyses by a group of laboratories on one or more homogeneous materials by the same method for the purpose of determining the performance characteristics of a method. The ability to evaluate the various parameters in a collaborative study depends upon the experimental design of the study. If a method is time-consuming, replicates and even duplicates may be sacrificed in favor of analyzing more commodities. When only a single determination per material per laboratory is conducted, only RSD_x can be calculated, not RSD_o; if multiple injections are specified, RSD_e can be estimated. RSD_e can also be estimated through a system-suitability test if one has been specified.

A system-suitability test is a quality control procedure that ensures adequate performance of the instrument and associated hardware, electronics, and recording components over the range of interest of the analyte. The test usually involves measuring the repeatability of examining replicate portions of the analyte solution and measuring the resolution of the analyte signal from that of any potentially interfering signal that might arise from an accompanying impurity or from an internal standard. The system-suitability test usually includes a minimum resolution value between crucial peaks in a chromatogram as well as a maximum RSD for area or peak height measurements from sequential injections. This value in turn is still not equal to RSD_e because the latter term includes the operation of transforming an instrument reading (peak height or area) into concentration through the analytical curve that includes the variability introduced by calibration.

Another approach to parameter estimation is use of an incomplete-block design: every laboratory does not analyze every material, but each laboratory obtains the same type of information. Because the number of analyses is smaller, the confidence interval associated with RSD_x is larger. Youden has suggested in the *Statistical Manual of the Association of Official Analytical Chemists (AOAC)* that RSD_o may be estimated by using the difference between the results for two very similar materials, conveniently designated as a *Youden pair* (1). Steiner, in another section of the same manual, implied that every material also should be analyzed in duplicate to provide a direct estimate of RSD_o. This disparity is intentional. In addition, the American Society for Testing and Materials (ASTM) has published guidelines (2) for conducting collaborative studies that refer to the AOAC manual — as does a recent British paper (3). Within-laboratory and within-instrument precision can always be estimated in a single laboratory by appropriate replication at different times without resorting to an expensive interlaboratory study design. There is, however, no way to obtain RSD_x without resorting to an interlaboratory study.

Estimation of the RSD_e parameter is appropriate as part of the system-suitability specification, which generally includes resolution or selectivity values for pairs of peaks. The calculation of resolution includes the peak width as well as the difference in capacity factors; thus, it provides more information about system suitability than does selectivity. It is easy to show that the same selectivity value for two peaks encompasses a broad range of resolution values, from baseline separation to severe overlap, depending upon peak widths. As peaks coalesce RSD_e may increase, and the measurements from unresolved peaks can be less consistent.

A number of experimental designs can be implemented in a single laboratory that may suggest how a method might perform in another laboratory. Youden has described a simple design, termed a "ruggedness test," that evaluates the relative effect of (typically) seven variables by performing a series of (typically) eight experiments that combine different levels of the variables in specified combinations (1). Evaluating the effect of each of the seven variables, one at a time, at two levels in combination with each of the other six variables, would require 2^7 (that is, 128) independent experiments. Multiple levels of three or four factors can be evaluated with Latin and Greco-Latin squares, respectively. The principle involved is similar to the incomplete-block design mentioned earlier and involves the use of symmetry. Standard statistics and experimental design texts discuss these designs in great detail.

RESULTS AND DISCUSSION

Horwitz and Albert have evaluated primary collaborative data from the studies of almost 1000 individual drug dosage forms and bulk products analyzed by partition (4), gas (5), and high performance liquid (6) chromatographic methods — as well as nonchromatographic methods — using a consistent statistical treatment throughout. These publications categorize methods by type and applicability by matrix and analyte, and they include numerical data for RSD_x, RSD_o, and concentration as well as number of determinations, laboratories, and outliers. Almost all of the data came from the *Journal of the AOAC (JAOAC)*. The AOAC requires the performance of a collaborative study as a prerequisite for adoption of analytical methods; approved methods are compiled at five-year intervals in the *Official Methods of Analysis of the AOAC*.

In dealing with this quantity of information, the computer becomes an indispensable tool for storing, analyzing, and displaying the information in an organized and consistent manner. An APL (7) program developed by Albert and Horwitz was used for computing the primary parameters, for outlier treatment, and for compiling the data base. That program is described in detail elsewhere (4). The evaluation, identification, and removal of outliers has been undergoing constant evolution as compilation continues; the final procedure will be the subject of a future publication. This chapter is more concerned with the intercomparison of the various chromatographic techniques and with the expected within- and between-laboratory variabilities. The authors have examined the available data for the common chromatographic techniques, using SAS (8) to explore possible trends in the data. The majority of the collaborative assays of drugs that have been published in *JAOAC* during the past decade have used a chromatographic separation step.

The important relationship noticed in the earlier publications on drug analysis was the dependence of RSD_x on concentration only, independent of analyte, matrix, and method. Figure 1 shows the linear regression lines of RSD_x/log concentration (but labeled linearly) for the major chromatographic method types: diatomaceous silica separation/spectrophotometric determination (DISP); gas-liquid (GLC); ion-exchange (IONX); and high performance liquid (HPLC). The negative logarithm is used as the concentration axis to avoid compressing the data at low concentrations and to place the 100% concentration point at the origin. The decrease in between-laboratory precision (that is, a larger RSD_x) with lower concentration is a well-known phenomenon. The lines are probably not significantly different from each other, but this observation has not yet been examined in detail. Table I presents this information in tabular form.

The maximum difference in RSD_x (absolute) is 0.5%–0.6% between the DISP and the IONX methods over all concentrations. From the figure it is apparent that the HPLC method has the steepest slope, perhaps reflecting to the maximum extent the greater influence of concentration on precision at the lower concentrations. In view of the relatively high variability of the individual studies that went into constructing these curves, however, these effects and interactions are probably of only secondary significance. In fact, Boyer and co-workers suggest, from their empirical examination of similar data for trace elements over the concentra-

TABLE I

Average Relative Standard Deviations (RSD_x) of Various Chromatographic Methods as a Function of Concentration (wt %)

Method type	RSD$_x$ at concentration					ΔRSD_x/10-fold concentration change
	100%	10%	1.0%	0.1%	0.01%	
DISP	1.7	2.1	2.5	2.8	3.2	0.38
HPLC	1.8	2.3	2.8	3.3	3.8	0.50
GLC	2.2	2.5	2.9	3.2	3.5	0.33
IONX	2.2	2.6	3.0	3.4	3.8	0.40

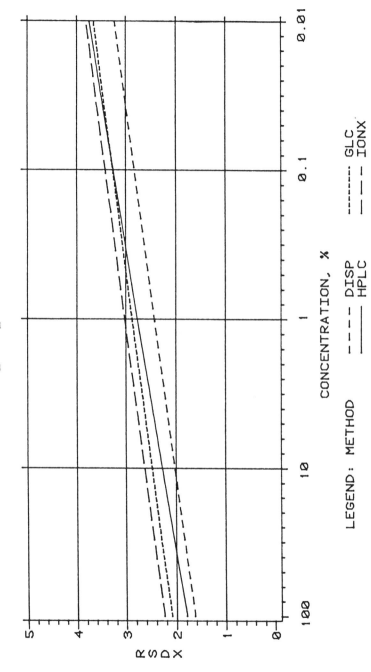

FIGURE 1. The calculated linear relationship between the between-laboratories relative standard deviation (RSD$_x$) and concentration (wt %) for the major classes of chromatographic methods.

tion range of 10% to parts-per-million, that the highest expected value of RSD_x is twice the average value found historically for a given concentration (9). On the basis of the authors' experience with pharmaceutical and other analytes, particularly pesticide formulations in the same concentration range as drugs, this factor of two appears suitable for drugs as well. Overall, the relative standard deviation (RSD) of chromatographic methods for bulk drugs and dosage forms is probably best represented by a straight line from $RSD_x = 2\%$ at 100% concentration to an $RSD_x = 3.5\%$ at 0.01% concentration — a change in slope of approximately 0.4 RSD_x units (% absolute) per 10-fold change in concentration. On the basis of the suggestion by Boyer and co-workers, occasional RSD_x values as large as 4% at 100% concentration and 7% at 0.01% concentration would be considered acceptable.

Table II shows the relative importance of the four chromatographic methods as a function of relative concentration, that is, the weight percentage of the analyte in the dosage form. The DISP method is used over the entire concentration range from 100% to 0.01%. The other methods are used more often for the higher-concentration preparations, although use of HPLC extends to dosage forms with the lowest analyte concentrations. The table also provides the frequency distribution of methods with concentration and associated RSD_x.

The design and statistical parameters for the chromatographic methods are summarized in Table III. A total of 103 drugs have been examined in these chromatographic studies, almost half by the DISP method. A few of the drug entities have been used in studies of several types of methods. Four hundred collaborative assays have been performed by an average of 10 laboratories for each, involving a total of 6345 individual determinations. The overall average RSD_0 is 1.4%, the RSD_x is 2.6%, and their ratio is 0.56 with an average outlier rate of 1.8% for the determinations. All of the averages have been weighted by the number of determi-

TABLE II

The Frequency Distribution of Number of Collaborative Assays (N) of Bulk and Dosage Form Drugs as a Function of Mean Concentration per Assay and Type of Chromatographic Method

Concentration (%)	DISP			HPLC			IONX			GLC		
	RSD_x (mean)	N	Conc (mean)	RSD_x (mean)	N	Conc (mean)	RSD_x (mean)	N	Conc (mean)	RSD_x (mean)	N	Conc (mean)
0.01–0.05	3.37	16	0.04	2.77	6	0.03	—	—	—	—	—	—
0.05–0.1	2.81	12	0.08	3.68	3	0.08	3.87	1	0.10	—	—	—
0.1–0.5	2.83	34	0.23	3.01	4	0.32	3.20	5	0.30	2.85	2	0.35
0.5–1	2.41	16	0.76	3.46	17	0.78	1.42	1	0.91	2.00	2	0.95
1–5	3.29	33	2.32	3.06	23	2.55	2.74	4	2.96	4.29	17	2.97
5–10	2.06	19	6.71	1.99	13	7.34	2.68	4	7.81	2.50	6	7.48
10–50	1.84	38	27.28	2.02	20	31.22	2.63	37	25.58	2.28	21	26.88
50–100	1.64	8	72.31	1.75	29	82.73	—	—	—	2.07	9	73.48

TABLE III

The Design and Statistical Parameters for the Collaborative Assays of Chromatographic
Methods Published in the *Journal of the AOAC* through May 1985

Method	Total number			Average number of labs	Average		Ratio $RSD_o/$ RSD_x	Outliers (%)
	Analytes	Assays	Detns		RSD_o	RSD_x		
DISP	49	176	2983	11.5	1.5	2.6	0.54	2.1
HPLC	21	115	1888	8.6	1.2	2.5	0.58	0.5
GLC	16	57	725	8.2	1.4	2.9	0.53	1.2
IONX	17	52	749	9.8	1.8	2.7	0.66	4.1
Total	103	400	6345					
Average*				10.0	1.4	2.6	0.56	1.8

* Averages weighted by determinations for each analyte.

nations for each analyte. The range of the ratio for RSD_o/RSD_x of 0.53–0.66 is practically the same as that quoted originally ($1/2$ to $2/3$) in a review of the variability of AOAC methods of analysis for pharmaceuticals (10). Although the RSDs by method do not appear to differ significantly from each other, the ratio RSD_o/RSD_x is somewhat higher for the IONX method than for the others, indicating greater within-laboratory variability for this method. The IONX method also has a significantly greater percentage of outliers than the other three types. These observations may reflect the application of this method to the analysis of the most difficult drug dosage forms with the largest number of components — as many as five — cases in which an incomplete separation may also affect the results of many of the other components.

Analyte recovery is not considered in this summary because in all collaborative studies where formulated materials were present analyte recovery was within a few percent of 100%.

Although not compiled in the statistics of these chromatographic methods, the instrumental error, RSD_e, should not be greater than 1% in order to avoid undue influence on RSD_o and RSD_x. Most system-suitability tests include this specification. In HPLC, an RSD for replicate injections of 0.5% can be obtained without difficulty, except at high detector sensitivity levels (low concentrations) for which there is significant baseline noise. The photomultiplier tube, one of the most commonly used light sensors in spectrophotometric and HPLC detectors, generally provides a high signal-to-noise ratio. The real problem of noise generation lies within the ultraviolet light source, especially with variable-wavelength detectors. The lamp simply does not provide enough energy outside the range of interest, that is, 220–350 nm. To compensate, detector gain must be increased, leading in turn to noise from stray light and possibly even from cosmic rays (shot noise). Detector noise is observed as extra jitter in the baseline; spikes, drift, or noise that comes and goes may indicate a system problem such as dirty or cracked cell windows or leaking fittings on the high-pressure side. To overcome such problems, other detectors may be used or the analyst may resort to shifting the analytical wavelength by derivatization and indirect determination. To minimize instrument error, a system-suitability test — with repetitive injections of a test mixture to attain an RSD_e of less than 1% — should be part of all HPLC method protocols offered for use in other laboratories.

Table IV presents a history of the precision (expressed as RSD_x) of chromatographic methods. Beginning in 1954, only the partition method was available;

TABLE IV

The Frequency Distribution of the Number of Assays (N) of Bulk and Dosage Form Drugs as a Function of Average Between-Laboratories Relative Standard Deviation (RSD_x) and Year of Publication

	Method							
	DISP		GLC		IONX		HPLC	
Year	RSD_x (mean)	N	RSD_x (mean)	N	RSD_x (mean)	N	RSD_x (mean)	N
1954	1.76	2	—	—	—	—	—	—
1955	2.12	4	—	—	—	—	—	—
1956	2.65	5	—	—	—	—	—	—
1958	5.98	7	—	—	—	—	—	—
1959	3.09	2	—	—	—	—	—	—
1960	3.79	5	—	—	—	—	—	—
1961	2.48	2	—	—	—	—	—	—
1962	2.64	1	—	—	—	—	—	—
1963	3.29	7	—	—	—	—	—	—
1965	3.78	6	—	—	—	—	—	—
1967	2.81	4	3.26	4	—	—	—	—
1968	1.69	13	—	—	—	—	—	—
1969	—	—	—	—	2.99	3	—	—
1970	2.44	12	2.17	5	2.45	16	—	—
1971	2.29	31	—	—	2.51	16	—	—
1972	1.94	15	4.03	18	—	—	—	—
1973	2.61	19	2.04	9	—	—	—	—
1974	2.34	8	—	—	3.21	13	—	—
1975	2.44	10	—	—	—	—	—	—
1976	1.51	4	1.77	4	—	—	—	—
1977	2.92	7	2.86	6	—	—	—	—
1978	—	—	2.34	5	2.54	4	3.91	3
1979	—	—	2.22	6	—	—	—	—
1980	2.61	6	—	—	—	—	—	—
1981	—	—	—	—	—	—	3.55	32
1982	—	—	—	—	—	—	3.51	2
1983	—	—	—	—	—	—	1.95	17
1984	—	—	—	—	—	—	2.11	46
1985	1.71	6	—	—	—	—	1.57	15

GLC was introduced in 1967, IONX in 1970, and HPLC in 1978. HPLC is now obviously displacing most other methods. With the first three types of methods there was little change in precision with time; with HPLC, however, there appears to be an improvement with time, possibly a result of better optimization of methods and the introduction of system-suitability tests.

Table V provides a complete listing of the statistical parameters by analyte and matrix of drugs analyzed by the HPLC method. There is little apparent effect of matrix for powders (bulk, tablets, capsules) or for true solutions (elixirs, tinctures, injections). Suspensions and ointments may show somewhat higher RSDs, as might be expected from sampling problems, but their number is too small to draw conclusions.

The analysis of this large amount of data is still in progress, and the results presented here are preliminary. Nevertheless, the statistical parameters from collaborative studies appear to be more or less constant with time, with a possible improvement to a constant limiting value after the initial introduction of a new technique such as HPLC. Low analyte concentration definitely results in lower precision, but HPLC compares favorably with other analytical methods in this respect.

Acknowledgment

The authors are greatly indebted to Michel Margosis for permitting use of the chromatographic portion of his drug data base of statistical parameters calculated from AOAC collaborative studies before their initial publication. They also thank Richard Albert for permitting use of his program, FLOW-CHART, before it has been described in the literature.

REFERENCES

(1) W.J. Youden and E.H. Steiner, *Statistical Manual of the AOAC* (Association of Official Analytical Chemists, Arlington, Virginia, 1975).

(2) ASTM Standard E-691-79, *Interlaboratory Testing* (American Society for Testing and Materials, Philadelphia, 1982).

(3) B.V. Fisher, *Anal. Proc.* **21,** 443 (1984).

(4) W. Horwitz and R. Albert, *J. Assoc. Off. Anal. Chem.* **67,** 81 (1984).

(5) W. Horwitz and R. Albert, *J. Assoc. Off. Anal. Chem.* **67,** 648 (1984).

(6) W. Horwitz and R. Albert, *J. Assoc. Off. Anal. Chem.* **68,** 191 (1985).

(7) *APL.SV Version 3,* implemented on an IBM 3081-D (IBM DP Division, White Plains, New York).

(8) *Statistical Analysis System (SAS), Version 82.4,* implemented on IBM 3081-D (SAS Institute, Cary, North Carolina).

(9) K.W. Boyer, W. Horwitz, and R. Albert, *Anal. Chem.* **57,** 454 (1985).

(10) W. Horwitz, *J. Assoc. Off. Anal. Chem.* **60,** 1355 (1977).

TABLE V

Average Concentration and Relative Standard Deviations Between- (RSD$_x$) and Within Laboratories (RSD$_o$) for Drugs Assayed by HPLC, Arranged by Dosage Form (N = Number of Assays per Item)

Analyte	N	Concentration	RSD$_o$	RSD$_x$
Bulk drug				
Ampicillin hydrate	1	86.7	0.96	1.53
Ampicillin anhydrous	1	99.3	1.15	1.60
Colchicine	1	97.3	0.91	1.33
Hydrocortisone	3	98.5	0.73	1.24
Methyldopa	1	100.1	0.98	0.98
Prednisolone	3	99.8	0.75	1.39
Tablets				
Allopurinol	3	41.7	0.59	2.10
Amitriptyline	3	14.6	0.91	2.02
Chlorothiazide	3	33.3	1.24	2.17
Chlorpheniramine maleate	3	1.10	—	4.30
Colchicine	3	0.798	1.42	2.35
Diazepam	4	4.07	0.65	0.95
Hydrochlorothiazide	3	5.73	0.94	1.79
Hydrocortisone	4	4.76	0.96	3.25
Methocarbamol	3	87.7	2.82	3.00
Methyldopa	9	57.8	1.12	1.69
Oxazepam	1	7.73	0.99	1.90
Prednisolone	3	2.35	1.25	2.28
Primidone	4	60.4	1.78	1.79
Pseudoephedrine	6	29.0	—	2.68
Sulfamethoxazole	3	81.1	1.15	1.42
Sulfisoxazole	2	76.0	0.48	1.68
Powders				
Oxazepam	5	7.91	1.70	2.46
Prednisolone	1	4.53	0.88	2.01
Sulfisoxazole	1	78.7	0.34	1.39
Solutions				
Amitriptyline	2	1.01	0.45	1.64
Ampicillin sodium	1	87.2	1.04	1.76
Chlorpheniramine maleate	3	0.0408	—	1.53
Colchicine	1	0.0507	0.65	2.13
Methocarbamol	3	10.2	0.66	1.71
Physostigmine salicylate	1	0.480	1.75	2.97
Pilocarpine	3	1.82	2.30	3.90
Pseudoephedrine	6	0.595	—	2.80
Sulfisoxazole	2	4.09	0.66	1.65
Triprolidine	6	0.844	—	4.32
Ointments				
Hydrocortisone	8	0.994	—	4.67
Physostigmine salicylate	1	0.240	1.42	4.06
Sulfisoxazole	1	0.415	0.51	1.44
Suspensions				
Probenecid	3	0.101	3.48	3.91
(Total) or average	(115)	27.8	1.20	2.49

THE STANDARDIZATION OF HPLC COLUMNS FOR DRUG ANALYSIS: ARE C_{18} COLUMNS INTERCHANGEABLE?

Donald J. Smith

Division of Drug Chemistry
Food and Drug Administration
Washington, DC 20204

The proliferation of column manufacturers producing similar types of columns has created a problem in the specification of generic columns for HPLC analysis. The attempt to find equivalent columns from different manufacturers has brought to light the fact that the common nomenclature used for columns is inadequate. A C_{18} *reversed-phase column* does not completely describe all the factors that play a role in the separation mechanism of that column. Unless a system comes close to describing the differences fully, it cannot be used to help select equivalent columns. Finding equivalent columns from various manufacturers is a major problem. An equally important problem is the lot-to-lot variation of columns from the same batch and batch-to-batch variation from the same manufacturer.

In order to eliminate the problem of column-to-column reproducibility, column production should be treated just like any other critical process. For example, the quality control system used for pharmaceuticals should be adapted to HPLC columns. Every batch of raw material — silica packing material — should be thoroughly tested, and there should be well-defined packing material specifications to aid in eliminating possible sources of variation.

The variation in column packings has resulted in a variety of approaches to the problem. Atwood and Goldstein (1) measured 24 different production lots of reversed-phase liquid chromatography column packing material (Vydac TP-201; The Separations Group, Hesperia, California) for variations in separation factor from lot-to-lot. Quality control ranges for separation factors and capacity ratios of lots to be used for packing Vydac TP-201 columns were calculated, and it was concluded that most of the measured column-to-column variation within lots was probably caused by differences in mobile-phase strength and laboratory temperature. Goldberg (2) has obtained data illustrating the vastly differing chromatographic properties of commercially available reversed-phase packings. Addressing the problem of interlaboratory collaborative studies, Wayne and co-workers (3) of the Association of Official Analytical Chemists (AOAC) stated that as long "as the given separation mode and basic mobile phase composition is maintained, the Associate Referee should allow flexibility in the use of alternative columns by providing a generic description and minimum performance standards, relative to retention time, elution order, and theoretical plates." They also recommended that no methods using other column types be adopted by the AOAC as official unless silica or C_{18} bonded silica has been demonstrated to be unacceptable.

Debesis and associates (4) recommend that the specificity test for column-to-column variability be repeated using at least three columns from three different batches from one column manufacturer. Kirkland (5) found that chromatographic performance of packing material is affected by particle size and shape, by specific surface area, by pore size and distributions, and by specific pore volume. De Stefano and co-workers (6) state that "for reproducible results in reversed-phase HPLC, the researcher must have control of several parameters: particle characteristics and control . . . [and] bonded phase chemistry." Bristow and Knox (7) recommend that the following data be obtained for a column:

- *operating conditions:* temperature of column, designation of packing, method of packing, composition of eluent, composition of test sample, detection method, and injection method
- *properties of eluent and solute(s):* viscosity of eluent and diffusion coefficients of solute(s) in eluent
- geometrical parameters: column bed length, column bore, and particle size
- chromatographic parameters.

What is necessary is a standardized procedure that will enhance the possibility for finding equivalent columns from different sources. It is not enough simply to specify the side chain; the column packing material must be described with greater specificity. An ideal standardized method for describing a column must include specifications for all the various components of HPLC columns.

A format for column specifications and performance characteristics and for describing packing material and HPLC columns in generic terms has been developed at the author's FDA laboratory. The complete definition of an HPLC column is required to ensure that developed analytical methods can be reproduced. An adequate description of an *equivalent* column will therefore allow the selection of columns having the best quality and stability characteristics.

The following criteria and format are considered significant for defining a column in generic terms as listed in Tables I–IV. Information about the basic particle, bonded-phase characteristics, and column specifications should be included to make the definition as useful as possible. All factors in Tables I–IV contribute information toward specifying an HPLC column completely. The manufacturer must supply the chromatographer with the majority of this information so that a decision can be made based on lot-to-lot and batch-to-batch variation.

TABLE I

Specifications for Basic Particles

1. Brand name: (source, if not manufacturerered):	
2. Chemical composition: (for example SiO_2)	
3. Particle shape:	
4. Particle size: Mean size (μm):	
Size distribution: (attach graphic distribution if available)	
5. Pore diameter: Mean (Å):	
Distribution:	
6. Pore volume (mL/g):	
7. Surface area (m^2/g):	
8. Maximum pressure limit (psi):	
9. Operating range: Temperature ($^\circ$C):	
pH:	

The basic type of particle selected is important in column stability and permeability. These are some of the parameters that generally affect the final products. The quality of the bonded phases is determined by the type of the particle and the chemical bonding methodology used. Column efficiency (N) and peak symmetry (A_s) reflect the quality of the column, while the capacity factors (k') and selectivity (α) determine the column's capabilities to retain and separate certain compounds. The methods of calculation used to obtain these performance factors are critical. The most efficient and reproducible method of calculation uses a computer with appropriate programs based on the formulas listed in Table V. As an alternative, manual measurements can be made (as drawn in Figure 1) and these distances used in calculations. Column manufacturers generally use one of these methods for quality control and in developing the performance data that accompanies each purchased column. Those performance factors can be used in determining the condition of column upon receipt and for periodic quality assurance checks.

Other considerations in selecting HPLC columns include the quality control information provided by the manufacturer, such as strict specifications for the bulk packing, identifying serial numbers, individual performance checks, and warranty included with each column. The above information is necessary to specify the

TABLE II

Specifications for Bonded Phases

1. Brand name:
2. Basic particle (refer to Table 1):
3. Bonded phase type:
4. Surface coverage (μmol/m^2):
5. Elemental analysis (% C):
6. Specific name of reagent used   for preparing bonded phase:
Purity of above reagent (%):

7. Specific name of reagent used
   for silanizing residual open sites,
   if such a reagent used:

Purity of above reagent (%):
8. Residual hydroxyl groups (%) and   method of determining that value:
9. Operating range: pH:
Temperature (°C):

quality and performance of a column uniquely and can also be used to describe adequately any manufacturer's columns or type of column. The test systems included in the specifications use model compounds and are designed to reflect the general performance of a column. In order to ensure that a column is suitable for a particular analytical procedure, HPLC methods should include a *system-suitability test,* which may require using specific compounds involved in the procedure. The responsibility for determining the method system-suitability test rests with the developer.

Acknowledgments

The author wishes to thank Milda Walter, Richard Thompson, Richard Krause, James Nelsen, and Steve Gonzales, all of the Food and Drug Administration, for helpful suggestions.

TABLE III

Specifications for Columns

1. Dimensions: Length (cm):	
i.d. (mm):	
2. Type(s) of endfittings:	
Frit pore size (μm):	
3. Packing type (refer to Tables I and II):	
System used to determine above value:	
4. Column efficiency (N/column):	
System used to determine above value:	
5. Peak asymmetry (skew):	
Method of calculation for above value:	
6. Column permeability:	
7. Selectivity: System used to test selectivity:	
Respective values:	
8. Reproducibility of column selectivity between columns (% range):	
9. Maximum operating pressure (psi):	

REFERENCES

(1) J.G. Atwood and J. Goldstein, *J. Chromatogr. Sci.* **18**, 650–654 (1980).

(2) A. Goldberg, *Anal. Chem.* **54**, 342–344 (1982).

(3) R.S. Wayne, *J. Assoc. Off. Anal. Chem.* **66**, 448–452 (1983).

(4) E. Debesis, J. Boehlert, T. Givand, and J. Sheridan, *Pharm. Technol.* **6**(9), 120–137 (1982).

(5) J.J. Kirkland, *J. Chromatogr.* **83**, 149–167 (1973).

(6) J.J. De Stefano, A.P. Goldberg, J.P. Larmann, and N.A. Parris, *Ind. Res. Dev.* **22**, 99–103, April (1980).

(7) P.A. Bristow and J.H. Knox, *Chromatographia* **10**, 279–289 (1977).

TABLE IV

Miscellaneous Information and Specifications

	Basic particle	Bonded phases
1. Packing batch size (average):		
2. Availability of bulk packing:		
3. Method of packing columns:		
Maximum packing pressure (psi):		
4. Frequency of quality control checks:		
5. Type of warranty:		
6. Other information you feel important and pertinent to your products:		

TABLE V

Equations for Calculating Performance Factors[*]

Factor	Computer method	Alternate manual method (see Figure 1)
Efficiency: $N =$	$2(h't_r/A)^2$	$16(t_r/W)^2$
Peak symmetry (skew): $A_s =$	$2(\tau/\sigma)^3/[1 + (\tau/\sigma)^{3/2}]$	b/a
Capacity factors: $k' =$	$(t_r - t_0)/t_0$	$(t_r - t_0)/t_0$
Selectivity: $\alpha =$	k'_2/k'_1	k'_2/k'_1

[*] Recorder chart speed $= 25$ mm/min.

$t_0 =$ elution distance of unretained component (mm).

$t_r =$ retention time (mm).

$a, b =$ peak half width at 10% peak height (mm).

$W =$ peak width at 10% peak height (mm).

$\tau =$ exponential function.

$\sigma =$ standard deviation.

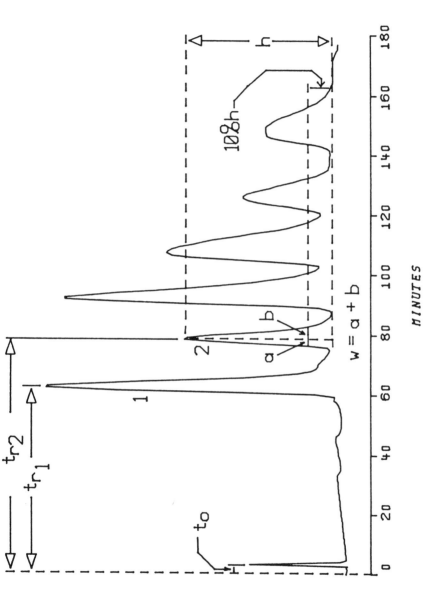

FIGURE 1. Manual measurements necessary to calculate performance factors.

PART FIVE

LC IN THE PHARMACEUTICAL INDUSTRY: CURRENT AND FUTURE TRENDS

TRENDS IN THE HPLC ANALYSIS OF MARKETED PHARMACEUTICALS: A REGULATORY VIEW

Eric B. Sheinin

Division of Drug Chemistry
Food and Drug Administration
Washington, DC 20204

Since 1965, New Drug Application (NDA) methodology has been subjected to validation in Food and Drug Administration (FDA) laboratories. Before that time, only a paper review process was used to determine the adequacy of these methods for regulatory purposes. From the inception of this program until December, 1981, satisfactory validation of the proposed methodology was one of the requirements for NDA approval. Since 1981, it has been possible, under certain circumstances, for an NDA to be approved before completion of the methods validation process. Nonetheless, FDA still expects any problems encountered during validation to be corrected in a timely manner.

Over the past 10 to 12 years, the use of high performance liquid chromatography (HPLC) has increased dramatically. Under ideal circumstances, one procedure would be sufficient to ensure the potency (or strength), purity, and identity of a sample. Quite often an HPLC method will come very close to meeting all of these requirements for both the new drug substance (NDS) and the dosage form. Other less-specific techniques such as nonaqueous titration or electronic spectrophotometry continue to play an important role for evaluation of the NDS when coupled with a chromatographic procedure for the detection and control of potential impurities.

Internal reports prepared during the last four years have summarized the number of NDAs validated over that time period. In 1981, 39 dosage form assays were validated; of those, 27 (69%) were HPLC procedures. In 1984, 24 of 26 (92%) dosage form assays relied on HPLC. A similarly strong trend has been noted in 1985. Procedures for impurities, while almost universally relying on chromatography, have included more thin-layer chromatography (TLC) procedures than HPLC procedures. Recent trends have shown a shift to HPLC, or a combination of HPLC with either TLC or gas-liquid chromatography (GLC). For the period

1981–1984, HPLC methods have increased from 6 of 36 (17%) in 1981 to 12 of 17 (71%) in 1984. This reliance on HPLC is expected to continue in the future.

Although HPLC has led to improved analytical procedures in the pharmaceutical industry as well as in many others, it has also led to an increase in the amount of time required to complete the methods validation process. The difficulties commonly encountered in transferring a method from research and development laboratories to quality control laboratories, even within the same organization, are multiplied when an organizationally distinct laboratory such as FDA's becomes involved.

The opportunity to participate in the validation of some 30 to 40 NDAs each year provides the author's laboratory with a broad view of the current status of HPLC in relation to practical applications. Many of the techniques discussed elsewhere in this book — the use of diode array detectors, chiral stationary phases, or microbore columns — have not yet appeared in NDAs. Most of the examples presented in this chapter are based on work carried out in the laboratories of the Center for Drugs and Biologics in which Type 1 (new chemical entities), Type 2 (new salts, esters, or other derivatives), and Type 4 (new combinations) drugs are validated (1). Other examples are based on the evaluation of proposed reference standards for the United States Pharmacopeial Convention, Inc. (USP). The future use and direction of HPLC in pharmaceutical analysis will be determined by the industry. As the newer techniques become more or less routine, their effect on regulatory methodology will grow. The examples presented below represent a mere introduction to the application of HPLC to the analysis of marketed pharmaceuticals. Many more NDAs are received each year than could be included in a review of this type.

EQUIPMENT

Columns

Columns may be specified in a variety of ways, but the one most commonly encountered is the specification of the brand name or the equivalent — for example, "μBondapak C_{18} or the equivalent." Specification of the brand name alone is next most common; generic terms are seen least often. The latter range from the classification adopted by the USP (2), for example, L-1, L-2, and so forth, to more descriptive nomenclature, such as "a reversed-phase, C_8 column." Regardless of how they are specified, equivalency, or, more appropriately, lack of equivalency, between columns of the same generic type causes considerable difficulty during validation. Any precolumn or guard column required for the method must be specified as well. A more detailed description of column equivalency is found elsewhere in this book.

Figure 1 shows three chromatograms obtained on "equivalent" CN columns; the one specified in the method was a μBondapak CN (Waters Associates, Milford,

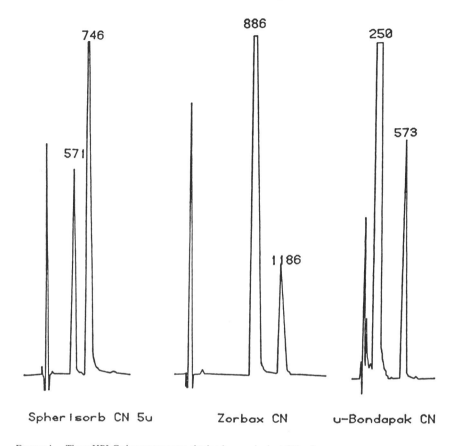

FIGURE 1.   Three HPLC chromatograms obtained on equivalent CN columns.

Massachusetts). No system-suitability test was included, but the chromatograms submitted with the NDA indicated that a high degree of resolution was attainable. Initially the specified column was used and retention times similar to those reported by the sponsor were obtained. The resolution, however, was not nearly as good. Another μBondapak CN column was not available, so two equivalent columns were tried. As shown in the figure, the order of elution of the two peaks was reversed when the Spherisorb-CN column (Phase Separations, Norwalk, Connecticut) was used. In addition, the peaks were narrower, with faster retention times than specified. The Zorbax-CN column (Du Pont Company, Wilmington, Delaware) yielded results that indicated it could probably be used with suitable modification of the mobile phase so that the desired resolution would be obtained. On this latter column, the elution order was identical to that given in the NDA.

Another example of column nonequivalency is illustrated in Figure 2. The official assay (3) for vitamin D includes a system-suitability test that uses the USP Vi-

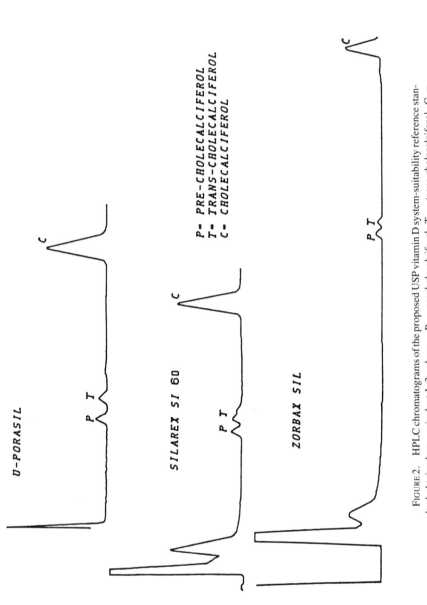

FIGURE 2. HPLC chromatograms of the proposed USP vitamin D system-suitability reference standard obtained on equivalent L-3 columns. P = precholecalciferol, T = *trans*-cholecalciferol, C = cholecalciferol.

tamin D System-Suitability Reference Standard. This standard contains cholecalciferol, precholecalciferol, and *trans*-cholecalciferol. Specifications are given for the relative standard deviation of the response, for the resolution, and for the expected retention times. A 25 cm × 4.6 mm stainless-steel column packed with chromatographic packing L-3 was specified. Representative chromatograms obtained on µPorasil (Waters Associates), Silarex SI-60 (Lion Technology, Dover, New Jersey), and Zorbax SIL (Du Pont Company) columns are shown in this figure. Because different flow rates were used with each column, the retention times were different. Even so, it is readily apparent that there is a difference in resolution from column to column. The system-suitability results are given in Table I. Each of the columns gave quite different resolutions and relative standard deviations. For this particular sample it appears that the µPorasil column was the optimum choice.

The use of a generic description does not necessarily mean that all columns within a particular group will provide equivalent results. This is shown by an example of an HPLC assay for both the NDS and the formulated tablet of an antiarrhythmic agent. A 25 cm × 3.9 mm stainless-steel column packed with silica, 100–10 µm, was specified. As part of the system-suitability test, the resolution factors were given for the active component and two related compounds that eluted together.

Initial attempts at validation were performed on a Nucleosil 100-10 column (Machery-Nagel, Düren, West Germany) purchased especially for this NDA. The first chromatogram in Figure 3 shows the system-suitability test mixture run at 1 mL/min with the specified mobile phase (methanol/ethyl ether/perchloric acid, 69:31:0.02). No separation was achieved, nor were these compounds retained on the column. Repeated injections to saturate the column did not improve the situa-

TABLE I

System-Suitability Test Parameters for USP Vitamin D System-Suitability Test Reference Standard

Column	Resolution $(k')^*$	Component	RSD	RRT[†]
Zorbax SIL	1.09 (1)	Precholecalciferol	1.69	0.65
(4 inj)	1.00 (4)	*trans*-Cholecalciferol	5.92	0.67
		Cholecalciferol	2.61	1.00
Silarex SI 60	0.86 (2)	Precholecalciferol	4.21	0.55
(6 inj)	1.00 (6)	*trans*-Cholecalciferol	6.70	0.59
		Cholecalciferol	1.54	1.00
µPorasil	1.33 (1)	Precholecalciferol	4.21	0.52
(5 inj)	1.54 (5)	*trans*-Cholecalciferol	4.83	0.55
		Cholecalciferol	0.77	1.00

* Number in parentheses is the number of the injection used to calculate resolution.
† RRT = relative retention time.

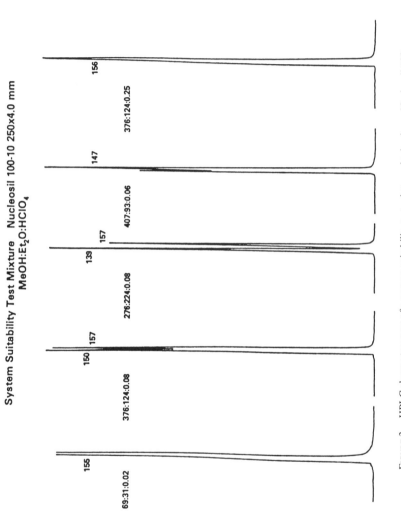

FIGURE 3. HPLC chromatograms of a system-suitability test mixture obtained on a Nucleosil 100-10 column with varying mobile phase compositions.

tion. The other chromatograms in this figure represent initial attempts to improve the chromatography via mobile phase modification. Increases in the methanol content, the ether content, or the acid content provided some improvement in resolution of the mixture into two peaks but failed to move them adequately from t_0. Substitution of a 10-μm LiChrospher Si-100 column (E. Merck, Darmstadt, West Germany), also purchased for this NDA, resulted in some improvement over the Nucleosil column but still was not totally satisfactory (Figure 4). Unsuccessful attempts at mobile phase modification are also shown; a representative chromatogram submitted by the applicant is included as well.

Figure 5 shows several typical chromatograms obtained on a new 10-μm LiChrosorb Si-100 column (E. Merck) supplied by the applicant. The resolution between the active component and the related compounds was 13.8; because the latter still elute at t_0, however, the use of the term *resolution* is not appropriate. Re-

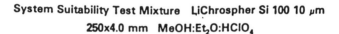

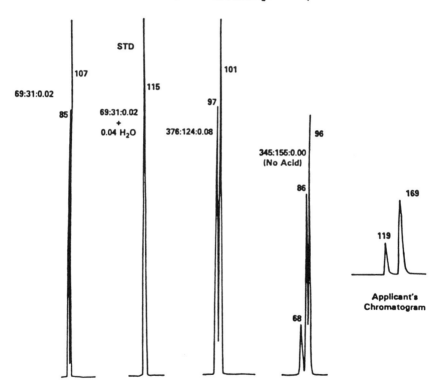

FIGURE 4.   HPLC chromatograms of a system-suitability test mixture obtained on a LiChrospher Si-100 10-μm column with varying mobile phase compositions.

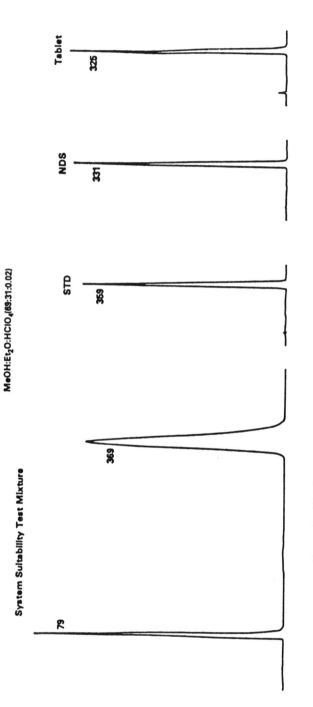

FIGURE 5. HPLC chromatograms of a system-suitability test mixture, reference standard, new drug substance, and dosage form obtained on a new LiChrosorb Si-100 10-μm column.

gardless, satisfactory chromatographic results were obtained for the system-suitability test mixture, the reference standard, the NDS, and the dosage form assay. Similar results also were obtained on a used LiChrosorb column supplied by the applicant. The only difference between the satisfactory and unsatisfactory columns is in particle shape: the former have irregular silica particles, the latter have spherical particles. Because the particle shape appeared to be critical in this instance, it was recommended that the method include the brand name of the desired column.

Detectors

Although a wide variety of detectors are available and new ones are continually being added to the market, the variable-wavelength UV detector continues to be specified in the great majority of NDAs. It has replaced the fixed-wavelength UV detector in popularity; because increased sensitivity results when the wavelength of maximum absorbance is used, fixed-wavelength detectors now are specified very infrequently. No specific examples will be presented because the use of UV detectors is illustrated throughout this chapter. Other detectors are used sparingly and will be discussed individually at this point.

Refractive index (RI) detectors are generally not as sensitive as the UV detectors but are quite useful for compounds that do not have acceptable UV properties. Figure 6 shows the chromatogram obtained for a sugar concentrate using a BA-X4 anion exchange resin (Benson Company, Reno, Nevada) with a mobile phase of 0.20 M boric acid, 0.12 M glycerol, and 0.025 M sodium acetate. The system-suitability test contained a requirement for a minimum resolution of 1.1 between galactose and lactulose. The other minor components of the concentrate and the dosage form are indicated on the figure as well. The order of elution shown here did not agree with the applicant's; the lactose and buffer peaks were reversed. Further, the FDA laboratory found that neither retention times nor peak heights were reproducible; this was attributed to the column and not to the detector. During the past six years, only one other NDA used an RI detector.

Another example of the use of an RI detector is presented in Figure 7. Proposed USP reference standards of lactose, lactulose, and galactose were examined by HPLC to determine their purity profiles. A μBondapak Carbohydrate column and a mobile phase of acetonitrile/water (70:30) were used. As seen from the three chromatograms in this figure, no additional peaks were observed. Each of these sugars would have been resolved from the other two if they were all present in a mixture. Conceivably, this column would have provided a viable alternative to the BA-X4 anion exchange resin used in the NDA discussed above.

The *fluorescence detector* is not as widely used as the UV detector, either, but unlike refractive index detectors it can be extremely sensitive. Figure 8 shows two chromatograms obtained for the reference standard and the sample with a fluorescence detector set at an excitation wavelength of 326 nm and an emission wave-

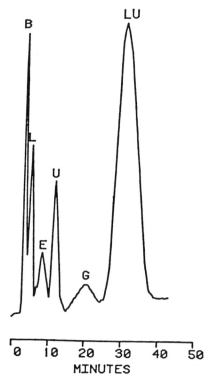

FIGURE 6.   HPLC chromatogram of sugar concentrate obtained with a refractive index detector. B = buffer, 3.40 min; L = lactose, 5.25 min; E = epilactose, 8.13 min; U = unknown, 11.42 min; G = galactose, 20.15 min; LU = lactulose, 30.23 min.

length of 436 nm. The active ingredient was the alkaloid methylergonivine male- ate; the response shown was obtained from an injection of 100 ng. This level was found to be at the upper end of the straight line obtained during verification of the linearity of the detector response. The sensitivity inherent in this method was con- siderably greater than shown in this figure; the actual emission specified in the NDA was given as a range of 418–700 nm, which is attained with an appropriate filter. FDA's detector is able to operate in this manner as well as with a specific emission wavelength; 436 nm was selected based on past experience with ergot al- kaloids. Additional work showed the wavelength of maximum emission to be 430 nm. The relative standard deviation for six reference standard injections was 0.07%, demonstrating the precision attainable with this detector under the proper conditions. During the past six years, this was the only instance in which a fluores- cence detector was specified for an NDA.

Another useful, but infrequent, detection system is the *electrochemical detec- tor.* In the past six years two NDAs have used such a system. The chromatograms

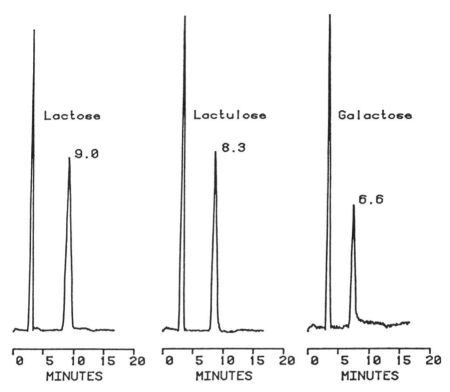

FIGURE 7.   HPLC chromatograms of proposed USP reference standards, lactose, lactulose, and galactose obtained with a refractive index detector.

presented in Figure 9 were obtained for a 1% lidocaine injection containing epinephrine 1:100,000. Three detectors, connected in series, were specified. Lidocaine HCl was detected at 254 nm, methylparaben was detected at 280 nm, and epinephrine was detected electrochemically at a potential of +0.90 V. Although the NDA method specified a 2-μL injection with a 10-μL syringe, FDA investigators believed that the use of a 10-μL loop would provide added precision. The NDA procedure was based on a stability-indicating method reported in the literature (4). Each of the components of interest could be detected with any of the specified detectors, but the applicant believed that an additional degree of specificity was achieved by using the three detectors. Adequate system-suitability tests were provided for each detector.

Ovens

The majority of methods submitted to FDA merely specify that they be run at ambient temperature. This specification is likely a reflection of the widespread use

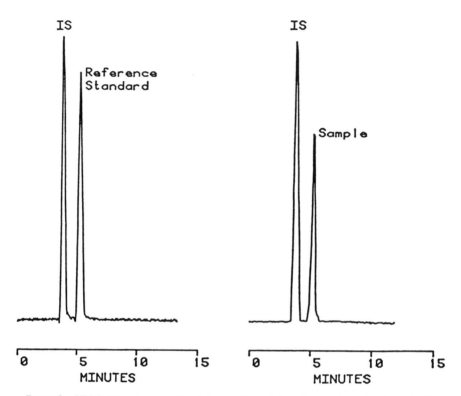

FIGURE 8.   HPLC chromatograms of methylergonivine maleate reference standard and sample obtained with a fluorescence detector.

of instruments that do not contain ovens or other means of controlling the column temperature, such as heating blocks. It also is a reflection of the ability precisely to control the temperature of the laboratory. All of FDA's instruments, however, are equipped with some means of controlling the column temperature; these include built-in ovens, stand-alone ovens, or heating blocks. In most instances the actual column temperature is not critical. It is important, however, to maintain a constant temperature in order to help avoid peak drift, which is necessitated by the almost universal use of data systems for the manipulation of chromatograms. Many laboratories, including FDA's, suffer temperature drift and thus require temperature control even for ambient operations. In other instances, temperatures above or below ambient are required to achieve resolution between various components or because of equilibria considerations.

The chromatogram shown in Figure 10 was obtained during the validation of an NDA for a new imaging agent and is representative of a series of procedures for the assay and the control of potential impurities. The peaks labeled 1–5 are isomers; together they constitute the active component. This chromatogram was run

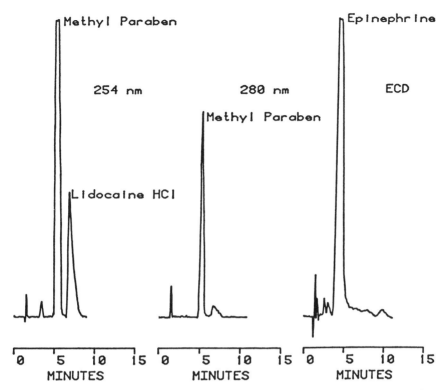

FIGURE 9.   HPLC chromatograms of a 1% lidocaine injection showing the simultaneous detection of lidocaine hydrochloride (254 nm), methylparaben (280 nm), and epinephrine (electrochemical detection).

at 15 °C; other procedures in the set of methods were run at either 15 °C or 20 °C. Both temperatures required the use of a circulating water bath operating somewhat below the desired temperature. Slight changes in mobile phase composition and column temperature are sufficient to afford separation of the various impurities from interfering peaks. A sixth isomer elutes in the series of impurity peaks and is quantified by a separate procedure. Because the conditions for each procedure are very similar, the analyst has minimal leeway in modifying the conditions should the system-suitability test specifications not be met. A Spherisorb S5 Phenyl column (Phase Separations) is specified for all procedures; failure to pass the system-suitability test requires replacement of the column. The only option for the analyst is to wash the column with dilute sulfuric acid. It was found that a single column would not pass the test for all procedures; at least two were required, making acceptability of the proposed methods for regulatory use questionable.

   Another example, Figure 11, was run at 70 °C because of the appearance of rotamer peaks at lower temperatures. The single peak broadens and eventually splits

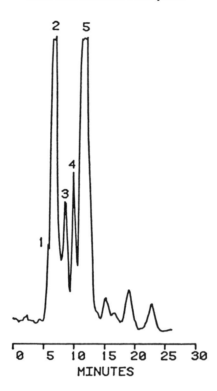

FIGURE 10.   HPLC chromatogram obtained at 15 °C for an imaging agent showing the separation of five isomers (1–5) and several related impurities.

into two poorly shaped peaks as the temperature is lowered. The series of chromatograms shown here were obtained with slightly varying mobile phase compositions; the k' values proved extremely sensitive to these changes. A PRP-1 column (Hamilton Company, Reno, Nevada) was used with a detector wavelength of 210 nm. Acetonitrile/phosphate buffer pH 6.8 (20:80) was the specified mobile phase. The system-suitability test included suggested values for k' of 4.00 and N (number of theoretical plates) of 1000, but no firm specifications were included.

The first chromatogram in this figure was obtained under the conditions described in the assay. The directions suggested changing the mobile phase composition in 2% increments if satisfactory resolution was not obtained. Increasing and decreasing the acetonitrile content did not prove successful in satisfying the system-suitability test parameters. Even so, satisfactory results for the assay were obtained by using a mobile phase with 18% acetonitrile. A typical chromatogram for the assay is included in the figure as well. FDA recommended that the applicant include actual specifications in the system-suitability test. Further, these specifications should give a more realistic indication of the suitability of the system for the intended use.

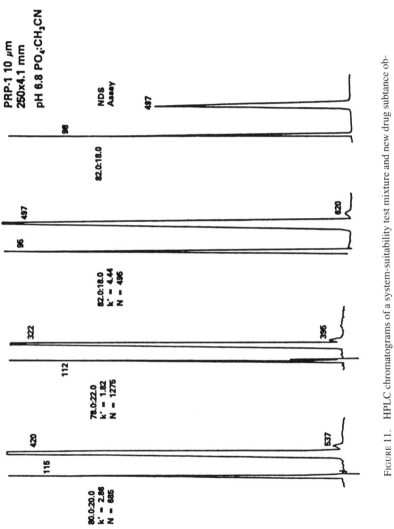

FIGURE 11. HPLC chromatograms of a system-suitability test mixture and new drug subtance obtained on a PRP-1 10-μm column with varying mobile phase compositions.

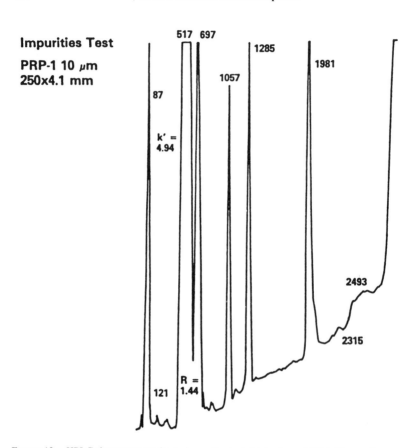

FIGURE 12.   HPLC chromatogram for an impurities test obtained on a PRP-1 10-μm column.

The impurities in the NDS are controlled by a gradient elution HPLC method; a representative chromatogram is shown in Figure 12. Again, although FDA was unable to meet the suggested values for k' (5.3) or resolution between the active component and its diastereomer (1.7), satisfactory results were obtained in comparison to those reported by the applicant.

The dosage form assay contained a similar type of system-suitability test. A LiChrosorb RP-8 column run at 80 °C was specified with a mobile phase of acetonitrile/phosphate buffer, pH 2.0 (32:68). An RP-8 guard column (Brownlee Labs, Santa Clara, California) between the injector and the analytical column was required because of the acidic mobile phase. Typical chromatograms are illustrated in Figure 13; minimum suggested k' values for the three marked peaks were 2, 4, and 10, respectively. The two small peaks are potential degradation products that are added to the standard for the system-suitability test. Under the assay conditions, the suggested k' specification was met for only one peak.

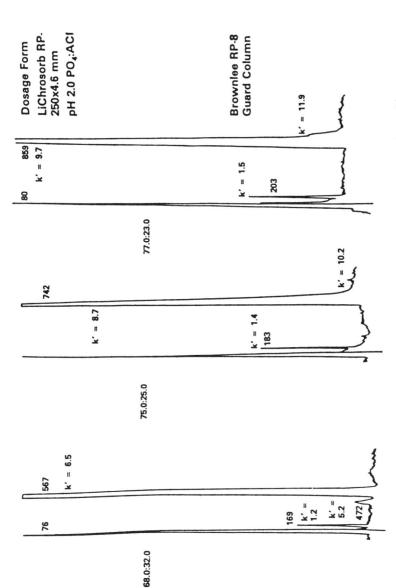

FIGURE 13. HPLC chromatograms of a dosage form obtained on a LiChrosorb RP-8 column with a guard column and varying mobile phase compositions.

Lowering the acetonitrile content to 25% and 23% resulted in some improvement, but the chromatography deteriorated. In addition, one of the degradates now eluted after the main peak rather than before. It was believed that part of the problem was the guard column. Accordingly, the system was replumbed to eliminate this portion of the system, and a precolumn before the injector was added. The precolumn was maintained at the same temperature as the analytical column. Representative chromatograms obtained with 32% and 23% acetonitrile are shown in Figure 14. Although the k' values for the latter system were higher, the former mobile phase gave better chromatography in terms of peak shape and resolution and was used for the assay. A representative chromatogram is included in the figure. Table II summarizes the conditions and the system-suitability test parameters for both the NDS and the dosage form. None of the mobile phase combinations used satisfied all of the specifications. The NDA required replacement of the column if these parameters were not met.

SYSTEM-SUITABILITY TESTS

As recently as 1978, the majority of HPLC methods received by the author's laboratory for validation was submitted without system-suitability tests. A great

TABLE II

System-Suitability Test Parameters for New Drug Substance and Dosage Form (Tablets)*

NDS (Isomer)

Mobile phase (ACN/PO₄, pH 6.8)	k'	N	T	RRT
20:80	2.86	685	0.83	1.27
22:78	1.82	1275	1.00	1.22
18:82	4.44	495	0.90	1.33
Specifications	4	1000	—	—

Dosage Form (Tablets)

	Mobile phase (ACN/PO₄, pH 2.0)	k' (impurity 1)	k' (impurity 2)	k' (drug)
Guard column	32:68	1.2	5.2	6.5
	25:75	1.4	10.2	8.7
	23:77	1.5	11.9	9.7
Precolumn	32:68	1.3	5.7	6.8
	23:77	2.2	18.3	15.6
Specifications		2	4	10

* k' = capacity factor; N = number of theoretical plates; T = tailing factor; RRT = relative retention time.

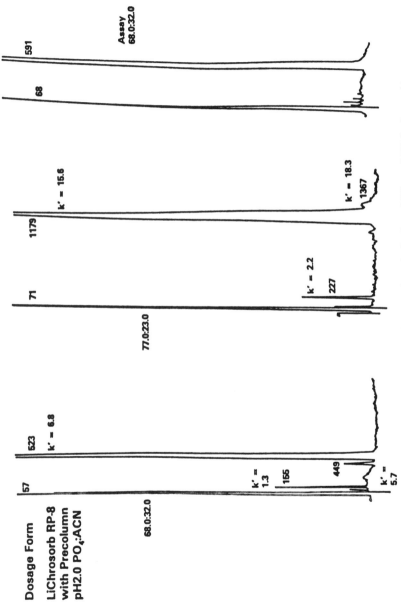

FIGURE 14. HPLC chromatograms of a dosage form obtained on a LiChrosorb RP-8 column with a precolumn and varying mobile phase compositions.

deal of time and effort were expended during the validation of these methods in an attempt to make them work when the system (or column) was unsuitable. Since then, the number of proposed methods containing system-suitability tests has increased dramatically. Part of the impetus for this trend is greater awareness and understanding on the part of the pharmaceutical industry of FDA's needs and of the validation process. The draft "Guideline on Methods Validation" published in 1984 (5) by the Center for Drugs and Biologics contains a section on system-suitability tests. The parameters acceptable to the center are defined in *USP* (6). Numerous reviews have been published on the utility of the various parameters, including a review by Roman in *Pharmacopeial Forum* (7). The specifications encountered in NDAs run the gamut from an explicit limit, such as "the resolution must be greater than 1.5 and the relative standard deviation must be less than 2.0%," to a mere suggestion, such as "the resolution is typically 2.5." From a regulatory point of view, the latter is unacceptable.

Specifications must be determined by the applicant during methods development and then are used by regulatory laboratories to verify the acceptability of the chromatographic systems; again, explicit values are required. Nonetheless, some freedom to modify the system and guidance in the method as to what modification would be appropriate is desirable. The number of tests to be included is at the discretion of the applicant, although the agency may make suggestions under certain circumstances. Parameters not given in *USP* (6) may also be included.

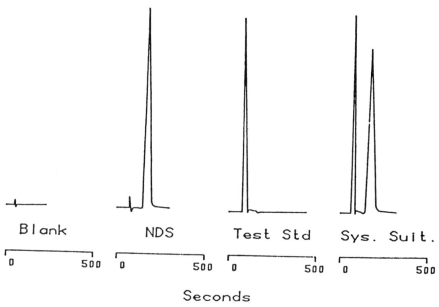

FIGURE 15. HPLC chromatograms of the system-suitability test solution for a new drug substance assay.

As a rule of thumb, for an internal standard method the minimum resolution between the internal standard and one or more active components must be specified. The maximum allowable relative standard deviation of the response is another desirable parameter. Further, the internal standard should be commercially available whenever feasible. If the method in question is used to control the level of impurities, the minimum resolution between the active component and the impurity that is most difficult to resolve must be given. In an external standard method, the maximum peak tailing factor and the maximum relative standard deviation may suffice (5).

The example illustrated in Figure 15 represents a poor choice for a system-suitability test. The chromatogram obtained for the system-suitability test mixture is shown on the right side of the figure. The NDS and the test standard were used to calculate the resolution. The required resolution did not appear to be a problem because there was adequate resolution. Closer examination of the chromatograms for the sample blank as well as for the test standard alone reveals, however, that the standard actually elutes at t_0. The expected retention time and the maximum tailing factor would be a much better choice of parameters. Other examples of system-suitability tests are given throughout this chapter.

TYPES OF SAMPLES ENCOUNTERED

Solids

Solids for oral use constitute the most common type of dosage form encountered. Many of the examples discussed above are tablets or capsules, so relatively little emphasis will be given here. Similar consideration is given to controlled-release capsules; for the most part, their analyses do not involve special or unusual procedures. The one drug that will be discussed here is of interest because various dosage forms using two different HPLC systems have been subjected to methods validation during a period of several years. These include tablets, capsules, controlled-release capsules, and an injectable.

Figure 16 shows representative chromatograms for these various dosage forms. The first three were all obtained on μBondapak C_{18} columns using a mobile phase of methanol/water (65:35). The first peak in each instance was the internal standard, ethyl paraben. The retention times and peak shapes varied from column to column, especially with the peak shapes shown in the first chromatogram. On this particular column, the specified mobile phase represented the optimum chromatography. Even with this relatively poor peak shape the analytical results were satisfactory, indicative of the ruggedness of the procedure. The final chromatogram was obtained for the injectable dosage form using a Partisil PXS ODS-2 column (Whatman, Clifton, New Jersey) and a mobile phase of methanol/water (70:30). The internal standard, eluting first, was tolualdehyde. The peak for the active

440   LC in Pharmaceutical Development

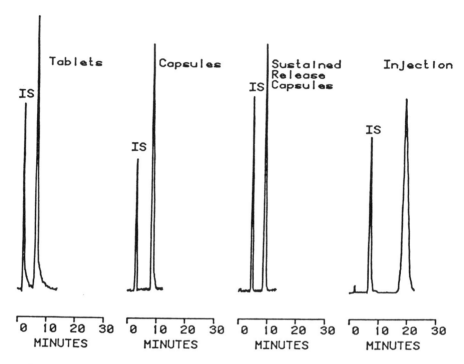

FIGURE 16.   HPLC chromatograms from the assay of four different dosage forms of the same active component.

component was rather broad but was still acceptable because the peak shape and the resolution from the internal standard were satisfactory.

Liquids

Liquid dosage forms lend themselves well to analysis by HPLC because the active ingredient is already in solution. Often, direct injection or simple dilution with water or the mobile phase provides adequate sample preparation. Relatively few NDAs of liquid dosage forms, however, are received for validation.

Figure 17 presents the chromatogram obtained when an elixir containing phenylpropanolamine HCl and brompheniramine maleate was diluted with the mobile phase and injected on a μBondapak C_{18} column. Dextromethorphan HBr was added as an internal standard. The sodium benzoate preservative could be quantified by the same procedure as well as by a separate GLC method. The mobile phase consisted of acetonitrile/water with octanesulfonic acid sodium salt and postassium nitrate (35:65) plus an additional 1.5 mL of acetic acid/L. The detector wavelength was changed from 254 nm to 280 nm at approximately 14.5 min to provide an adequate response for the dextromethorphan HBr. The retention times

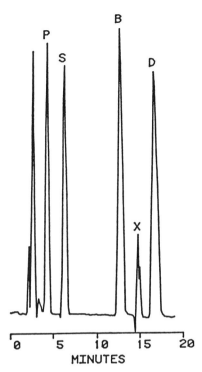

FIGURE 17.   HPLC chromatogram of an elixir containing phenylpropanolamine hydrochloride and bromapheniramine maleate obtained on a μBondapak C$_{18}$ column. P = phenylpropanolamine HCl, 4.30 min; S = sodium benzoate, 6.25 min; B = bromapheniramine maleate, 12.55 min; X = wavelength change from 254 nm to 280 nm, 14.52 min; D = dextromethorphan HBr, 16.45 min.

on FDA's system were somewhat longer than those specified in the method; the change in wavelength was called for at approximately 8.2 min. Because the peak shapes and the resolution between all of the components were acceptable, no attempts at modification of the mobile phase to decrease the retention times were considered. One must, however, question the rationale for selecting dextromethorphan HBr as the internal standard; a better choice would be a compound that did not require a change in the wavelength during the analysis.

Injectables in Oil

The analysis of oil-based dosage forms can present problems because of interference of the oil with the chromatography or because of its effect on the column. The formulation presented in Figure 18 was an oil-based injectable containing testosterone enanthate. The assay was run on a μBondapak C$_{18}$ column with a mobile phase of acetonitrile/methanol/water (6:3:1). Sample preparation involved dilu-

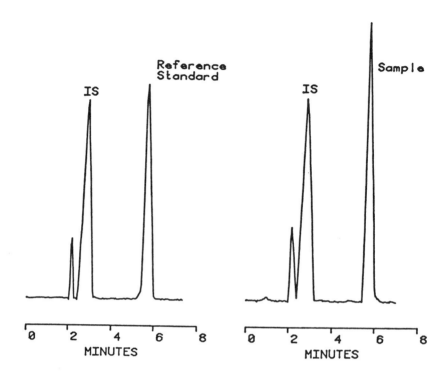

FIGURE 18.   HPLC chromatograms of the reference standard and sample obtained on a μBondapak C₁₈ column during the assay of an oil-based testosterone enanthate injectable.

tion with methanol/chloroform (1:1). Initially, the column passed the system-suitability test, but before the validation was completed the column had to be replaced, presumably because of the oil in the formulation.

Problems of this type often may be avoided by a preliminary clean-up procedure — for example, use of a Celite column (8) — or by including a guard column in the HPLC system. Previous work in the author's laboratory has shown that columns such as Zorbax ODS and Spherisorb ODS separate the common esters of testosterone, free testosterone, isomers of testosterone, and the preservatives present in the formulations. The bulk of the oil was removed before injection by reversed-phase partition chromatography on silanized purified diatomaceous earth.

Ointments and Creams

Ointments and creams, like oils, present special problems as a result of the cleanup required to isolate the active ingredient from the excipients. The chromatogram in Figure 19 was obtained after cleanup with a flocculating solution of methanol/water/glacial acetic acid (300:700:1) and tetrahydrofuran. The active

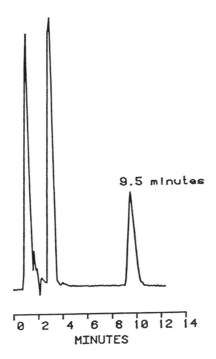

FIGURE 19.   HPLC chromatogram of a hydrocortisone butyrate cream obtained on a μBondapak C₁₈ column.

ingredient, an ester of hydrocortisone, was then injected onto a μBondapak C₁₈ column using a mobile phase of acetonitrile/water/acetic acid (400:600:5). At a flow rate of 2 mL/min, the retention time was approximately 9.5 min and no interferences were noted. Without a cleanup procedure, the useful lifetime of the column would have been reduced to the point of rendering the assay useless. This serves to illustrate the fact that no technique, including HPLC, can solve all analytical problems by itself; other basic techniques of chemistry will continue to play an important role in pharmaceutical analysis. During development, special considerations must be given to confirming the quantitative extraction of the active component from dosage forms of this type.

Suppositories

Another dosage form that can present unusual problems during analysis is the suppository. Extraction of the active component, a prostaglandin, from one formulation and subsequent analysis led to the chromatogram shown in Figure 20. Sample preparation involved warming the sample in acetonitrile (internal standard solution), cooling at −20 °C overnight, warming the sample to room temperature, and filter-

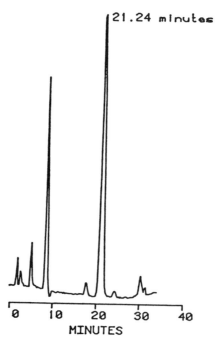

FIGURE 20.   HPLC chromatogram of a suppository containing a prostaglandin obtained on a Li-Chrosorb RP-18 column with a Brownlee RP-8 precolumn.

ing. Finally, a concentrated solution was eluted from a SEP-PAK C_{18} cartridge (Waters Associates) with acetonitrile/water (80:20) and injected onto the analytical column. Analysis was carried out on a 5-μm LiChrosorb RP-18 column with a 3-cm Brownlee RP-8 precolumn. The mobile phase consisted of acetonitrile/methanol/water (40:160:75). As seen in the figure, separation of the three potential impurities from the prostaglandin was more than adequate. Once the extraction procedure was completed, the chromatography was routine. The original method resulted in an overloaded column. Dilution of the sample (1:25) and the use of a larger loop improved the chromatography to the point shown in the figure.

Transdermal Preparations

One example of the use of HPLC to assay the novel dosage form, transdermal skin patches, is presented in Figure 21. Formulations of this type are designed to administer a drug at a continuous rate for an extended period of time. They also present problems in the extraction of the active component from the dosage form. As was true for suppositories and ointments, once the extraction conditions are devised the chromatography becomes more or less routine. In the present case, the extraction of scopolamine HBr involved removal of the polyester backing and

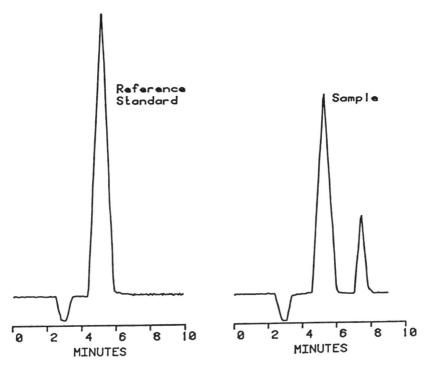

FIGURE 21. HPLC chromatogram of the reference standard and sample obtained during the analysis of a transdermal skin patch.

treatment with chloroform at 60 °C for 30 min. Intermittent shaking ensured quantitative extraction and separation of the patch into layers. Chromatograms of both the reference standard and the dosage form are shown in the figure. The additional peak eluting at about 7.5 min in the dosage form is well separated from the scopolamine peak.

Fat Emulsions

Figure 22 presents several representative chromatograms obtained during the analysis of a fat emulsion containing phosphatidylethanolamine (PE), phosphatidylcholine (PC), and lysophosphatidylcholine (LPC) as the active components; glycerin and soy oil were added to the dosage form and were controlled by HPLC. The method specified a refractive index detector, but, because of equipment considerations, FDA investigators attempted to use a UV detector operating at low wavelengths. This proved to be satisfactory for the glycerin and soy oil, but the sensitivity for the PE peak was such that it was barely detectable. Each procedure involved a separate extraction and a different column. For the soy oil content, a 1:1 mixture of hexane/isopropanol was used for the extraction and the concen-

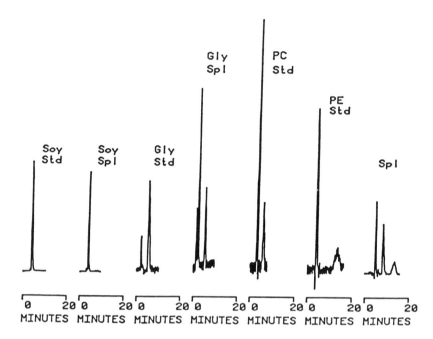

FIGURE 22.   HPLC chromatograms of the various components of a fat emulsion.

trated residue was dissolved in a 99:1 mixture of the same solvents. The assay was performed on a Zorbax SIL column with determination at 215 nm rather than with a refractive index detector as discussed above.

Extraction of glycerin from the fat emulsion involved a different scheme. A mixture of acetone/methanol/glacial acetic acid (30:10:1) was added to a 10 mL aliquot of the sample followed by 150 mL of petroleum ether. The residue from the aqueous layer was dissolved in water, neutralized with dilute NaOH, and injected onto a μBondapak Carbohydrate column. A mobile phase of water/acetonitrile (5:95) was used to elute the glycerin. The variable-wavelength detector was operated at 190 nm for this assay. Although the decreased sensitivity at this wavelength resulted in a noisy chromatogram, peak shape was adequate and reproducible peak heights were obtained.

The three active components were also separated from the raw egg phosphatide by an HPLC method. When the LPC was determined at 190 nm after extraction with chloroform/methanol (2:1), the baseline was found to be too noisy to allow quantification. The PC and the PE, however, were quantified on a μBondapak-NH$_2$ column with a mobile phase of chloroform/methanol-water (25:1)/acetic acid (500:500:4). The elution order for PC and PE was reversed compared to the chromatogram submitted by the applicant and the PE peak was very broad, as is seen in the figure.

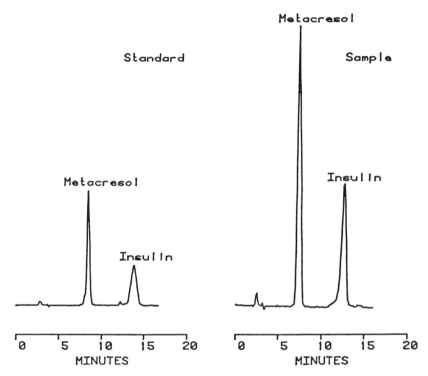

FIGURE 23.   HPLC chromatograms of human insulin reference standard and sample obtained on a Zorbax TMS column.

Other Examples of Interest

The current official monograph for insulin (9) contains both a bioassay and an HPLC assay. Previously (10), only a bioassay was included in the monograph as the quantitative procedure. Similar changes are occurring for many other large, biologically active molecules. Insulin is of special interest because it is a widely used pharmaceutical and because FDA's laboratory has validated many of the proposed HPLC procedures and developed other HPLC procedures (11). The chromatogram in Figure 23 was obtained during the analysis of biosynthetic human insulin and shows the separation of the insulin peak from metacresol for the reference standard and the dosage form. Similar separations are achievable for pork and beef insulin and the desamido-insulins. Polymeric forms of insulin elute from the column with much longer retention times. In the present case, a Zorbax TMS column and a mobile phase of 0.1 M, pH 2.0 $(NH_4)H_2PO_4$/acetonitrile (74:26) were used. The retention times and the resolution showed remarkable dependence on relatively small changes in acetonitrile content.

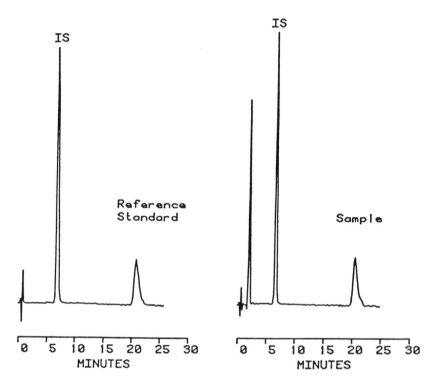

FIGURE 24.   HPLC chromatogram of a nine amino acid peptide reference standard and sample obtained on a Bio-Sil ODS-53 column.

An NDA for a small peptide containing nine amino acids used HPLC as the dosage form assay. Representative chromatograms of the reference standard sample and the dosage form are presented in Figure 24. The system used was a 15 cm × 4 mm Bio-Sil ODS-53 column (Bio-Rad Laboratories, Richmond, California) with a mobile phase containing acetonitrile/phosphate buffer, pH 7.0 (23:77). This mobile phase was used to control the level of impurities in the NDS as well. It was found that premixing the mobile phase before use resulted in strong background absorbance that in turn led to a noisy baseline, to problems in equilibrating the column, and to the adsorption on the column of a strongly UV-absorbing material. The latter necessitated cleanup of the column with methanol. These problems were alleviated by mixing the components of the mobile phase via the ternary proportioning valve in the chromatograph. Alternatively, if the mobile phase must be premixed then the premixing should be done immediately before use. Under the former conditions, no problems were noted with the chromatography.

Another NDA for a peptide (as the acetate salt) with 10 amino acids contained HPLC assays for both the NDS and the dosage form. The method specified a

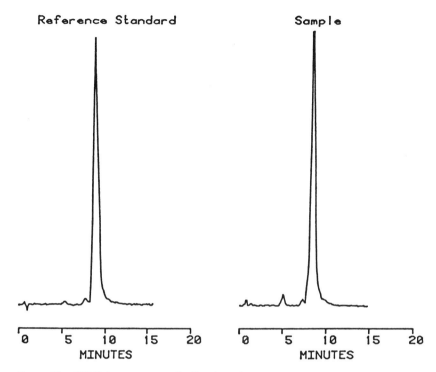

FIGURE 25.   HPLC chromatograms of a 10 amino acid peptide reference standard and sample obtained on a Hypersil ODS column.

Hypersil ODS column (Shandon Southern Instruments, Sewickley, Pennsylvania) and a mobile phase consisting of acetonitrile/phosphate buffer, pH 6.5 (18:82). Representative chromatograms of the reference standard and the sample are shown in Figure 25. The small peaks eluting at about 5.2 and 7.5 min are impurities that are controlled in the NDS by a TLC procedure. Satisfactory results were also obtained with a μBondapak C_{18} column.

IMPURITIES

The use of HPLC to control the presence of impurities in the NDS and the dosage form has increased dramatically over the past four years, as was discussed above. Not all of the NDAs validated in FDA's laboratory contain methodology to control impurities in the dosage form; furthermore, those that do have methods of this type often use procedures other than HPLC. Still, the use of HPLC is increasing. Many of the NDAs discussed above contained methods to control dosage

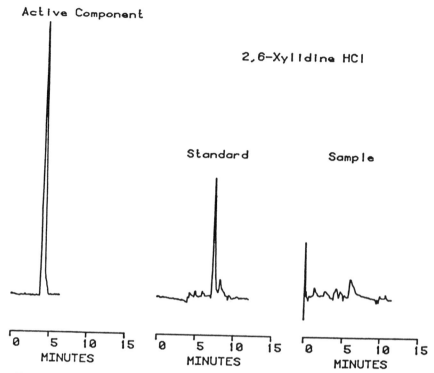

FIGURE 26.   HPLC chromatograms showing the simultaneous detection of 2,6-xylidine hydrochloride with an electrochemical detector and the active component with a UV detector at 254 nm after separation on a μBondapak C$_{18}$ column.

form impurities by HPLC. In order to present a broader perspective, different products will be discussed here.

Dosage form impurities can arise from various sources; most commonly they are degradation products that can form during either the formulation of the material or the storage of the product. Additionally, any synthesis or by-product impurity in the NDS will still be present in the dosage form. The finished product, however, is not usually examined for this type of impurity unless it also can arise as a degradation product. Some methods are more complete and do include limits for NDS impurities in the dosage form.

Figure 26 illustrates the determination of a degradation product, 2,6-xylidine HCl, that can arise during storage. The active ingredient is determined at 254 nm with a UV detector; the impurity is determined electrochemically at levels as low as 0.05%. The chromatographic system included a μBondapak C$_{18}$ column with a mobile phase of 0.01 M octanesulfonic acid, sodium salt/acetic acid/methanol (48:2:50); the electrochemical detector was operated at a potential of +0.90 V.

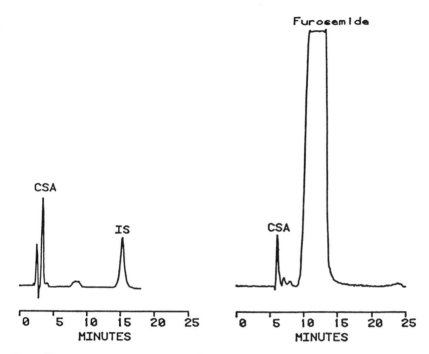

FIGURE 27. HPLC chromatograms of two furosemide formulations showing the detection of 4-chloro-5-sulfamoylanthranilic acid (CSA).

Because the impurity is determined simultaneously with the assay, two recorders are required. Both chromatograms are shown in the figure. The UV signal was passed through the HPLC data system, and a strip chart recorder was used for the output from the electrochemical detector. Only trace amounts of the degradation product were found in the dosage form.

Occasionally FDA is required to validate methodology for generic products marketed by different companies; generally the methodology is different for each NDA. In one such example, presented in Figure 27, two different formulations of furosemide were subjected to method validation. The NDA for the first product, furosemide tablets, contained a procedure to control the level of 4-chloro-5-sulfamoylanthranilic acid (CSA) and two trace impurities. The first chromatogram was obtained on a LiChrosorb RP-18 column using a mobile phase of PIC A/acetonitrile (70:30). For the second product, furosemide injection, the levels of CSA and other trace impurities were also controlled by HPLC. In this instance a μBondapak C_{18} column and a mobile phase of 1% acetic acid in water/methanol (50:50) were used. In both systems, the CSA eluted before furosemide and was well resolved. As an alternative to HPLC, *USP XXI* (12) determines the presence of CSA in each of these dosage forms by a colorimetric procedure.

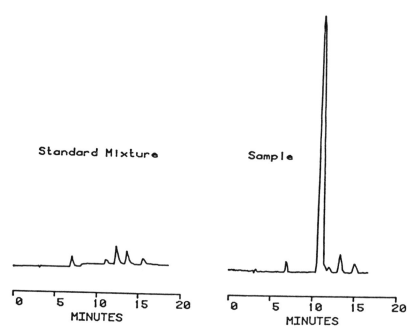

FIGURE 28.   HPLC chromatograms showing the control of impurities in a derivative of vitamin A obtained on a Zorbax SIL column.

One final example of an impurity procedure is shown in Figure 28. Several possible isomers of the active component, a derivative of vitamin A, were controlled by an HPLC method using a Zorbax SIL column and a mobile phase of hexane/ethyl acetate/acetic acid (970:30:1). When a standard solution was run, adequate resolution was obtained between all of the components of the mixture except for two isomers that coelute at about 12.5 min. When the dosage form was assayed, the peak from these two compounds was not completely resolved from the active component. Nonetheless, the data system was still capable of accurately determining their level in the capsules.

MISCELLANEOUS

NDAs for solid dosage forms often contain content uniformity and dissolution methods that are subjected to validation in our laboratory. When HPLC is used for the assay, the content uniformity procedure commonly uses the same system. In fact, many assays are based on the average of 10 (or more) tablets or capsules examined for content uniformity. Much less common is the use of HPLC to determine the extent of dissolution. Nearly all of the dissolution tests validated over the

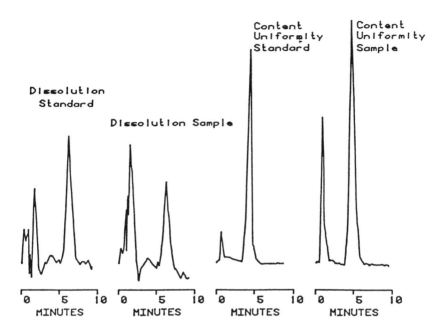

FIGURE 29. Chromatograms demonstrating the use of HPLC for dissolution testing and content uniformity procedures obtained on a μBondapak Phenyl/Corasil II column.

past six years used spectrophotometry as the quantitative step. Figure 29, however, shows the chromatogram obtained when HPLC was used instead. The samples, 0.5 mg and 1.0 mg antihypertensive tablets, were dissolved in 900 mL of water at 37 °C using the USP paddle method. The mobile phase consisted of acetonitrile/0.2% ammonium carbonate (70:30) on a μBondapak Phenyl/Corasil II column. A 50-μL sample loop was used; the detector setting was 0.005 AUFS. Peak heights were measured because of poor reproducibility of peak area measurements. The same chromatographic system was used for the content uniformity procedure except that a 20-μL sample loop was used and the detector setting was 0.02 AUFS. The baseline was much more stable than in the dissolution test, and peak areas were used to quantify the active component. The dosage form assay used the average of the values obtained in the content uniformity procedure.

One final example, Figure 30, presents the content uniformity procedure for an oral contraceptive containing norethindrone and mestranol. A Spherisorb S5-C_6 column with a mobile phase of acetonitrile/water (50:50) was used. Both active ingredients were quantified simultaneously, the results were reproducible, and system-suitability specifications were met. The internal standard, progesterone, eluted between the two components of the drug. As was true in the previous example, the average of 10 individual tablet values was taken as the assay result.

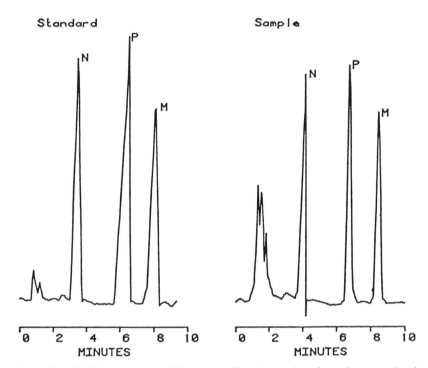

FIGURE 30.   HPLC chromatograms of the content uniformity procedure of an oral contraceptive obtained on a Spherisorb S5-C$_6$ column. N = norethindrone; P = progesterone; M = mestranol.

CONCLUSION

The use of HPLC to control the quality of marketed pharmaceuticals has grown rapidly during the past 10 years. Additionally, the variety of samples analyzed — as well as the types of procedures performed — has increased during this time period. HPLC is certainly the dominant technique in the field of pharmaceutical analysis; its versatility and practicality have been demonstrated in this chapter. Innovations and state-of-the-art applications are presented elsewhere in this book, but these uses have yet to appear in the NDAs received for validation in FDA's laboratories. During the next few years this situation is expected to change as these techniques become established as more or less routine.

One technique that certainly will have an effect on NDA methodology is the use of a diode array detector to check for impurities hidden under the peak of the active component in bulk drug substances and in dosage forms. Another is the resolution of optical isomers on chiral stationary phases. This technique is especially important in the pharmaceutical industry because the majority of active components exist as stereoisomers with one form being more desirable than the other. Nevertheless, HPLC has allowed the pharmaceutical industry to make great strides in controlling and monitoring the quality of its products. As mentioned above, however, the use of HPLC also has been responsible for an increase in the length of time required for FDA laboratories to validate NDA methodology.

REFERENCES

(1) Center for Drugs and Biologics Staff Manual 4820.3, August 31, 1976.
(2) *The United States Pharmacopeia XXI* (U.S. Pharmacopeial Convention, Rockville, Maryland, 1985), pp. 1230–1231.
(3) *Ibid.*, pp. 1215–1219.
(4) S.M. Waraszkiewicz, E.A. Milano, and R. DiRubio, *J. Pharm. Sci.* **70** (11), 1215–1218 (1981).
(5) "Draft Guidelines for Submission of Samples and Supportive Analytical Data for Methods Validation in New Drug Applications," *Fed. Reg.* **49** (89), 19412 (May 7, 1984).
(6) *The United States Pharmacopeia XXI* (U.S. Pharmacopeial Convention, Rockville, Maryland, 1985), pp. 1229–1230.
(7) R. Roman, *Pharmacopeial Forum* **8** (4), 2237–2238 (1982).
(8) E. Smith and O.H. Riggleman, "Liquid Chromatographic Determination of Steroids and Steroid Esters in Oil Injections III — Analysis of Testosterone Esters in Oil Injections," Abstract I, paper presented at 36th National Meeting of the Academy of Pharmaceutical Sciences, Montreal, Canada, 1984.
(9) *The United States Pharmacopeia XXI* (U.S. Pharmacopeial Convention, Rockville, Maryland, 1985), pp. 534–541.
(10) *The United States Pharmacopeia XX* (U.S. Pharmacopeial Convention, Rockville, Maryland, 1980), pp. 900–901.
(11) D.J. Smith, R.M. Venable, and J. Collins, *J. Chromatogr. Sci.* **23**, 81–88 (1985).
(12) *The United States Pharmacopeia XXI* (U.S. Pharmacopeial Convention, Rockville, Maryland, 1985), p. 453.

TRENDS IN THE USE OF LC IN THE PHARMACEUTICAL INDUSTRY

W. John Lough

Physical and Analytical Services Unit
Beecham Pharmaceuticals
Harlow, Essex CM19 5AD
United Kingdom

In other parts of this book, modern liquid chromatography methods for the analysis of drug substances and advances in the use of LC in individual areas of the pharmaceutical industry have been described. In this chapter, the use of LC and its importance over all the stages of pharmaceutical development will be examined. To some extent, this will constitute an overview, bringing together the topics already described and discussing their relative importance. More important, however, the *trends* taking place in the use of LC in the pharmaceutical industry will be highlighted.

Trends of any kind may most readily be identified and understood by knowing the factors that govern them. The main factors that govern trends in the use of LC in the pharmaceutical industry include:

• advances in the LC technique itself
• changes in the pharmaceutical industry
• the demands placed on LC by the specific needs of pharmaceutical applications.

RECENT ADVANCES IN LC

There are many sources of up-to-date information on advances in the LC technique itself. Indeed, taken together, the new LC methods described in the opening chapters of this book give a good reflection of recent advances. Despite the many advances in LC over recent years, however, the overall pattern of the new development of the technique can be described fairly simply.

By the late 1970s, column technology had reached the stage at which efficiencies on the order of 20,000 theoretical plates could regularly be obtained when using conventional 250 mm × 4.6 mm columns packed with 5-μm spherical particles. Technological limitations were such that there was little point and little prospect in producing columns containing particle sizes of less than 3 μm, so that it was no longer possible as it was in the past to obtain substantial improvements in efficiency by using yet smaller particle sizes.

The well-known equation for chromatographic resolution follows:

$$R_s = \frac{1}{4}\underbrace{\left(\frac{\Delta k'}{k'}\right)}\quad \underbrace{\left(\frac{k'}{1+k'}\right)}\quad \underbrace{\left(N^{1/2}\right)}$$

relative difference in retention, that is, selectivity factor	degree of retention factor	column efficiency factor

where

$$k' = \frac{t_r - t_0}{t_0}$$

$$N = \text{column efficiency} = 5.54\left(\frac{t_r}{W_{1/2}ht}\right)^2$$

From this equation it can be shown that it is easier to improve resolution by improving selectivity rather than by improving efficiency. The middle and late 1970s saw the initiation of a period during which chromatographers sought to improve upon the selectivities available from those inherited from classical chromatography in normal-phase LC systems and from reversed-phase systems using alkyl-bonded phases. The effect of secondary chemical equilibria (1), whether modified by additives in the mobile phase or by changes in bonded phases, has been exploited to improve selectivity.

The first such development was ion-pair chromatography. This technique has become widely adopted (2) and has been comprehensively reviewed (3). This was followed by ligand-exchange chromatography; this technique was more limited in its applicability, but the improvements in selectivity obtainable over other systems were dramatic. The work of Karger, who first announced his results on enantiomeric separations of dansyl amino acids (4) in Baden-Baden in 1978, exemplified the selectivity of ligand exchange. Activity in this important area continued with the use of different metals, different chiral ligands, and studies of the mechanism. By 1983, and again in Baden-Baden, Karger was able to announce considerably enhanced resolutions of dansyl amino acid enantiomers obtained by placing spacers between the chiral ligands in the bonded phase (5).

The separation of enantiomers by a chiral chromatographic system usually represents one of the most severe tests available in achieving sufficient selectivity to obtain a separation. Another of the most significant pieces of work in enhancing selectivity by the use of secondary chemical equilibria took place in this area when Pirkle used charge-transfer interactions to achieve enantiomeric separations. His broad-spectrum chiral phase has been particularly useful in the separation of enantiomers of electron-rich aromatic molecules (6).

One of the more recent manifestations of this emphasis of making the most out of selectivity was the intense activity between 1980 and 1983 on the optimization of LC mobile phases. Rather than making use of different mechanisms such as ion-pairing, these optimization techniques were typically applied to conventional reversed-phase or normal-phase systems. The optimal ratio of one or more organic solvents with either water or n-hexane required to achieve a desired separation was determined. Some of the most sophisticated techniques incorporated optimization of pH and other ionic effects (7). The variety of methods for quantifying the feature of the chromatogram to be maximized and the variety of methods for arriving at the optimal solvent system to achieve this have been reviewed (8); some of the methods have been directly compared (9).

There is still much important work being carried out on improving selectivity for difficult separation problems, but since 1981 and 1982 much of the focus of LC research has again been on column technology and, in particular, on low-dispersion chromatography. Essentially this is chromatography in which there is very little dilution of the injected band of solute. Bracketed in this category are fast (or high speed) (10), microbore (11), and capillary LC (12). Only in capillary LC is there a continuation of the search for very high plate numbers.

Because of the high back pressures involved, 3-μm particles were not packed in 25-cm columns to produce higher efficiencies than those packed with 5-μm particles. Similar efficiencies could be obtained with shorter — for example, 15-cm — columns. With very short (5-cm, for example) columns packed with 3-μm particles, efficiencies on the order of 5000–7000 theoretical plates are achievable. Even though, for these small particles, efficiency sometimes falls off more rapidly with flow rate than is true for larger particles (13), it is still possible to obtain very difficult separations in very short analysis times. This so-called fast LC technique is now widely used, and Erni (14) has already used it for analysis times far shorter than will be required in the foreseeable future.

Microbore LC is a more recent technique and involves the reduction of column diameter (from 4.6 mm to 2 mm or 1 mm) rather than particle size. The advantages of doing this are to give a higher mass sensitivity of detection, better compatability with LC/MS interfaces of the moving belt or direct liquid inlet type, and to reduce solvent consumption, which not only reduces solvent costs but also allows the use of expensive or exotic solvents. The coupling of microbore columns in series to produce high efficiencies (at the expense of analysis time) has been cited as an advantage (15), but recently there have been examples of this coupling also being done with conventional columns (16). Nonetheless, in column-coupling experiments the advantage of reduced solvent consumption, not to mention packing material, would come into greater play.

In capillary LC, the columns are usually fused-silica capillary tubes. These may be very fine (10–50 μm or slightly wider) open-tubular capillaries packed with conventional LC packing materials. The technique marks a continuation of the quest for very high efficiencies. These efficiencies are obtained by using long capillaries — on the order of 30 m — but this is very much at the expense of analysis time. Research into capillary LC, particularly excellent work by Novotny (17) and Japanese investigators (18), has been progressing since the early 1980s, but it is still very much a research technique.

Discussion of these latest techniques, such as microbore and capillary LC, leads this brief synopsis of the general direction of recent LC advances to the subject of LC equipment. As yet, there is no commercially available equipment for capillary LC. Commercially available equipment for microbore LC has been criticized in general because there is too much extracolumn dispersion to make the most of the columns, particularly when those of 1-mm rather than 2-mm i.d. are being used (19). It is surely only natural, however, that commercially available equipment should lag behind relatively new techniques. In general, new pumping, injection, and detection systems have kept pace in terms of reproducibility and reliability with what has been required of them. In the case of detectors, they are now more sensitive and many different types are available. The most significant advances in instrumentation, however, have been in automation brought about by the application of computers, usually microprocessors. This has involved the automation not only of pumps, injection systems, and, more recently, detectors, but also of all aspects of data acquisition and handling. As will be seen later, some of the LC advances that have been described have application in only a limited number of areas of LC usage by the pharmaceutical industry, but the large increase in automation has been universally beneficial.

CHANGES IN THE PHARMACEUTICAL INDUSTRY

The main changes in the. pharmaceutical industry that have affected LC use are two: the increase in information required by regulatory authorities and the need for the industry to be more cost effective.

Regulatory authorities must satisfy themselves not only with the safety of a new drug, but also with its quality and efficacy (20). The testing of drug safety involves not only toxicology studies, but also studies of drug metabolism and pharmacokinetics. LC is used extensively in the identification of metabolites and in the monitoring of drug and metabolite concentrations in biological samples from toxicology studies, clinical trials, or purely pharmacokinetic studies. Pharmokinetics is a very significant area of LC use and continues to grow.

The information required on drug quality has been summed up by Sullman (20). In her paper on international regulatory requirements with respect to the pharmaceutical aspects of new drug development, she states that, in satisfying themselves as to the quality of a new drug, regulatory authorities must receive satisfactory answers to a number of questions:

• Is the method of synthesis of the active ingredient producing drug of consistent and acceptable purity (and impurity)?

- Is the active ingredient what it is said to be? That is, is there proof of structure?
- Is the method of manufacture of the final formulation producing a consistent and acceptable end product?
- Is the active drug available?
- Are the methods of sampling and the analytical methods suitable for obtaining answers to these questions?
- Is the product going to be stable over the proposed shelf life?
- Does the drug/formulation that is proposed for marketing reflect the drug/formulation used for preclinical testing and in the clinical trials?

LC may be involved in the replies to all of these questions — other than the second, regarding proof of structure. For instance, individual companies have been deliberating (21) on the best way they see fit to comply with questions concerning system suitability and are trying to ensure a unified approach to the subject from all the LC services in that company. In addition, more effort is being made to optimize methods as early as possible so that a large number of modifications do not need to be documented and so that information on the quality of several batches of a drug is obtained using the same method throughout.

The need for the pharmaceutical industry to be more cost effective has also resulted in an increase in the use of LC. Several examples of this will be discussed later, but one need only think, for instance, of assays of drugs in biological samples. Complex extraction procedures followed by a spectrophotometric measurement are being replaced by LC methods involving minimal sample preparation (22). These LC methods are not only much more rapid, but yield far more information.

With increasing use of LC has come the associated need for increased sample throughput, which may be achieved by the use of fast (or high speed) LC, but, as will be discussed later, use of this type of LC is not as widespread in the pharmaceutical industry as it possibly could be. Higher sample throughputs have been achieved largely by a much greater emphasis on automation, which allows not only unattended and overnight runs but also frees the operator for other tasks, thus helping to reduce overall labor costs.

It was noted previously that equipment manufacturers have occasionally been criticized for not keeping up with developments in column technology. Such criticisms could not, however, be leveled against the manufacturers with respect to automation. The automation of pumps, injection systems, detectors, and data handling referred to earlier has been of paramount importance to the use of LC in the pharmaceutical industry. The increase in automation has been *the* major current trend in LC in the pharmaceutical industry over the past five to six years, and this trend will continue for some time to come.

DEMANDS PLACED ON LC BY PHARMACEUTICAL APPLICATIONS

The third factor in determining current trends in LC in the pharmaceutical industry is the specific requirements demanded of LC from each pharmaceutical application. These requirements are obviously very different for each area of pharmaceutical development. As such, the requirements for each area and the consequent trends in LC use in that area will be described on an individual basis. There are, however, some general observations that can be made.

Other than in the very earliest stages of research, the development of an analytical method must take into account the fact that the method will probably be used in laboratories other than the laboratory carrying out that development work. These other laboratories may be on other sites and in some cases may even be in other countries. The methods must therefore be very robust; that is, the method must still work when the performance of the LC system is not quite at its optimum, including such factors as slightly higher baseline noise than usual, slight loss in column efficiency, or slight loss in resolution caused by difficulties in reproducing exactly the desired mobile phase. Thus, the system must be as simple as possible.

One aspect of this requirement is *equipment*. The method must be geared to the least well-equipped of the laboratories likely to use it. Although equipment manufacturers would no doubt like it otherwise, it is not possible even in the most successful of pharmceutical companies for every laboratory using LC to be equipped with state-of-the-art instrumentation. Just as there is a time lag between new LC advances and the development of suitable equipment, there is a time lag, often of longer duration, between the development of new equipment and its widespread use in the pharmaceutical industry. The more obviously relevant a new LC advance is to a particular pharmaceutical application, the more rapidly will it be adopted. In the following discussion of what is required from LC in the individual areas of pharmaceutical development, the more relevant of recent LC advances will be identified.

Research Chemistry

During the course of the organic chemistry involved in the synthesis of novel compounds for pharmacological testing, by far the main use for LC is to monitor the progress of reactions and to check the purity of reaction products. An LC system has become as much an integral part of an organic chemistry laboratory as, for instance, a rotary evaporator or a thin-layer chromatography (TLC) setup. No great sophistication is required to monitor reaction progress, and at this stage ultrahigh purity is not quite so critical. Therefore, for some time now the standard LC system for the typical organic chemistry laboratory in the pharmaceutical industry has consisted of a reciprocating dual-piston pump, a loop injector, a conventional 250 mm × 4.6 mm stainless-steel column usually packed with a 5- or 10-μm ODS silica, a variable-wavelength UV detector, and a strip-chart recorder — or, of late, an inexpensive single-channel integrator/plotter. If more sophisticated instrumentation is required, there is always a specialist analytical laboratory available. In very few companies, however, would the specialist analytical laboratory be responsible for all of the LC work in research chemistry. It is far more practical to allow such laboratories to concentrate on problems for which detailed expertise is required and to allow the organic chemists to have basic systems for instantaneous follow-up to their reactions.

This situation should also apply in the future, because there is no genuine need for changing this arrangement. One slight change is that the reciprocating dual-piston pump might be replaced in some cases by a programmable ternary (or quaternary) gradient pumping system. Ternary gradients are seldom if ever used in this area of LC application, but such pumps allow very rapid switching of mobile phases. Even this capability is not usually necessary, because a synthetic chemistry program will involve work on a series of compounds that have some common structural features and consequently are not dramatically different in their chromatographic properties. Moreover, most ternary gradient pumping systems are far too expensive to warrant use in this environment. Only a few such pumping system are inexpensive enough to be applicable; of these, some are cheaper but less versatile, having only binary gradient capability.

In this area of pharmaceutical development there is also some movement toward the use of disposable cartridge columns in cartridge holders instead of stainless-steel columns. Such an approach would be of dubious value in a specialist analytical laboratory with its own column-packing equipment and its need to have as many columns as is practical with different packing materials available. Because of their ease of use, however, these disposable cartridges may have a place in the LC system of research organic chemists.

Packed stainless-steel columns have an advantage in analytical LC if there is any intention of repeating the work on a preparative scale. Despite the introduction many years ago of the Prep 500 system (Waters Associates, Milford, Massachusetts), which used radially compressed cartridges, most of the preparative LC in research chemistry in the pharmaceutical industry is carried out with packed stainless-steel columns. Often the packing materials used in cartridges are not available loose for packing in preparative or semipreparative stainless-steel columns, so the direct scaling-up of conditions used for the analytical-scale separation with a cartridge may not be possible.

Preparative LC is, in fact, the other major LC method in use in the area of research chemistry. In recent years significant advances have been made in the theory and practice of preparative LC. For instance, Kraak and Poppe (23) have studied the maximization of sample throughput, and Gariel and colleagues (24) have used a nonlinear optimization procedure to achieve a production rate that is often 10-fold greater than that achieved by simply scaling up analytical conditions. Much of this work, however, has not found application in the early stages of drug research.

The preparative applications are usually one-off problems for which time is not a major consideration. For final products, purity is the main priority and, in most cases, only sufficient material for pharmacological testing (10–500 mg) is required. Another factor involved is that in most companies this work is carried out by the organic chemists themselves and not by specialist chromatographers. The simplest approach is therefore the one most often preferred.

An analytical separation is developed so that good resolution is obtained with retention indices as low as possible. This approach is done for two reasons: to minimize solvent consumption in the subsequent preparative work and to minimize overloading of the LC stationary phase, as follows:

$$k' = \frac{\text{quantity of solute in stationary zone}}{\text{quantity of solute in mobile zone}}$$

The amount of sample injected is then gradually increased step by step until the necessary resolution is just maintained. Repeat injections are then made until the desired amount of pure compound is obtained. Despite its lack of sophistication, this approach is very popular. A typical example is the work carried out in the author's analytical unit (Figure 1). The impurity, B, arises from the loss of one molecule of water from the major component, which is a peptide with 18 amino acids (25).

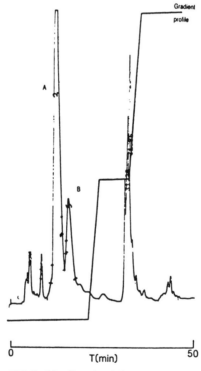

FIGURE 1.   Preparative HPLC of β-cell tropin. Column: LiChroprep RP8, 250 mm × 16 mm (E. Merck, Darmstadt, FRG); mobile phase: acetonitrile/water/trifluoracetic acid (32:68:0.1) followed by gradient step to clean up column; detection: UV 274 nm; flow rate: 8.0 mL min⁻¹. Injection volume: 2.0 mL; sample load: 35 mg. A = main peak, B = major impurity.

Just as a very high proportion of analytical LC in the pharmaceutical industry is carried out under reversed-phase conditions, a high proportion of the preparative LC is now also carried out under these conditions. There is still a strong tendency, however, to look for normal-phase conditions whenever possible. Beyond the obvious advantages of less expensive packing material and easier isolation of the compound from the mobile phase, normal-phase LC can sometimes offer much greater selectivity and resolution between closely related compounds than are obtainable in the reversed-phase mode.

This was the case in the author's company's work on isomeric benzamides (26) with nitrogen-containing cyclic side chains (Figure 2). These types of compound can show central nervous system (CNS) activity and/or activity as gastric-motility enhancers. Separation of equatorial and axial isomers can therefore often be crucial in obtaining exactly the desired profile of drug activity. It was found on the one hand that good isomer separations were rarely obtainable under reversed-phase conditions. On the other hand, a much improved separation could always be obtained by using an aminopropyl-silica column in the normal-phase mode; even better separations were obtained on silica (Figure 3). An amine had to be added to the mobile phase to prevent peak tailing of these strongly basic compounds; because of the use of the relatively volatile sec-butylamine, this did not cause problems in the isolation of the solid compound from collected mobile phase fractions.

Because of the laboriousness of this approach to preparative LC, there is now a trend to the use of automatic repeat injectors and peak-sensing fraction collectors. Until recently, fraction collecting usually had to be done on a do-it-yourself basis, but today there is equipment especially designed for unattended and overnight operation.

When preparative LC of synthetic intermediates is being carried out, larger quantities (500 mg to 5 g) may be required. Even in this range — unless the separation is very easy — repeat injections on 250 mm × 22 mm i.d. columns packed with materials with particle sizes between 5 μm and 20 μm tend to be used rather than systems containing larger and wider columns. There is, however, much more emphasis on maximizing sample throughput, often including some form of recycling, and less emphasis on maintaining good resolution throughout.

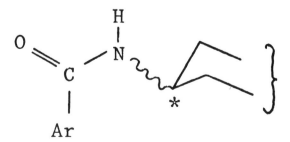

FIGURE 2.   A general benzamide structure; note the axial/equatorial isomerism at C*.

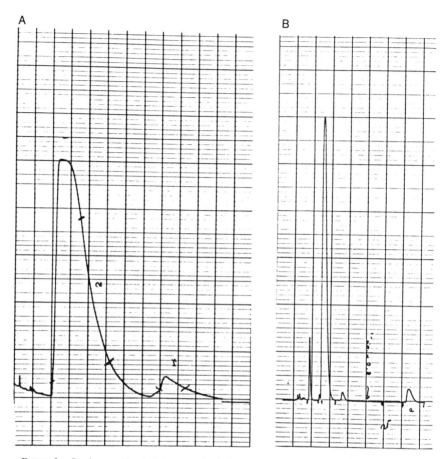

FIGURE 3.   Semipreparative LC of saturated solutions of benzamide in various mobile phases. (A) 500-μL injection ($\sim$10 mg) on APS-Hypersil, 250 mm $\times$ 7 mm (Shandon Southern Instruments, Sewickley, Pennsylvania); mobile phase: methylene chloride; flow rate: 5.0 mL min^{-1}. (B) 200-μL injection ($\sim$ 5 mg) on Hypersil, 250 mm $\times$ 7 mm; mobile phase: 0.005M sec-butylamine in acetonitrile; flow rate: 4.0 mL min^{-1} changing to 8.0 mL min^{-1} before elution of last peak.

The other LC application areas in research chemistry include its use in structure-activity relationships. Reversed-phase LC retention indices are used in structure-activity relationships (27) in which drug activity is governed by its ease of transport to the receptor site in the body. Often a correlation is established between retention data and partition coefficients (28) so that retention data can be used to determine unknown partition coefficients that are then used in the structure-activity relationship (or, for example, to predict solubility). In other cases a direct correlation between retention indices and activity can be made (29).

Work of this nature contributes to the cost-effectiveness of research in that chemists can tell when they have made a compound with the optimum hydrophobicity for a certain series of compounds and therefore have no need to produce any further analogues. Because the determination of retention indices is much more rapid than the determination of partition coefficients, there ought to be a very strong trend towards their use in structure-activity work — more so because many of the earlier problems in obtaining good correlations have now been solved. Unger (30) has used an amine in the mobile phase to screen residual silanol groups in the ODS silica, which give unwanted interactions, and Tomlinson (28) has developed a mathematical treatment to allow the correlation of the retention indices and partition coefficients of the un-ionized form of basic drugs. This type of LC work has, however, largely been superseded by the use of molecular graphics, despite the fact that the two methods are complementary. In molecular graphics the study is not so much on transport properties as on the details of the steric and electronic interactions at the receptor site. A reasonable number of compounds must be synthesized and tested to build up information on the nature of the receptor site, but once this has been done the technique has the considerable advantage that new drug structures can be studied without first having been synthesized.

Enantiomeric separations are very important in research chemistry. Even though there have been many publications in the literature on the direct resolution of enantiomers by LC on a chiral stationary phase or with a chiral additive in the mobile phase, there is still a strong dependance in organic research in the pharmaceutical industry on the formation of diastereomers for analytical separations and on fractional crystallization of a salt formed with an optically active species of opposite charge for preparative work. The problem is that, because of the very precise nature of the requirements for the interactions between the solute and the chiral phase to achieve separation, most chiral chromatographic systems are applicable only to a limited class of compounds. Drug research chemists cannot often afford the luxury of experimenting with various chiral LC systems that have not already been shown to work for the particular type of compound. They must be able to turn to a method on which they can depend to work fairly quickly.

One phase that has more general application (6) is Pirkle's broad-spectrum chiral stationary phase (Figure 4). As such, it has become fairly widely adopted, but even this phase has considerable limitations in its use. When the phase (Pirkle

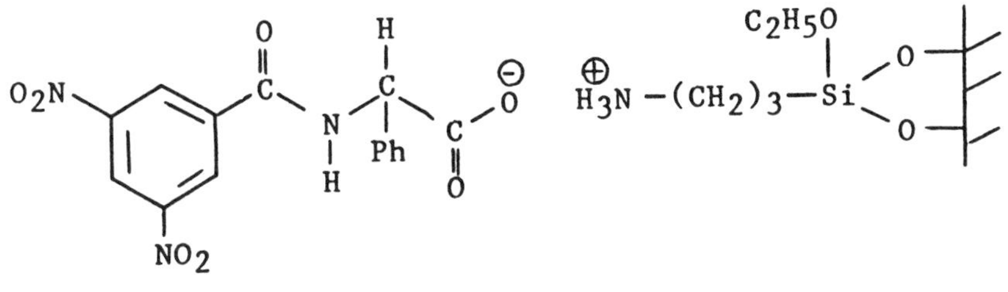

FIGURE 4.   The structure of Pirkle Type 1A stationary phase.

Type 1A, Regis Chemical Co., Morton Grove, Illinois) was first introduced commercially, it was tested at the author's unit for almost all of the different types of racemic compounds involved in the company's research. Enantiomeric separations
were obtained for only a small portion of the compounds, that is, for the classes of
compounds shown in Figure 5.

These results agree with the examples given by Pirkle and others. For successful
enantiomeric separation, a naphthalene ring (or higher polynuclear aromatic system) in close proximity to the chiral center is required. It was not possible to elute
some strongly basic compounds from the column without stripping the stationary
phase. This would not apply, however, to the covalent form of the Pirkle phase.
With the Type 1A phase, difficulties were also encountered with carboxylic acids
and ketonic compounds not eluting from the column, which was presumably
caused by ionic interactions and imine formation, respectively, with residual amino groups.

Wainer and co-workers (31) have extended the scope of the phase by using
achiral reagents containing a naphthalene ring for derivatization before injection.
This represents no real advance on diastereomer formation followed by LC on an
achiral (cheaper) phase. The technique, however, may still gain widespread use in
the future, because enantiomeric separation of the derivatives can be fairly reli-
ably obtained.

At present, the only other type of chiral LC system in general use in research organic chemistry is ligand-exchange LC, using a chiral ligand either as a bonded
phase or as a mobile phase additive. Such systems are used almost exclusively for
separations of amino-acid enantiomers (5), often as part of an amino-acid analysis
of a peptide. They have also been used (32) for enantiomeric separations of those
amino-acid derivatives that are side chains for penicillins.

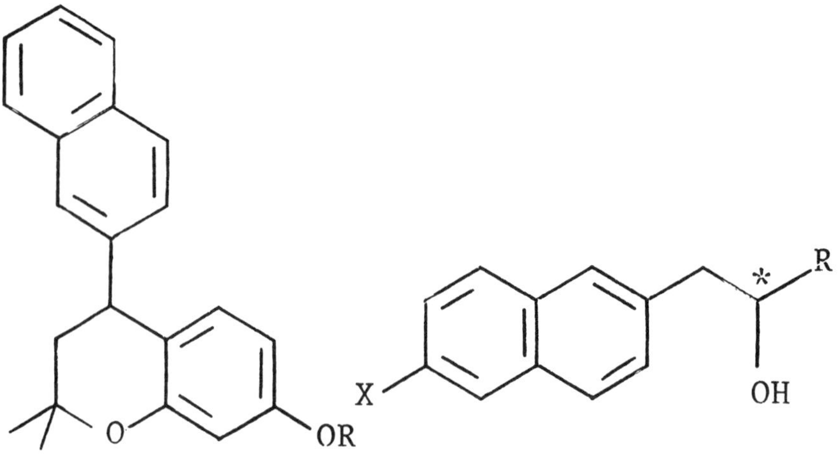

FIGURE 5.   Drug classes for which enantiomeric resolution on Pirkle Type 1A phase was possible.

In the field of amino-acid analysis in general, there is a very strong trend away from dedicated amino-acid analyzers and toward amino-acid analysis on conventional LC systems, whether it be by ligand-exchange LC or by reversed-phase LC of amino-acid derivatives prepared by reaction with, for instance, phenylthiohydantoin (33), phenylisothiocyanate (34), or p-N,N-dimethylaminophenyl isothiocyanate (35). Although amino-acid analysis is important in research chemistry for synthetic peptides, its main area of application is in research biochemistry.

Research Biochemistry

In the past, the research biochemistry area of pharmaceutical development involved the study of the mechanism of disease, its effect on the biochemistry of the body, and, from these, the development of a biochemical assay to screen potential drug substances for activity against that disease. Today, the scope of biochemistry in pharmaceutical research is much wider, largely because of a massive expansion in work involving biopolymers, which embraces both the use of biopolymers as medicines and the study of biotechnological processes for producing drugs. Of all the forms of biotechnology, genetic engineering holds the greatest potential (36). In the health field it is already being used for the synthesis of hormones, antibiotics, insulin, and interferons.

The role of LC in assays for small-molecule drug substances in biological systems has been firmly established for a few years now. Much work has, for instance, been published on LC systems for the analysis of the prostanoid content of biological samples (37). Even more work has been carried out on LC assays for catecholamines and their metabolites (38). There is not much scope left for further replacement of more traditional biochemical assays with LC assays; the trend is now toward the refinement of LC methods already in use. For instance, there is an increase in use of improved electrochemical detectors such as dual-electrode models. In this type of detector, improved sensitivity and stability are obtained by the use of two fully porous electrodes in series.

The very large potential for growth in the LC of biopolymers has been responsible for much development work on column technology and support materials. This development work, in turn, has opened up many new possibilities in the LC of biopolymers. Much has been done in improving the mechanical properties of polymeric packing materials, but perhaps more significant has been the introduction of wide-pore ($\sim$ 300 Å as opposed to the more conventional $\sim$ 100 Å) silica-based materials (39). The narrow spread of pore sizes and the use of the 5-μm silica-based particles has greatly enhanced size-exclusion chromatography (40), but possibly the greatest advantage of the wider pore sizes is that ion-exchange and reversed-phase LC of peptides, proteins, and polynucleotides can now be carried out in high-efficiency columns. In chromatographic supports, the bulk of the sites of interaction lie within the pores. The solute molecules must therefore be able to enter the pores for retention to take place. Size-exclusion, ion-exchange, and reversed-phase LC are now the main LC methods for biopolymers but affinity chromatography (41) and hydrophobic-interaction chromatography (42) (similar to reversed-phase LC but using aqueous mobile phases in order to prevent denaturation and controlling hydrophobicity with salt concentration) have also benefited from the new packing materials. Wide-pore supports can also be used for chromatofocusing (43), but in this technique a narrow band width is obtained by the focusing effect of a pH gradient and is not so dependent on the use of small particles.

In analytical-scale LC of biopolymers, there are often only very limited quantities of materials available. It has been said that this will inspire the market for microbore LC (19). In view of the higher mass sensitivities achievable with microbore compared to conventional LC, this ought to be the case. This change, however, has yet to take place to any great extent.

Preparative LC of biopolymers is important not only in the research stage, but also in the production of biopolymers as medicines. This aspect will become more and more important. Preparative LC is feasible as a means of production because the amounts of material required usually are less than is required for small-molecule drugs. It is also necessary because proteins are easily denatured and the purification of all biopolymers requires much milder conditions than methods such as the crystallization techniques used for traditional drug substances. There has therefore been considerable effort put into preparative LC of biopolymers, and the wide-pore materials have been invaluable in this area. Much of the interest has, however, been in ensuring high mass recovery. This has involved the use of very short-chain alkyl bonded phases and polar bonded phases (44). Deactivation procedures have also been developed to reduce interactions with residual surface silanols (45).

Changing methods in amino-acid analysis and the overall growth of this area of LC work in research biochemistry were mentioned earlier. This growth has been very much enhanced by the availability of very pure polypeptides and proteins from preparative LC.

Chemical Development

In the chemical development stage of the progress of a potential drug substance toward the market, its chemical synthesis is scaled up and modified. Batches of compound prepared from these scaled-up syntheses are used in toxicology studies in animals and then, later in the development process, in drug metabolism and pharmacokinetic studies and in clinical trials.

As mentioned earlier, LC is heavily involved in the provision of data on batch-to-batch drug quality (20). There is a strong trend to the provision of more of this data and an emphasis on its being very well documented. There is an equally strong emphasis on the documentation not only of the methods used to provide the drug quality data, but also of the tests carried out to validate them.

Drug companies must give careful consideration to what approach they should adopt to matters such as *validation tests,* that is, tests done during method development to show that the method is scientifically sound and adequate for the assay; *acceptance criteria,* that is, criteria the validation test results must meet for the validation to be acceptable; and *system-suitability tests,* that is, test(s) done each time the system is set up for a particular analysis to ensure that it meets the criteria required for a reliable assay. The approach adopted by the quality control department of Hoffmann–La Roche has been set out by Debesis and colleagues (21). On method validation, they describe the test, acceptance criteria, and documentation on each of six aspects of the method, including specificity, linearity, precision, accuracy, sensitivity, and ruggedness. For system-suitability tests of LC systems for a particular method, they measure repeatability (relative standard deviation of peak height, peak area, or response ratio upon repeat injections of the same solution) and resolution of the main compound peak from a critical impurity peak.

Other companies will probably not take exactly the same approach, but this view from Roche provides an excellent focal point around which deliberations can be based. For instance, there is an implication that a one-point calibration is made each time a method is carried out. In the author's laboratory at Harlow, at least a three-point calibration is carried out for analyses of final products, which provides a check against nonlinearity arising from the LC system (as opposed to from the method). It is not unknown for this to occur, for instance, when there is a combination of higher-than-usual baseline noise and a nonoptimal setting of electronic integration parameters. A system-suitability test very similar to that used by Roche is carried out in the author's laboratory on an informal basis, but much greater emphasis is placed on completing the method and then letting the confidence limits that can be placed on the result be the judge of the system. This, of course, gives rise to a massive increase in the expenditure of operating time if the system turns out to be unsuitable. This problem can be totally offset by the use of fast LC, for which the total LC operating time is minimal in comparison with the rest of the overall analytical process (sample submission, sample preparation, and reporting).

As well as harmonizing their approach to LC method validation, some companies are harmonizing their approach to LC methods. In at least one company, a list of *preferred* or *first-choice* packing materials for each type of stationary phase has been drawn up. In addition, each LC method is developed on a specific batch of a packing material, and enough columns containing that batch are packed so that every time the method is carried out — no matter in which lab or at which site — it can be carried out with this same batch of packing material. By operating with such policies, significant savings can be made in the bulk purchase of a limited range of packing materials, and problems arising from a lack of column-to-column robustness (it is widely recognized that different batches of the same packing material sometimes exhibit different selectivities) of an LC method are effectively eliminated.

There is, however, another school of thought. There are also benefits to be obtained from holding a very wide range of packing materials and different batches of each packing material. In this situation, there ought to be a much greater chance of a new column yielding an impurity peak that hitherto had been unobserved. Also, with respect to method validation, it might be better to test for column-to-column ruggedness rather than temporarily eliminate it as a potential problem because the method may be tested by the regulatory authorities; there may be practical problems in supplying regulatory authorities in every country with an LC column containing the appropriate batch of packing material. In addition, other workers may not be able to reproduce the method if it is published.

Increasing emphasis on rigorous LC method validation, especially with respect to specificity (resolution of main peak from potential impurity peaks), is one of the reasons for a greater interest in mobile phase optimization procedures. Although many such procedures have been compared (9), there seems to be no strong trend toward the use of one over the others. The fact that there are so many different procedures is partly attributable to the fact that most of them were developed by instrument manufacturers. Each instrument manufacturer needed its own optimization procedure for incorporation into the software available with its equipment. There are analysts in the pharmaceutical industry who are authorities on optimization procedures and therefore have distinct preferences, so much so that the type of optimization procedure available can be the deciding factor in the choice of new equipment. In general, however, there is a tendency for most analysts simply to adopt the procedure associated with the equipment they already have.

The author's laboratory contains mostly equipment from a company that provides software for unattended optimization by a simplex procedure (46). This method is one of the less sophisticated and has been criticized for its sluggishness and inability to cope well with the existence of many local optima (47). Nonetheless, it can be useful for problems in which the number of components is small and the number of factors limited. It is therefore important to recognize these situations.

Despite the obvious current and future need for optimization procedures, it has only been necessary at the author's company to attempt to use them on two occasions. On both occasions there were a large number of components, but there was a separation problem only for a critical few of them. On one occasion (Figure 6), a limited number of factors were studied (ratio of aqueous to various organic mobile-phase components), and it soon became apparent that no improvement in resolution could be obtained. Increases in selectivity were accompanied by unacceptable increases in retention and peak tailing. The problem might eventually have been solved by the introduction of factors such as pH changes or the addition of an amine additive to the mobile phase, but this was not necessary. The problem was more easily solved by the use of a different type of stationary phase. The other occasion (Figure 7) involved not a chemical development problem, but work on a prostaglandin separation carried out in conjunction with research biochemists.

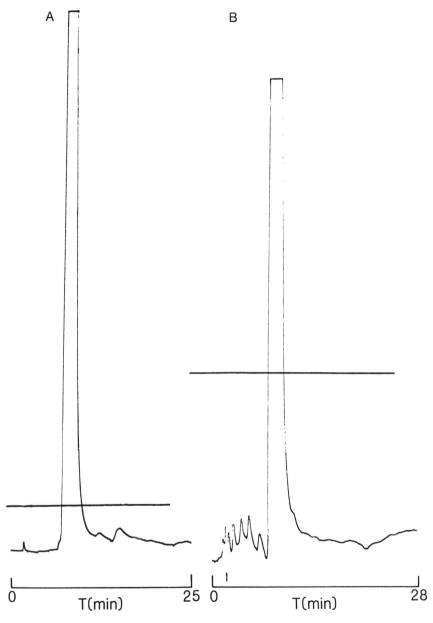

FIGURE 6.   BRL 26441A. Optimized separations of main peak from impurity (each at 0.1% by weight) peaks. (A) On Zorbax-CN (Du Pont Company, Wilmington, Delaware); coelution with main peak or detection problems from broad asymmetric impurity peaks; (B) on Partisil-SCX (Whatman, Clifton, New Jersey).

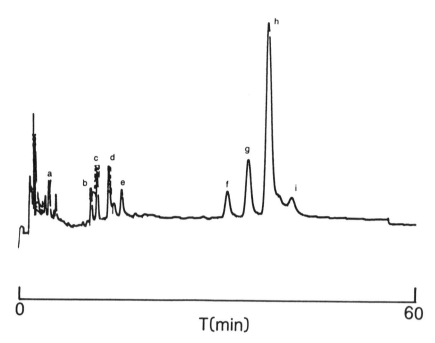

FIGURE 7. Separation of mixture of prostanoids on ODS-Hypersil, 250 mm × 5 mm. Mobile phase: acetonitrile/tetrahydrofuran/water/acetic acid (31.5:1.9:66.6:0.01); flow rate: 2.0 mL min⁻¹; detection: UV 210 nm. (a) 6-keto $PG_{F2\alpha}$; (b) TX_{B2}; (c) $PG_{F2\alpha}$; (d) PG_{E2}; (e) PG_{E1}; (f) PG_{A1}; (g) PG_{A2}; (h) PG_{B1}; (i) PG_{B2}.

The problem was solved simply by the addition of tetrahydrofuran to the mobile phase. Implementation of the simplex procedure resulted in very little further improvement.

The unattended operation of these optimization procedures is but one facet of the major trend to automation in the use of LC in the pharmaceutical industry. Automation is particularly important in LC for the chemical development phase. Automation is much more than loading up an autosampler with, for instance, 100 samples of the same compound and having the results calculated automatically. In the chemical development phase it is more common to encounter a large number of samples — but of several different types of compounds. In the past, this might have been dealt with by a laboratory having a relatively large number of isocratic LC systems, all set up with different conditions. Today, a single typical LC system (Figure 8) can easily cope with rapid, unattended (and, when necessary, overnight) switching between many different types of conditions. This stems from the introduction of features such as programmable autosamplers, automatic solvent selection, automatic column selection, and automatic wavelength drive for UV detectors. Despite this, analytical laboratories in the pharmaceutical industry do

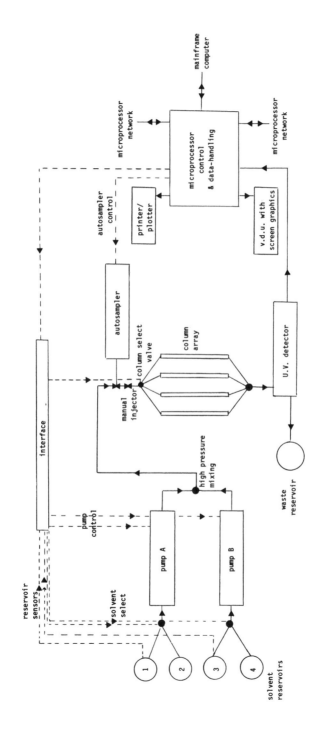

FIGURE 8. A modern, automated LC system.

not have fewer chromatographers and fewer LC systems. This is, of course, because of the increase in the LC workload.

On the data handling side, the increase in speed of calculations and results reporting is only one benefit. The availability of reintegration in conjunction with screen replotting facilities has allowed very accurate integration to be carried out without the necessity for repeat injections until the many electronic integration parameters such as peak width, noise, and skim ratio have all been set to their optimal values, which is particularly important for small peaks and/or partially resolved peaks and/or peaks on a sloping baseline. The replotting of chromatograms is also a valuable time-saving device in obtaining the best chromatograms for presentation purposes. This is not just a superficial advantage for the improvement in quality of reports: more and more LC work must be presented to regulatory authorities, and it is very important that the work is presented well.

A major growth area in the pharmaceutical industry as a whole has been the building up of computer networks to allow rapid transfer of data between different laboratories on a site and between different sites. This process has also been taking place within analytical units, where microprocessors in individual laboratories (chromatography, spectroscopy, physical methods) are all being linked to a central computer that houses laboratory management software. The many advantages in laboratory management and sample tracking of such systems have been well documented (48). For LC, there has been the advantage that raw data can be stored in the short term in the memory of the central computer and in the long term on hard disks. It is still wise to retain some hard copies of chromatograms for good laboratory practice (GLP) and regulatory requirements, but this development means that there is less need for archives filled with reams of chromatograms and that, if necessary, raw data can be recalled for reprocessing many years after it is first obtained.

Although the scale of the syntheses in chemical development is much larger than in the earlier stages of research, there is still a need for preparative LC. The custom-built preparative LC systems that were introduced in the mid-1970s come into their own for preparative work in the 5–100 g range, provided sufficient resolution can be obtained. Very serious attempts have been made to extend the range of preparative LC to production scale (49,50). Nonetheless, many companies still take the view that the technique would be far too expensive to contemplate in that environment. One of the major aims of a chemical development program is to develop a process that can be directly transferred to manufacturing plant. Because of this, the role of preparative LC in chemical development is much less than it otherwise might be.

Drug Metabolism and Pharmacokinetics

Drug metabolism and pharmacokinetics essentially are concerned with the fate of a drug once it enters the body of an animal or human. Obviously, animals are used in the early stages of these studies and then, later, healthy volunteers must be found.

This area of pharmaceutical development is undoubtedly one of the most difficult and challenging for a chromatographer. The problems of distinguishing between the drug, its metabolite(s), and any endogenous material in the biological samples (urine, plasma, and others) are compounded by the very low levels of the drug and its metabolite(s) present. With more-active drugs being made, this latter feature is becoming a greater problem. Also, in pharmacokinetic studies in which the concentration of drug and metabolites is monitored with time, it is usual to take large numbers of samples from a large number of patients (animals or volunteers) in order to obtain statistically significant results. High sample throughput is therefore especially important.

As with the LC of biopolymers, much research has been directed toward improvements in this area because of the difficult problems that had to be overcome. Many of the recent advances in LC methods have had a significant effect on the use of LC in drug metabolism and pharmacokinetic studies in the pharmaceutical industry. The most important of these new methods with respect to this area are listed below, along with a description of the effect they have had:

- *solid-state extraction techniques and the automation thereof,* providing major reductions in time taken for sample preparation
- *improved detector technology,* providing a great increase in sensitivity, especially from use of improved electrochemical detectors and on-line derivatization to allow electrochemical or fluorescence detection
- *diode-array detection,* which is especially useful in distinguishing between drug-related peaks and endogenous peaks and is also useful in metabolite identification and the monitoring of peaks with widely differing λ_{max} values in one chromatographic run
- *column switching,* which greatly increases the resolving power of the LC system in tackling the complex mixture involved
- *ion-pair LC,* which allows the retention of ionic species to be altered in a predictable way, so that the desired selectivity with respect to nonionic species and ionic species of opposite charge can easily be obtained
- *computer-assisted mobile phase optimization,* which minimizes time taken to develop mobile phase conditions to separate all peaks in the complex mixture of drug, metabolites, and endogenous compounds and allows rapid adjustment of conditions when an unexpected endogenous peak interferes with the assay.

LC/MS will be very useful in the future but has not, as yet, been used in pharmaceutical development on a widespread basis. Low-dispersion chromatography techniques offer improvements in mass sensitivity, which are especially useful in sample-limited situations. The application of fast and microbore LC in this area, however, has not been as prevalent as might have been expected. This may perhaps be because of the requirements of low extracolumn dispersion placed on the LC systems used and, in the case of microbore, because of the need to have pumps capable of delivering very low flow rates in a reproducible and pulse-free fashion. Despite not giving such an increase in mass sensitivity as microbore LC, the use of fast LC is more commonly found. This is mainly because of its very short analysis times, but also because it was introduced a longer time ago and is used more easily with conventional equipment.

With all of these new developments, it would seem that, in principle, it should now be a quick and easy matter to develop and set up a rapid, high-sensitivity method for the determination of a drug and its metabolites in a biological fluid. Indeed, a scheme has been proposed that would apply to all basic (pK$_a$ > 7) drugs (51). For reasons of economy, it is ideal to approach method development in such an organized and structured way as this — that is, using one or a very limited number of outline methods and making only minor alterations in conditions in order to obtain the method for each different drug.

This is very rarely possible, however, because of a number of practical considerations. Not the least is that a particular drug may be especially suited to assay by a technique or method not incorporated in the general scheme. More often, though, the difficulty arises from equipment availability. Analytical laboratories in drug metabolism departments are geared for dealing with very large sample numbers, so that a method may need to be carried out on a system that is not in use at the time, whether or not it is the best system for doing the job. Also, as mentioned earlier, the method must be compatible with the level of equipment available in other laboratories that are to use it.

Some of the equipment that is available for use at the time may not even be LC equipment. There is still a large body of drug metabolism work carried out using gas chromatography. It is only recently that it has been possible (with fluorescence and electrochemical detection) to achieve sensitivities with LC that have been obtained for a longer time by GC with, for instance, flame-ionization or electron-capture detection. There is still therefore an abundance of GC systems and GC expertise in drug metabolism departments. A GC method may also be preferred in cases in which GC methodology must be developed in order to identify metabolites by GC/MS. In general, with drugs becoming more and more active and, consequently, lower drug levels being found in the body, GC will continue to be used for some time to come for those assays that require very high sensitivity.

However simple they are designed to be, reaction systems for derivatization within an LC system represent an unnecessary complication. Especially in this environment, assays must at all times be kept as simple as possible. This is why solid-state extraction procedures are so popular. Like the solvent extraction procedures used in the past, they give considerable sample cleanup; moreover, extractions can be carried out while other samples are being chromatographed. On the one hand, because of the considerable sample cleanup, fairly simple LC conditions are required. On the other hand, when direct injection of urine or plasma following only protein precipitation is used, a complex LC system of high resolving power is more likely to be necessary. In addition, solid-state extractions can now be carried out automatically and rapidly so that they represent a minimal increase in time over direct injection. The increasing use of automated solid-state extraction procedures is therefore perhaps the strongest trend in the use of LC in drug metabolism and pharmacokinetic studies.

Formulation

Formulation is normally associated with the final marketed pharmaceutical preparation of a drug, but formulation development work begins much earlier, running in parallel with chemical development from its initial stages. Also, regulation of drug safety testing has led to a growth in the analysis of the formulations used in toxicology studies.

With respect to these formulations, the regulatory authorities must be satisfied that the *identity* of the drug in the formulation has been confirmed, that the *concentration* of the drug is as stated in the study protocol, and that the *homogeneity* of the formulation has been established (52). As such, formulations have been analyzed by LC rather than UV spectroscopy for many years now. LC retention time is a much better confirmation of drug identity than a UV spectrum because the UV absorbances of compounds closely related to the drug (analogues, potential impurities, degradation products) can be almost identical to that of the drug itself. In addition, some of the formulations under study can contain more than one active component, and LC is needed to enable the simultaneous determination of all of the active components.

The experience of the analytical unit at Beecham has been with solutions and solid suspensions in liquids that are typically used in short-term toxicology studies. Surprisingly, in some cases analysis on these formulations is still carried out by UV spectroscopy (53). The unit's aim was to establish a general method that was as universal as possible. Because some formulations contained more than one active component and, for some formulations, stability data were required, it was decided that, if possible, LC should be used to analyze all of these dosage solutions. The major problem was that LC was slower than UV determinations.

Erni has pointed out that speed in LC can be used advantageously if the analysis time is limited by the chromatographic time, such as occurs in cases in which simple sample preparation possibilities exist or no sample preparation is required (54). In the author's case, it was found that, in general, sufficient resolution along with very short analysis times could be obtained on short columns *even* with conventional equipment. It was necessary to demonstrate this for conventional equipment, because not all of the LC systems in the laboratory are equipped with detectors having low-volume flow cells and short time constants. It has not been necessary to use such detectors with low injection and connecting-tubing volume to achieve even shorter analysis times.

For many of the samples, however, analysis time was not limited by chromatographic time. Aqueous methyl cellulose was frequently used as a carrier. In analyzing such formulations by reversed-phase LC, Ng found it necessary to precipitate methyl cellulose by sodium chloride treatment (55); otherwise blockage of the LC column occurred when the sample solution was injected. In work at the author's company (56), it was found that no blockage occurred on direct injection if a polar bonded phase was used as the stationary phase in order to allow the use of mobile phases containing a very high proportion ($\geq 80\%$) of aqueous component.

It was also found that the dilution of samples to give solutions of the appropriate concentration for injection onto the LC system was either time-consuming by weighing or by use of volumetric glassware, or was not sufficiently accurate by manual use of microsyringes. Dilution with the Microlab 1000 diluter/dispenser (Hamilton Company, Reno, Nevada) was evaluated for aqueous methyl cellulose solutions and found to be very rapid, accurate, and precise.

In longer-term toxicology studies, animal diet is more often used as the drug carrier. Stavrou has elaborated on the analysis of drugs in animal diet (52), dwelling particularly on the difficulties of obtaining a representative sample of the formulation and of developing a suitable extraction procedure.

The factors mentioned in the context of formulations for toxicology studies are also applicable to the LC analysis of the capsules, tablets, and injectables used in volunteer studies, in clinical trials, and, of course, for the eventual marketed product. Erni's point about the advantageous use of speed in LC is relevant to this area of pharmaceutical development — especially for content uniformity (homogeneity), dissolution rate, and stability tests — because very large numbers of samples are involved. These tests represent perhaps the first area of pharmaceutical development in which a low-dispersion chromatographic technique has had a genuine, widespread effect.

As for chemical development, though, a major influence of trends in LC use in this area has been the influence of regulatory authorities. Of the seven questions on drug quality on which Sullman states the regulatory authorities must be satisfied (20), four are directly concerned with matters relating to formulation — that is, questions on the consistency and acceptability of the final formulation, on the availability of active drug from the formulation, on the stability of the drug in the formulation, and on the necessity for the drug/formulation that is proposed for marketing to reflect the drug/formulation used for preclinical testing and clinical trials.

As was true in chemical development and other areas, the validity of LC methods used in preparing replies to questions from regulatory bodies must be tested. This is not an identical process to the validation test for an LC method for the drug substance itself: in the case of an LC method for a formulation, additional information must be given on the extraction of the drug from the formulation (percentage recovery). In some cases, even when using LC conditions identical to those being used for the drug substance itself, it may still be necessary to recheck accuracy, precision, and linearity. It is conceivable, although often unlikely, that cochromatography with an excipient might affect this.

The hydrophobicity (or lipophilicity) of a drug is among the physical properties that play an important part in deciding the type of formulation in which it will be presented. The body of work on the correlation of reversed-phase LC retention indices with partition coefficients has application in the field of formulation as well as in structure-activity relationships, and there is a trend for this application to increase. Tomlinson in particular has been active in this area (28).

Production

LC plays a role in production in the monitoring of the processes taking place in the plant (by checking the purity of intermediates) and in assuring the quality of the final product. All that has been said on the influence of regulatory questions on drug quality in the area of chemical development also applies to this area and probably more so.

The key to LC in monitoring a process is always being able to have rapid turnaround times so that any necessary changes in the process can be made as soon as possible. The samples are likely to come in threes and fours at any time of night or day rather than in batches of hundreds at any one time. There is thus no need for any trend away from the LC laboratory with several basic LC systems; indeed, they need not even be equipped with autosamplers. Quick turnaround time is achieved by having methods that are as simple and as similar as possible. In this way, hardly any time at all is lost in setting up the appropriate conditions for any samples that arrive.

The sample-load situation in the quality assurance of final products is entirely different. Here, very large numbers of samples of the same material are received on a regular basis. By this stage all the methods — LC and others — for the material have been developed. Again, there seems to be very little movement away from a well-established pattern. Usually equipment for isocratic LC is used. The main features of the LC systems used are autosamplers and microprocessors (handling up to four detector outputs) for the automatic calculation of purity figures. Most of these systems are now interfaced to a mainframe computer so that the large volume of generated results can be transfered directly to the data base therein. In some cases, there is also networking of all of the microprocessors, and there are also alternative systems in which a large number of detectors (up to 32) are all handled directly by the mainframe computer. Such systems are less flexible than the former and are becoming less popular.

EXTENT OF ADOPTION OF NEW LC TECHNIQUES

The most important of the recent developments in LC with respect to its use in the pharmaceutical industry can be judged by a combination of the overall degree to which they have been used and the importance of their use in individual areas. By this token, the most important of recent LC developments as far as the pharmaceutical industry is concerned could be said to be:

• automation of the control of LC systems
• computerized data handling
• the use of solid-state extraction procedures for the automated cleanup of samples of drugs in biological fluids
• the development of improved packing materials for the LC of biopolymers.

From the previous discussion of trends in LC use in individual areas of pharmaceutical development, it might also have been noticed that some of the recent LC advances, while having their uses, have yet to exert the kind of influence that might have been expected.

The applicability of fast LC has already been mentioned. It is a fact that many of the analytical procedures in the pharmaceutical industry are *not* limited by operating time. In these cases, there is not sufficient advantage in the use of fast LC to warrant the rapid replacement of conventional detectors (and integrators) with new equipment more compatible with the technique. As a result of the length of time it takes for a drug to advance from discovery to market, some of the analytical methods for drugs now approaching the production stage may have been initially developed several years ago. Therefore, even in cases in which the advantages of fast LC are demonstrable, the advantages may not be sufficient to justify the modification of the methods to incorporate the use of short columns containing 3-μm packing materials, especially given that many 10-μm and 5-μm stationary phases are not available in the 3-μm size.

This is not a problem for microbore LC, for which exactly the same packing materials can be used. The fact that microbore LC places much greater constraints on the equipment that can be used, however, has worked against this technique. The microbore LC systems purchased by LC users in the pharmaceutical industry have so far mostly been obtained with a specific application in mind; in most cases this would be for LC/MS. There has been little published evidence of the advantage of microbore LC in allowing the use of expensive stationary phases or of exotic mobile phases actually being taken advantage of for pharmaceutical applications. For routine work, the advantages of higher mass sensitivity and reduced solvent costs have not warranted a wholesale change from conventional pumps and detectors to microbore-compatible components. This is especially true given that it has been claimed recently that the gains in sensitivity obtainable in practice are much less than those in theory (57) and that the additional savings in solvent costs in moving from 2-mm to 1-mm microbore columns over the lifetime of an LC system are small compared to the cost of the LC system (58). In addition, even in the best equipment available today designed especially for microbore LC, extracolumn dispersion is such that significant losses in efficiency for early-eluting peaks still cannot be avoided.

Despite all of this, there can be little doubt that microbore LC will eventually be adopted throughout the pharmaceutical industry. Many of the new LC pumps being brought onto the market today for work at conventional ($\cong$ 2.0 mL min^{-1}) flow rates also have microbore flow rate (down to 10 μL min^{-1}) capability. Almost all new UV detectors have very fast time constants and may be used with low-volume flow cells. Thus, as old LC equipment in the pharmaceutical industry is gradually replaced, the situation will arise that most LC systems will be microbore-compatible. There have been statements (59) that 2-mm i.d. columns will become the norm, but there is little reason why 1-mm i.d. columns should not also be used.

Aside from the certainty that equipment will improve with respect to extracolumn dispersion, there are two other factors to consider with respect to 1-mm LC columns in pharmaceutical analyses:

- a major objective in the method development of a pharmaceutical assay is to obtain conditions of such a selectivity that the method is very robust and does not rely on very high efficiency for the achievement of separation
- in most pharmaceutical LC assays, particularly for drugs in biological fluids, the main peak(s) are not early eluting.

The role of diode-array detectors in the area of drug metabolism has been noted. They have also been used in checking the purity of drug substances themselves and can conveniently be used to scan for impurities at several different wavelengths during one chromatographic run. This is most important when compounds are first synthesized or are synthesized by a different route — that is, when previous knowledge of the amount and nature of impurities is limited. It could be said that every analytical laboratory should have one, but, especially at current prices, there is certainly no need for *more* than one. Indeed, in the chemical development or production environment they are more of a convenience than a necessity. For instance, similar information could be obtained in an acceptable time by the use of fast LC with several chromatographic runs at different wavelengths using a conventional variable-wavelength UV detector. Despite this, if diode-array detectors become available with very low-volume flow cells and their prices are substantially reduced, they could become the rule rather than the exception.

LC/MS is a technique that *will* find widespread application in the pharmaceutical industry. The technique has still to be fully developed, but even at this stage great interest is being shown in it.

INFLUENCE OF OTHER TECHNIQUES

Trends in LC usage cannot fully be discussed in isolation from other analytical techniques. After all, a major reason for the increase in LC's use has been its ability to replace other techniques. The story is not one-sided, though. LC has been well established in all areas of pharmaceutical development for some time now and, not surprisingly, there have been developments in other techniques that have allowed those techniques to rival LC for some pharmaceutical applications.

For instance, in the area of research organic chemistry, a chemist may occasionally choose to run a high-field nuclear magnetic resonance (NMR) spectrum of a sample, rather than do an LC separation, in cases such as when:

- *it is known that the sample is fairly pure and contains a* limited *number of unknown impurities* — structural information on the impurities is obtained, as well as on the purity of the compound and the molar percentage of the impurities from the integral
- *an isomer ratio is required for a new compound when no samples of each isomer*

have been prepared — if the isomer ratio was determined by LC (assuming the same UV extinction coefficient for each isomer), it would still not be known which isomer was which

• *the enantiomeric composition of a fairly pure compound is required* — it has been the experience of the analytical unit at the author's company that high-field (270 MHz) Fourier-transform ¹H NMR techniques using chiral-shift reagents — particularly (+)- or (−)-2,2,2-trifluoro-1-(9-anthryl)ethanol (Figure 9), which operates via a charge-transfer mechanism — can be more reliable to produce a result than can the use of LC with diastereomer formation or the use of commercially available chiral stationary phases.

In the past the resolution of low-field NMR instruments (60–100 MHz) and the high spectral acquisition times required resulted in these instruments not being able to compete with LC in analyses of the types of samples mentioned above. Today, on a high-field spectrometer operating at 200 MHz or more, it is often possible to determine *and* obtain structural information on an impurity in as few as 10 min if there is as little as 500 µg of the impurity present in the sample solution. Given the high cost of high-field NMR instruments (about 10 times the cost of a typical gradient LC system), this may seem to be an expensive method of tackling the problem. If there is already such an instrument in-house for the purpose of elucidating very difficult organic chemistry structural problems, however, using it to tackle simpler problems can be seen as merely maximizing its use. Because of the very short acquisition times required for the types of sample mentioned above, many such samples can be run in a small amount of instrument time. The enantiomeric resolutions require more instrument time but can often avoid the need for many hours of LC method development time.

In the production area, some companies are beginning to use near-infrared reflectance analysis (NIRA) — for example, the Infraanalyser (Technicon Industrial Systems, Tarrytown, New York) — for purity determinations of bulk intermediates and final products. This technique is very rapid but has its disadvantages. It

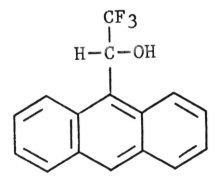

FIGURE 9.   The structure of (+)- or (−)-2,2,2-trifluoro-1-(9-anthryl)ethanol.

can only be used if the physical form of the compound being analyzed is stable. The calibration correlation of NIRA measurements with purity (as determined by LC and other techniques) must be very robust and must be based on at *least* 50 samples.

In other areas, there are always other techniques that can do an assay similar to LC [the Chromatatron automated preparative TLC system (Harrison Research, Palo Alto, California) in preparative work; various bioassays in drug absorption studies, in toxicology studies, and in clinical trials; and differential scanning calorimetry in purity determinations in which no reference material is available]. On the whole, however, there is no sign as yet of the use of alternative techniques on a large enough scale to cause a reduction in LC use.

Other techniques also influence LC use by their combination with LC to give hyphenated techniques. The future value of LC/MS to the pharmaceutical industry has already been mentioned; LC/NMR and LC/FT-IR will also be of value, but probably to a lesser extent.

FUTURE TRENDS

The trends of increasing LC use, increasing documentation of LC methods and LC results, and increasing automation in LC set in motion by changes in the pharmaceutical industry will continue (but perhaps with some leveling off) unless the industry takes a wholly unexpected turn. It is fairly certain that interest in biopolymers will continue to grow, with a concomitant increase in LC use in that area.

In some areas in the pharmaceutical industry, trends have more or less taken their course. LC has been adopted to satisfy a need and has established itself, and there is no desperate need as yet for further major improvement. The most significant future changes should therefore take place in areas in which there is a well-defined need for improvement.

The important trend toward automation, however, has not by any means taken its full course. In LC as in other methods, sample preparation is still the time-limiting step. Developments such as the use of diluter/dispensers and automated solid-stage extraction procedures have been mentioned; real breakthroughs could come with the use of robotics.

In research organic chemistry, chiral stationary phases that are applicable to a wide range of classes of compounds have yet to be found. The use of β-cyclodextrin inclusion complexes (60) has shown some promise, but it is the work of Hermansson (61) on an α_1-acid glycoprotein bonded phase that could herald a new era. Chromatographic separations on protein bonded phases could also have interesting implications in structure-activity relationships.

Such phases will, however, find no application in preparative enantiomeric separations, an area in which there is great need for improvement. On this scale, only simple chiral phases can be used, and there is little chance of finding a universally applicable simple chiral phase. An alternative approach might be to tailor-make a simple chiral system for each separate preparative enantiomeric resolution using molecular graphics to predict the optimal simple chiral phase, recrystallization salt, or easily removable derivatizing agent for each problem.

In the area of drug metabolism, high sample numbers are likely to get higher. Even with the many improvements in LC assays in biological fluids, the techniques still are not quick enough to cope easily with the ever-increasing sample numbers. Also, in toxicology studies and clinical trials in which, say, only information on drug absorption is required, they are more specific and give more information than is necessary. For such studies the developing trend toward the use of very rapid bioassays, such as radioimmunoassays and radiolabeled ligand-displacement assays for which virtually no sample preparation is required, may well accelerate.

Finally, it is worth touching on a possible future trend that might have intriguing prospects for pharmaceutical chromatographers. They have been content with computerized method optimization, safe in the thought that their vast wealth of chromatographic knowledge and experience is still necessary for true method development. In synthetic organic chemistry, however, synthetic routes can be designed by computer, drawing from information on chemical functional group reactivity and synthetic reagents stored in a large data base (62). Such an approach could be adopted for LC method development with a data base being built up on the behavior of as many possible types of molecules in as many possible types of LC conditions. The day may yet come when the robotic organic chemist types in the structure of a compound at a terminal and, within seconds, an LC method has been developed and the sample analyzed!

Acknowledgments

The author would like to acknowledge A.E. Bird, S. Bratt, M. Brightwell, B. Catherwood, J. Clarke, B. Davies, R.S. Oliver, and D.M. Smith for valuable discussions and V. de Biasi, G. Harper, and J. Humphries for their involvement in the LC work used to illustrate this chapter.

REFERENCES

(1) E. Tomlinson, *Chem. & Ind.* 687 (1981).
(2) I.S. Lurie and S.M. Demchuk, *J. Liq. Chromatogr.* **4,** 357 (1981).
(3) E. Tomlinson, *J. Pharm. Biomed. Anal.* **1,** 11 (1983).
(4) W. Lindner, J.N. Le Page, G. Davies, D.E. Seitz, and B.L. Karger, *J. Chromatogr.* **185,** 323 (1979).
(5) B. Feibush, M.J. Cohen, and B.L. Karger, *J. Chromatogr.* **282,** 3 (1983).
(6) W.H. Pirkle, J.M. Finn, J.L. Schreiner, and B.C. Hamper, *J. Am. Chem. Soc.* **103,** 3964 (1981).
(7) J.L. Glajch, J.J. Kirkland, and L.R. Synder, *J. Chromatogr.* **238,** 269 (1982).

(8) J.L. Glajch and J.J. Kirkland, *Anal. Chem.* **54**, 2593 (1982).

(9) J. Rafel, *J. Chromatogr.* **282**, 287 (1983).

(10) J.L. Di Cesare, M.W. Dong, and L.S. Ettre, *Chromatographia* **14**, 257 (1981).

(11) C. Eckers, K.K. Cuddy, and J.D. Henion, *J. Liq. Chromatogr.* **6**, 2383 (1983).

(12) P. Kucera and G. Guiochon, *J. Chromatogr.* **283**, 1 (1984).

(13) N.H.C. Cooke, B.G. Archer, K. Olsen, and A. Berick, *Anal. Chem.* **54**, 2277 (1982).

(14) F. Erni, *J. Chromatogr.* **282**, 371 (1983).

(15) R.P.W. Scott and P. Kucera, *J. Chromatogr.* **169**, 51 (1979).

(16) I. Halasz and G. Maldener, *Anal. Chem.* **55**, 1842 (1983).

(17) V.L. McGuffin and M. Novotny, *Anal. Chem.* **55**, 580 (1983).

(18) D. Ishii and T. Takeuchi, *J. Chromatogr. Sci.* **18**, 462 (1980).

(19) S.A. Borman, *Anal. Chem.* **56**, 1031A (1984).

(20) S.F. Sullman, *Pharmacopeial Forum* **10**, 4611 (1984).

(21) E. Debesis, J.P. Boehlert, T.E. Givand, and J.C. Sheridan, *Pharm. Technol.* **6**(9), 120 (1982).

(22) K. Shimizu, S. Amagaya, and Y. Ogihara, *J. Chromatogr.* (Biomed. Applns.) **272**, 170 (1983).

(23) A.W.J. de Jong, J.C. Smit, H. Poppe, and J.C. Kraak, *Anal. Proc. Chem. Soc.* **17**, 508 (1980).

(24) P. Gariel, C. Durieux, and R. Rosset, *Sep. Sci. Technol.* **18**, 441 (1983).

(25) J. Humphries, E.F. Nurse, S.J. Dunmore, A. Beloff-Chain, G.W. Taylor, and H.R. Morris, *Biochem. Biophys. Res. Commun.* **114**, 763 (1983).

(26) V. de Biasi and W.J. Lough, manuscript in preparation.

(27) R. Kaliszan, *J. Chromatogr. Sci.* **22**, 362 (1984).

(28) T.L. Hafkenscheid and E. Tomlinson, *J. Chromatogr.* **292**, 305 (1984).

(29) K. Jinno, *Anal. Lett.* **17**, 183 (1984).

(30) S.H. Unger and G.C. Chiang, *J. Med. Chem.* **24**, 262 (1981).

(31) I.W. Wainer, T.D. Doyle, and W.M. Adams, *J. Pharm. Sci.* **73**, 1162 (1984).

(32) A.M. Marshall, personal communication.

(33) T. Takeuchi, M. Yamazaki, and D. Ishii, *HRC & CC, J. High Resolut. Chromatogr. Chromatogr. Commun.* **7**, 101 (1984).

(34) R.L. Heinrikson and S.C. Meredith, *Anal. Biochem.* **136**, 65 (1984).

(35) T.J. Mahachi, R.M. Carlson, and D.P. Poe, *J. Chromatogr.* **298**, 279 (1984).

(36) D.C. Warren, *Anal. Chem.* **56**, 1529A (1984).

(37) R. Freixa, J. Casas, R. Rosello, and E. Gelpi, *HRC & CC, J. High Resolut. Chromatogr. Chromatogr. Commun.* **7**, 156 (1984).

(38) C. Kim, C. Campanelli, and J.M. Khanna, *J. Chromatogr.* **282**, 151 (1983).

(39) "Biopolymer Separations," Shandon Southern Products Ltd., Cheshire, U.K.

(40) M.G. Styring, C. Price, and C. Booth, *J. Chromatogr.* **319**, 115 (1985).

(41) B.C. Stark and W.Y. Chooi, *J. Chromatogr.* **319**, 367 (1985).

(42) D.L. Gooding, N.M. Schmuck, and K.M. Gooding, *J. Chromatogr.* **296**, 107 (1984).

(43) G. Wagner and F.E. Regnier, *Anal. Biochem.* **126**, 37 (1982).

(44) P. Roumeliotis and K.K. Unger, *J. Chromatogr.* **218**, 535 (1981).

(45) L.A. Witting, D.J. Gisch, R. Ludwig, and R. Eksteen, *J. Chromatogr.* **296**, 97 (1984).

(46) S.L. Morgan and S.N. Deming, *Anal. Chem.* **46**, 1170 (1974).

(47) R.E. Majors, H.G. Barth, and C.H. Lochmüller, *Anal. Chem.* **56**, 300R (1984).

(48) A.L. Robinson, *Science* **220**, 180 (1983).

(49) C.K. Shih, C.M. Snavely, T.E. Molnar, J.L. Meyer, W.B. Caldwell, and E.L. Paul, *Chem. Eng. Prog.* 53 (1983).

(50) T.H. Maugh II, *Science* **216**, 159 (1982).

(51) D.L. Massart and M. de Smet, poster presented at Seventh International Symposium on Column Liquid Chromatography, Baden-Baden, Federal Republic of Germany, May 3–6, 1983.

(52) A. Stavrou, *Anal. Proc. Chem. Soc.* **20**, 202 (1983).

(53) Minutes of Chemical Aspects of Toxicology Discussion Group, meeting, Ware, Hertfordshire, United Kingdom, March 1985.

(54) J.C. Gfeller, R. Haas, J.M. Troendle, and F. Erni, *J. Chromatogr.* **294,** 247 (1984).

(55) L.L. Ng, *Anal. Chem.* **53,** 1142 (1981).

(56) V. de Biasi, M.B. Evans, W.J. Lough, and D.K. Williams, manuscript in preparation.

(57) R. Gill, *Anal. Proc. Chem. Soc.* **21,** 436 (1984).

(58) B. Glatz, paper presented at Hewlett-Packard Analytical Symposium, Stratford-upon-Avon, England, June 11–13, 1984.

(59) Comment by R. Brownlee reported in *Chem. Eng. News* **61,** 31 (1983).

(60) W.L. Hinze, T.E. Riehl, D.W. Armstrong, W. DeMond, A. Alak, and T. Ward, *Anal. Chem.* **57,** 237 (1985).

(61) J. Hermansson, *J. Chromatogr.* **208,** 67 (1984).

(62) A.P. Johnson, *Chem. Br.* **21,** 59 (1985).

INDEX